One-of-a-Kind Production

AF181532

Prof. Yiliu Tu
Department of Mechanical and
Manufacturing Engineering
University of Calgary
2500 University Drive NW
Calgary, AB T2N 1N4
Canada
e-mail: paultu@ucalgary.ca

Dr. Paul Dean
CIM Systems Ltd.
142 Royal Oak Point NW
Calgary, AB T3B 5C6
Canada
e-mail: pdean@cimsys.org

ISBN 978-1-4471-6174-5 ISBN 978-1-84996-531-6 (eBook)
DOI 10.1007/978-1-84996-531-6
Springer London Dordrecht Heidelberg New York

British Library Cataloguing in Publication Data
A catalogue record for this book is available from the British Library

© Springer-Verlag London Limited 2011
Softcover re-print of the Hardcover 1st edition 2011
Apart from any fair dealing for the purposes of research or private study, or criticism or review, as permitted under the Copyright, Designs and Patents Act 1988, this publication may only be reproduced, stored or transmitted, in any form or by any means, with the prior permission in writing of the publishers, or in the case of reprographic reproduction in accordance with the terms of licenses issued by the Copyright Licensing Agency. Enquiries concerning reproduction outside those terms should be sent to the publishers.
The use of registered names, trademarks, etc., in this publication does not imply, even in the absence of a specific statement, that such names are exempt from the relevant laws and regulations and therefore free for general use.
The publisher makes no representation, express or implied, with regard to the accuracy of the information contained in this book and cannot accept any legal responsibility or liability for any errors or omissions that may be made.

Cover design: eStudio Calamar S.L.

Printed on acid-free paper

Springer is part of Springer Science+Business Media (www.springer.com)

Yiliu Tu · Paul Dean

One-of-a-Kind Production

Springer

Yiliu Tu · Paul Dean

One-of-a-Kind Production

Springer

Prof. Yiliu Tu
Department of Mechanical and
Manufacturing Engineering
University of Calgary
2500 University Drive NW
Calgary, AB T2N 1N4
Canada
e-mail: paultu@ucalgary.ca

Dr. Paul Dean
CIM Systems Ltd.
142 Royal Oak Point NW
Calgary, AB T3B 5C6
Canada
e-mail: pdean@cimsys.org

ISBN 978-1-4471-6174-5 ISBN 978-1-84996-531-6 (eBook)
DOI 10.1007/978-1-84996-531-6
Springer London Dordrecht Heidelberg New York

British Library Cataloguing in Publication Data
A catalogue record for this book is available from the British Library

© Springer-Verlag London Limited 2011
Softcover re-print of the Hardcover 1st edition 2011
Apart from any fair dealing for the purposes of research or private study, or criticism or review, as permitted under the Copyright, Designs and Patents Act 1988, this publication may only be reproduced, stored or transmitted, in any form or by any means, with the prior permission in writing of the publishers, or in the case of reprographic reproduction in accordance with the terms of licenses issued by the Copyright Licensing Agency. Enquiries concerning reproduction outside those terms should be sent to the publishers.
The use of registered names, trademarks, etc., in this publication does not imply, even in the absence of a specific statement, that such names are exempt from the relevant laws and regulations and therefore free for general use.
The publisher makes no representation, express or implied, with regard to the accuracy of the information contained in this book and cannot accept any legal responsibility or liability for any errors or omissions that may be made.

Cover design: eStudio Calamar S.L.

Printed on acid-free paper

Springer is part of Springer Science+Business Media (www.springer.com)

Preface

My research on one-of-a-kind production (OKP) can be traced back to the early 1990s when I carried out my Ph.D. study under supervision of Professor Hans Holm in Aalborg University of Denmark. Not very long after I started the research, I quickly realized that I mixed up the concepts. The interesting point is that I am not the only one who mixes up the concepts as I am often asked by people about the differences between OKP and one-off or job shop production. After they hear my answers, the following questions are usually raised: What is difference between an OKP system and a flexible manufacturing system (FMS)? What is new in OKP research compared or contrasted with the research in computer integrated manufacturing (CIM) and mass customization? These questions have motivated me to write a book on OKP for many years.

However, to clarify these questions is not our major motivation to write this book or developing the OKP system which enables companies to mass-produce customized products. Manufacturing business in developed countries, like Canada and USA, have followed a trend of moving manufacturing from these countries to Latin America or East Asia where there are favorable competitive conditions, e.g. large number of well-educated labours, low wages, etc. Some people in the countries like Canada and USA believe that manufacturing businesses in these countries need to be restructured to make high-tech products and that the low-tech products should be made overseas. However, the question here is: What products are considered as high-tech products? Cars used to considering high-tech products. To design a new model of car takes many years and a huge investment. However, to set up a manufacturing plant and to train people to make parts and assembly parts into cars along a flow line do not require the same amount of time. Therefore, we see more and more manufacturing plants being moved out of the countries, whereas we do not see enough new manufacturing plants open in these countries to avoid the large number of layoffs by these "restructuring" businesses. On the other hand, we do see some new demands for local manufacturing businesses. If your windows in your house or building need replacing you would most likely prefer talking to a local manufacturer to make and replace the windows rather than ordering the windows from a manufacturer in a far and remote area, particularly during the cold winter in Calgary, Alberta, Canada. Another case could be an oil/gas production company. After a well is drilled and the oil/gas is explored, the company would like to order and install the production equipment in the shortest time possible since the waiting time or lead-time of making and installing the equipment would cause the company to lose a significant amount of revenue. We can provide more cases where customers would like to deal with local manufacturing companies since they can express their requirements better to and get their orders fulfilled in a shorter lead-time from local manufacturing companies. Furthermore, it has been well recognized among researchers and practitioners in manufacturing that market demand for individualized products is increasing dramatically whereas demand for standard products is decreasing. The competitive priorities of customization, volume flexibility and lead-time are becoming more and more appealing to various manufacturing businesses. Therefore, the manufacturing companies that are able to rapidly and efficiently make customized products without a batch size limitation (i.e. they can make batches of one) will gain significant competitive advantages over the others. We refer these companies as OKP companies.

In general, the cost of a customized product can be roughly decomposed into two parts, constant part and variable part. The constant part is the cost of materials and components which is subject to international prices. This part of cost will not be affected by where and by whom the product is made. The variable part of cost includes the work efforts for engineering and producing the product. The engineering efforts include product design, production management (including quality assurance and knowledge management), process planning and production scheduling and control. The production effort mainly requires the use of shop floor labor force and production equipment. From our research we find that the engineering efforts usually take the major part of the variable cost in an OKP company. Therefore, to improve the efficiency of engineering efforts can dramatically reduce the cost and lead-time of a customized product. This is in fact our major motivation to carry out our research on OKP and write this book to report our research findings.

To improve the efficiency of engineering efforts to fulfill a customer order in an OKP company, a computer software system is needed to aid and integrate the engineering processes, such as product design, production management, process planning, and production scheduling and control. From our research, we have developed such a software system, named OKP Management and Control Software (OKPMCS) system© for which the copyright has been registered with Canadian Intellectual Property Office. Normally, the material requirement planning (MRP I) is used as the fundamental technique or driving engine for the computer aided production management and control software systems, such as MRP II (manufacturing resource planning) and enterprise resource planning (ERP) systems. However, the MRP I technique can result in technical hardships and inefficiencies when a company is making customized products with large variations and small volumes, which is often the case in OKP companies. This has limited the applications of these computer software systems in OKP companies.

The OKPMCS system does not follow the MRP I principle. Instead, it uses a novel data modeling technique called product production structure (PPS) as the driving engine. In addition to the PPS technique, this software system also includes or facilitates the application of other novel concepts, methods, computation techniques and system structure for improving the efficiency of engineering processes in an OKP company. For readers to easily read this book and understand these concepts, methods, computation techniques and system structure, we offer the readers *one year free of charge to use this software system* for either commercial or educational purposes.

This book consists of eight chapters. In Chap. 1, we define the OKP system and concept in comparing and contrasting it with the tradition approaches for improving production efficiency in manufacturing. The emphasis of Chap. 1 is placed on the common features and differences between the OKP system and traditional production systems so that readers can have a clear understanding of the OKP system and its important application meaning to manufacturing companies. Chapter 2 presents the framework of the OKP Management and Control Software (OKPMCS) system©. This chapter is for readers, who have some basic knowledge on computer programming, to appreciate the technical merits and novel features of the OKPMCS system in comparing with commercial MRP II and ERP systems. In Chap. 3, we present the PPS technique and a concurrent engineering method for rapid development of a customized product under the help of the OKPMCS system. Chapter 4 presents the novel computer aided cost estimating, quoting and tracking methods. In Chaps. 5 and 6, we discuss the methods for automatically generating product design and production data, which is particularly important for OKP companies to improve their efficiency of engineering processes. However, to read these two chapters, a reader has to have a basic knowledge of computer programming since the data generation method has been developed through application of computer modeling and programming techniques. The architectural design and the major data classes of information generation module (IGM) are specified in real-time process algebra (RTPA) in Appendix D so that readers can further understand this computer technology which is a novel and key technology for OKP companies to achieve both high customization and high production efficiency. In Chap. 7, the heuristic and computation methods for adaptive OKP scheduling and control are discussed. The efficiency and feasibility of these production scheduling and control heuristic and methods have been tested in industries. Finally in Chap. 8, we briefly summarize the functionalities, novel features and application of the OKPMCS system. The detailed user manual and installation manual of the OKPMCS system can be freely downloaded from our webpage according to the guidelines as outlined in Appendix A.

It is hereby, I would like to acknowledge the funding supports from the Public Good Science & Technology (PGS&T) Foundation in New Zealand and the Natural Sciences and Engineering Research Council of Canada (NSERC). I also would like to acknowledge the contribution to Chap. 4 made by Dr. Shane Xie in The University of Auckland (New Zealand) during his Ph.D. study under my supervision in University of Canterbury (New Zealand) between 1998 and 2002, and the contribution to Chap. 7 made by Dr. Mike Li during his Ph.D. study under my supervision between 2006 and 2010 in University of Calgary (Canada).

Finally, we would like to thank the anonymous referees for their valuable comments for improving the quality of this book and the editors who have made great efforts to publish this book. We also appreciate in advance for readers to give us your comments and opinions so that we can improve our research in the future.

Calgary, Alberta, October 16, 2010 Yiliu Tu

Contents

1 Introduction to the One-of-a-Kind Production System 1
1.1 Traditional Approaches to Manufacturing 3
1.2 Computer Integrated Manufacturing Systems 3
1.3 Mass Customization 4
1.3.1 Current Status of Mass Customization Systems 6
1.3.2 IT Support for Mass Customization 9
1.3.3 Information Systems and Modelling Techniques for Mass Customization 11
1.4 Risks and Disadvantages of Mass Customization 15
1.5 The One-of-a-Kind Production (OKP) System 16
1.6 The Requirements or Constraints for Achieving Mass Customization in an OKP Company 20
1.7 The Novelty of the OKP System 25
1.8 Chapter Summary 29
References 29

2 Framework of Computer Aided One-of-a-Kind Production System 33
2.1 Framework of the OKP Management and Control Software System 33
2.1.1 Information Technology Patterns 35
2.1.2 Real-Time Process Algebra (RTPA) Methodology 36
2.1.3 UML Class Diagrams and RTPA 37
2.1.4 Software Development Methodology 39
2.2 Modelling of Mass Production and One-of-a-Kind Production . . . 40
2.3 One-of-a-Kind Production (OKP) Strategy 41
2.4 Analysis of Requirements 43
2.5 Data Modelling Module 47

2.5.1 BOM Template . . . 47
2.5.2 BOO Template . . . 48
2.5.3 BOI Template . . . 52
2.5.4 Classes . . . 53
2.6 Information Generation Module . . . 56
2.7 Shop Floor Control Module . . . 56
2.8 Summary of OKP Management and Control Solutions . . . 57
2.9 Chapter Summary . . . 59
References . . . 60

3 Rapid Product Development in One-of-a-Kind Production . . . 61
3.1 Historical Review . . . 62
3.2 Stepwise Concurrent Product Design and Process Planning Management Module . . . 66
3.3 Incremental Design and Process Planning (IDPP) Module . . . 67
3.4 Basic Concepts, Methods and Control Structure for Production Planning and Control in OKP . . . 69
3.4.1 Product Production Structure (PPS) . . . 70
3.4.2 Production Planning and Control . . . 74
3.4.3 Production Planning and Control Structure . . . 76
3.4.4 Data Modelling for Managing OKP . . . 78
3.5 Chapter Summary . . . 78
References . . . 79

4 Cost Estimate and Control in One-of-a-Kind Production . . . 83
4.1 Review of PD Cost Estimate and Control Techniques . . . 86
4.2 Cost Index Structure . . . 88
4.3 Cost Estimate . . . 91
4.3.1 Generative Cost Estimate Method . . . 91
4.3.2 Variant Cost Estimate Method . . . 93
4.4 Cost Estimation and Optimization Procedure . . . 93
4.4.1 Stage 1: Cost Estimate . . . 93
4.4.2 Stage 2: Cost Optimization . . . 95
4.5 Some Considerations on Cost Estimation . . . 95
4.6 Cost Control . . . 97
4.7 Case Study . . . 98
4.7.1 Case Scenario and System Description . . . 99
4.7.2 Cost Estimate for Making the Case Product . . . 101
4.7.3 Cost Tracing and Control for Making Electric Generator ACSG500 . . . 109
4.7.4 Section Summary . . . 110
4.8 Chapter Conclusions . . . 111
References . . . 113

5 Knowledge Mapping 117
5.1 Scope of Knowledge Mapping Process 118
5.2 Graphical Interface 122
5.2.1 Processing BOM 122
5.2.2 Processing BOO and Establishing the PPS for a Customized Product 126
5.2.3 Processing BOI 135
5.3 Data Structure and Classes Specification 137
5.4 Chapter Summary 142

6 Information Generation 143
6.1 IGM Architecture 144
6.1.1 Input Definitions 145
6.1.2 Output Definitions 149
6.1.3 Process Definitions 151
6.2 Input Demand Specification 151
6.3 User Interface 152
6.4 Creation of Configuration Model 153
6.5 Basic Processes of Information Generation 158
6.5.1 Server Initialization 158
6.5.2 Processing Requests 161
6.5.3 Import Demand Items 161
6.5.4 Generation of Bill of Operation (BOO) 165
6.6 Chapter Summary 167
References 167

7 Adaptive Scheduling and Control of One-of-a-Kind Production . . . 169
7.1 Review of Flow Shop Scheduling 172
7.2 State Space (SS) Heuristic 176
7.2.1 Two Concepts in the SS Heuristic 176
7.2.2 The Steps to Achieve the SS Heuristic 179
7.2.3 The Computational Complexity of the SS Heuristic 181
7.3 Adaptive Production Scheduling and Control 182
7.4 Heuristics Comparison Through Case Studies 185
7.4.1 Case Studies on Taillard's Benchmarks 186
7.4.2 Case Studies on Operator Absence or Machine Breakdown 188
7.4.3 Industrial Case Study 191
7.4.4 An Extension of SS Heuristic 192
7.5 eGantt Chart Module for OKP Scheduling and Control 193
7.5.1 Resource Levelling 194
7.5.2 Resource Allocation under Limited Number of Resource 196

7.5.3 Minimum Cost Project Crashing Heuristic 200
7.5.4 eGantt Chart Module 203
7.6 Chapter Summary 210
References 212

8 One-of-a-Kind Production Management and Control Software System© 215
8.1 Introduction 216
8.2 PPSMaint Module 217
8.2.1 PPS Graphic Interface 217
8.2.2 Enterprise Production Data (EPD) Maintenance and Management. 219
8.2.3 Cost Tracing and Monitoring 219
8.2.4 Production Scheduling and Control 220
8.2.5 On-line Production Progress Monitoring and Control . . . 223
8.2.6 Miscellaneous Functions 223
8.3 PPSManager Module 224
8.3.1 Managing Outsourcing Operations 225
8.3.2 Managing Subcontracting Operations. 226
8.3.3 Managing Shop Floor Operations 226
8.3.4 Managing Work Centres and Work-Teams. 228
8.4 PPSShopFloor Module 229
8.5 Managing Complicated Projects or Products 231
8.5.1 Forward Scheduling—Bottom-Up Scheduling. 233
8.5.2 Backward Scheduling—Top-Down Scheduling. 235
8.6 Chapter Summary 236
References 237

Appendix A: Acessing the CIM Web Site to Download the OKP Management and Control Software System© 239

Appendix B: Relational Database Specification 243

Appendix C: RTPA Notation System 249

Appendix D: RTPA Specification of Architectural Design of The Information Generation Module (IGM) and its Major Classes 253

Appendix E: Industrial Case Study 285

Index 299

Chapter 1
Introduction to the One-of-a-Kind Production System

Jason Myers, President of the Canadian Manufacturers and Exporters (CME), in a speech on October 2008 said:

> In order to compete and grow in light of these new global market realities, manufacturers in Canada need to focus on the rapid development and commercialization of new products. They need to adopt flexible, automated, integrated, and reconfigurable technologies and production systems.

Nowadays demand for customized products has been increasing. In addition to customization, customers also expect the products to be delivered in a shorter time frame and at nearly the same cost as mass-produced items with a relative high quality. This market trend requires manufacturing companies to be able to rapidly and cheaply produce customized products with a required quality level. This requirement is somehow controversial with conventional understanding, i.e. a trade-off between customization and low cost through highly efficient production mode, such as mass production. On the other hand, this market trend also provides the manufacturing companies with new opportunities, particularly for these small or medium sized enterprises (SMEs) in developed countries like Canada. The typical examples are customized window and door manufacturers, structural steel construction companies, special industrial equipment manufacturers, etc. In these manufacturing businesses, the customers normally require the orders to be specified with their specific requirements and to be filled within a short lead-time. The high customization and short lead-time drive the customers to place their orders with a local company instead of an overseas competitor. However, the high cost of producing a customized product through a traditional production mode, e.g. job shop, may turn the customers away from these local companies. Therefore, to keep the high customization, short lead-time and at meanwhile to effectively reduce the production cost become the high priority strategy for manufacturing companies to effectively compete with overseas large and cheap producers. This is particularly true for SMEs in the developed countries, such as Canada, USA, etc.

In October 2006, Canadian Manufacturers & Exporters (CME) organized a conference in Calgary, Alberta to discuss technological strategies to improve the

Y. Tu and P. Dean, *One-of-a-Kind Production*,
DOI: 10.1007/978-1-84996-531-6_1, © Springer-Verlag London Limited 2011

competitiveness of Canadian manufacturing companies. At this conference, many companies indicated that both manufacturing resource planning (MRP II) and enterprise resource planning (ERP) systems were ineffective for their production needs. A survey conducted by us between January 2006 and February 2007 to investigate the technological status of manufacturing companies in Canada supports this observation. We have received valid responses from 78 companies in Canada. From this survey, we can see that 95% (74/78) of the companies have used automated facilities (e.g. computer numerical controlled (CNC) machines) in their production, whereas only 5% (4/78) of the companies have used an MRP II or ERP system effectively in their production. The main reason is that the MRP II or ERP systems have been broadly developed to meet the common needs of various manufacturing companies; and, they do not meet the specific needs of one-of-a-kind production (OKP) companies. According to Wortmann et al. (1997) and Tu (1996a), an OKP company is an advanced manufacturing company that produces customized products within a product domain. The characteristics of OKP can be briefly summarized as, high customization, great uncertainties in production control, complicated and dynamic supply chains, and dynamic production systems. Most of the SMEs in Canada fit this definition. They are competing with large manufacturing companies (often conglomerates) by offering customized products with shorter lead-times and higher quality. However, the major weakness of those SMEs is high production cost, resulting in the movement of production facilities overseas as Canadian manufacturing businesses restructure.

A better strategy for the manufacturing sector to survive and grow is to strengthen OKP while reducing cost. As to our belief, this strategy will eventually lead manufacturing companies, particularly those SMEs to realize the mass customization, which requires a manufacturing company to achieve both customization and economics of scale (Pine 1999). Because customization is in conflict with economics of scale in traditional manufacturing modes, most manufacturing companies today cannot achieve mass customization. As indicated by Svensson and Barford (2002), the main strategy in achieving mass customization is the application of flexible automation (e.g. CNC equipment) and information technologies.

This book presents the concepts, methods, and computer and computation techniques for design and implementation of a production system which is able to rapidly and efficiently produce OKP products. This production system is refereed as OKP system in this book. This OKP system has been studied and developed by Dr. Tu's research groups in the past two decades through research and industrial implementations. Before we fully define and describe the OKP system, we would like to briefly review the traditional production modes in the following sections. From the review, we aim at drawing the differences between the OKP system and the traditional production modes. These differences draw the fundamental requirements for the research and development (R&D) of the concepts, methods, and computer and computation techniques for design and implementation of the OKP system in manufacturing companies, particularly in SMEs.

1.1 Traditional Approaches to Manufacturing

As Pine (1993) pointed out, the benefits of mass production come from economies of scale. These include buying large amounts of raw materials at discounted prices, and making large quantities of standard products to gain efficiencies in production and reductions in waste and hence in unit product cost.

The type of information system framework adopted by mass or batch production companies includes the use of stock numbers for their standard products or modules that are all listed in their catalogues. Customers place orders by choosing products or compositions of standard modules from these catalogues. Orders are then entered into the system. The company selects and delivers ordered items from the warehouse of finished goods. The information system in a mass or batch production company is mainly used for demand forecast, inventory management, supply chain management and financial reports. Typical computer aided production management and control systems are called MRP II (manufacturing resource planning) or ERP (enterprise resource planning) systems. However, Jiao et al. (2006) find a fundamental logical error exists in all kinds of MRP II and ERP systems. That is the separation of bill of materials (BOMs) and bill of operations (BOOs), which leads to a limited success when implementing these systems in manufacturing companies. This is particularly true for most of SMEs or OKP companies where the high customization results in frequent changes of product design, process planning and production routines.

Within the production process in a mass or batch-production company, some of the information is used to drive shop floor equipment and so the use of CNC (computer numerical control) and robotic equipment has evolved. Often these pieces of automation equipment are used in isolation, which produces large inventories of sub-components. Sales analyses and marketing studies are conducted to determine future trends in demand for changes to existing products or the development of new products. These design processes have promoted the development of knowledge-based CAD systems to address problems that designers often deal with in isolation without collaboration/consulting with customers.

1.2 Computer Integrated Manufacturing Systems

Computer integrated manufacturing (CIM) systems can be achieved by the use of information systems, computer controlled manufacturing equipment (e.g. CNC equipment) and material handling systems (e.g. automatically guided vehicles, robots and reconfigurable conveyors). Dedicated manufacturing systems, e.g. mass or batch production systems, are characterized by rigid equipment designed specifically for a product or a restricted family of products. Typically, the standard or repetitive production operations are automated by using dedicated machines or limited use of computer-controlled manufacturing and material-handling equipment. They provide high production rates and low flexibility, and are expensive to change.

One of the key technologies in CIM systems is so-called Flexible Manufacturing Systems (FMSs). According to Matta and Semeraro (2005), FMSs are defined by CECIMO (Commit Europeenne de Cooperation des Industries de la Machine Outil) as automated manufacturing systems that are capable, with minimal human action, of producing any part of a predefined product family. FMSs are generally adopted for production at small to medium volumes, in variable lot sizes that differ in their compositions. System flexibility is usually limited to the family of parts on which the system was conceived. FMSs are characterized by its use of computer-controlled equipment and low volumes or small batch sizes. However, the FMSs are not successfully implemented in manufacturing companies due to its complexity, low reliability and high investment. This is particularly true for SMEs.

The one-of-a-kind production (OKP) system uses some of the methods from FMSs in order to achieve economies of scale and production system flexibility. From this point of view, an OKP system can be viewed as an FMS with a lot size of one. However, it has a much lower automation level than an FMS has. OKP companies often adopt automated and highly efficient manufacturing processes, e.g. flow production lines, CNC machining centres, etc., whereas one-off or job-shop companies normally adopt high flexible but low efficient manufacturing processes, e.g. universal and functional equipment and job-shop processes. However, in OKP companies, the automatic material handling systems are rarely used due to the high investment and low flexibility of the current automatic material handling systems. In OKP, it normally employs one-piece flow technology. To use the current automatic material handling systems in a one-piece flow production factory will be complicated and very expensive.

Fujimoto and Ahmed (2003) discussed the need to consider both process and product in order to provide customized items at a reasonable cost using economies of scale. The variety of product design and production process in OKP is created by differences in basic functions (thermal properties of windows), adaptability requirements (different size and shape of windows), optional functions (windows that open), and non-functional requirements (welded frame or screwed frame). The variety affects efficient management of OKP in the way of high inventories, feeding complexities, excessive capital investment, and change in assembly sequence and complexity in line balancing. If we assume that the FMS would have been designed with these considerations in mind, the information requirements are not the same as in conventional manufacturing modes.

1.3 Mass Customization

Svensson and Barford (2002) discussed the trend toward mass customization and identified that cost is a significant factor in the manufacturing process and it is related to the amount of customization required in products. As mass production has developed since the 1900s, the unit product cost of mass production has dropped dramatically. As products became available, customers started to demand better quality, so in the 1980s there was a push for higher quality products, resulting in rising costs. Costs further increased as customers started to demand more customization in the 1990s. As mass producers tried to introduce more variety into their products, costs increased because of higher inventories and increased setup times. Then the concept of merging mass production with customization was introduced. The phrase of mass customization has been promoted as the goal of reducing costs while maintaining high customization and quality.

Kotha (1996) wrote about the case of the National Industrial Bicycle Company (NIBC) and their venture into mass customization. The paper highlighted the advantages of selecting a product suitable for customization and being the first to market with a customized process. MacCarthy et al. (2003) defined the mass customization model as having six fundamental processes: order taking, product development and design, product validation and manufacturing engineering, order fulfilment management, order fulfilment realization, and post-order processing. McCarthy (2004a) defines mass customization as

> The capability to manufacture a relatively high volume of product options for a relatively large market (or collection of niche markets) that demands customization, without tradeoffs in cost, delivery and quality.

According to Selladurai (2003), mass customization is the integration of mass production principles with processes that produce custom products. Selladurai (2003) believes that mass customization is a trend for manufacturing companies in future. Selladurai (2003) also identifies a number of strategies required to be successful with mass customization. Selladurai (2003) mentions effective and rapid integration, specifically the need to quickly record customer needs and specifications, to change these into customized product designs and processes and integrate these to produce the final product when the customer requires it. Selladurai (2003) also emphasizes automating as many tasks as possible and integrating these tasks with the communications network and databases that provide the necessary information flow to make the process work effectively. Selladurai (2003) concludes that growth in the use of mass customization (MC) by SMEs would continue as customers demand ever-increasing customization. According to Frutos and Borenstein (2004):

> An MC system is highly dependent on well designed information systems that provide direct links among the main agents involved in the customization process, namely customer, company, and supplier.

They proposed a framework in which the customer could interact quickly with the company to obtain required customized products. The framework is based on an Internet application that would allow the customer to specify options and parameters of the required product. Their framework involves customers selecting from an online database of products and options. The system will then determine the feasibility of manufacturing and have the product made and delivered. At the same time, any new product variations can be added to the catalogue. Interaction with suppliers is important during both the design and production phases.

1.3.1 Current Status of Mass Customization Systems

McCarthy (2004a) in his editorial to the special issue of Production Planning and Control on mass customization says:

> However, despite the strong interest in configurationally research in the business and operations strategy area, there are few works that develop and propose models for understanding how to mass customize.

The benefits of mass customization (MC) have been identified by many researchers, and the characteristics of MC have been defined by others. Several researchers have proposed strategies to achieve mass customization (Kotler 1989; Pine 1993; Agrawal et al. 2001; Berman 2002; Tseng and Piller 2003). Such strategies include different manufacturing processes (Kotha 1996; Feitzinger and Lee 1997), infrastructures for ERP, the Web and e-Commerce, and data structures like generic bill of materials (GBOMs), generic product production structures (GPPSs) and generic bill of operations (GBOOs). McCarthy (2004b) maintains that the outstanding challenge is to design and operate integrated systems. McCarthy emphasizes one of the major requirements of the system as the ability to handle increasing variety and uncertainty in product information without affecting lead-time, cost or quality. As a second major requirement, the system should have the ability to transform and assemble materials to produce custom products (Ahlstrom and Westbrook 1999; Salvador et al. 2001; Tu and Xie 2001; Duray 2002). He maintains that the adoption of mass customization is inhibited by the lack of explicit studies on how to design and operate manufacturing systems accordingly. McCarthy (2004a) states that

> It is a system design problem that involves searching a space of strategic configurations, whose elements are different combinations of manufacturing capabilities

In the editorial, McCarthy (2004a) identifies that although the essence of mass customization is relatively fixed, it will mean different things for different companies at an operational level because of the number of complex factors that vary from company to company. No single solution can be applied and this suggests that a number of configurations will be needed to solve the problem.

McCarthy (2004a) continues his editorial by examining the contributions made to this problem by the eleven papers in the journal. Together, he claims, they make contributions that help address the "How" question by providing rich, and theoretically grounded descriptions of how manufacturing systems should be configured to achieve certain levels of mass customization. Comstock et al. (2004) research a Swedish mobile telephone company and try to move from mass production to mass customization. From their conclusions it is evident that the market

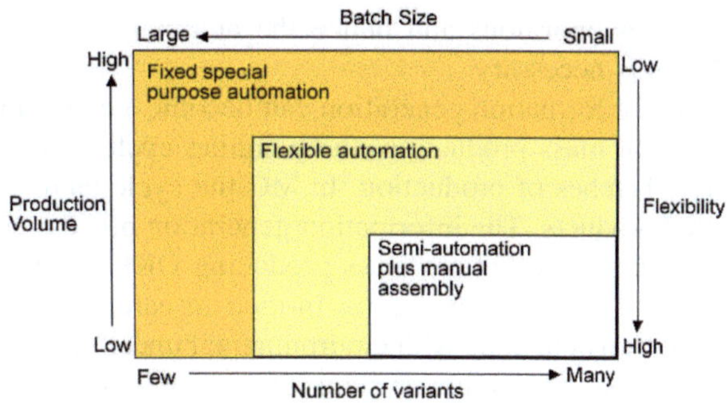

Fig. 1.1 Assembly option

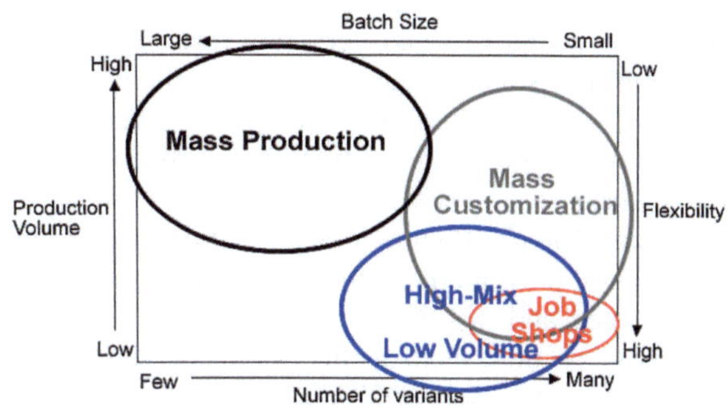

Fig. 1.2 Manufacturing strategies

place does not demand MC but rather some small variations in standard products that are ordered in large quantities. However, their work does confirm that to move to MC a company needs a simplified, customized, and automated data handling process that focuses on reducing lead-time and errors. They also found that MC initiatives in companies with mass production traditions can be successfully implemented by modifying their products, processes and systems appropriately.

Figure 1.1 is a modification of Rampersad (1993) and Heilala and Voho (2000), which addresses the choice of assembly principles as it relates to flexibility, number of variants, production volume, and batch size. What is not quantified is the amount of data required to run these systems and the constraints to be placed on the operation.

In reviewing Fig. 1.2 from the perspective of manufacturing strategies such as mass production (MP), job shops (JS), high mix/low volume (HL), and mass customization (MC), it is evident that there are overlaps among these strategies.

Considering data requirements for information to support the implementation of these strategies, it implies that as a company approaches MC, there is a significant increase for information required to manage and control the production. The increase in information volume comes about the fact that each custom product requires a different (or modified) version of the product family's bill of materials (BOMs) and bill of operations (BOOs). As a company increases its throughput of producing those custom products, the task of producing the BOMs and BOOs becomes enormous and hence the automation of data generation of BOMs and BOOs is necessary.

The information generation can become a constraint when the planning cycle is short. In mass production, the planning cycle tends to be long with long runs of large batches of production. In MC the cycle tends to be short with small runs of OKP products. The information generation process can become a critical issue for the companies that are mass-producing OKP products.

Tu et al. (2004) recognize in their research that OKP company managers have to find ways to cope with environmental uncertainty and increasing product variety through a production system that is able to produce OKP products without affecting lead-time, cost or quality. We also recognize that this system can be achievable with flexible manufacturing technology and information systems. However, we have found very few research papers with theoretical frameworks and empirical studies. This is one of the main motivations for publishing this book in which we provide not only a theoretical framework with a rigorous specification to achieve mass customization, but also case studies based on the use of the framework.

Tu and Xie (2001) employs an empirical study to establish that re-engineering setup times, preventative maintenance and cellular manufacturing have a statistical significance and positive impact on mass customization capability. In the case studies to be described in this book, the setup times are effectively reduced by employing automated CNC machines that are driven by a computer aided production control system. The benefits of cellular manufacturing are achieved by using flow production lines dedicated to a family of products requiring similar processes. The application of flow production lines makes material handling more efficient and reduced waiting times.

Salvador et al. (2004) reviews mass customization from the total supply chain aspect. They point out that MC comes with a "price"—higher inventories, higher manufacturing costs and longer delivery times unless specific steps are taken to address these issues. In this book, we will use case studies to validate that using the techniques and computer system presented in this book it is possible to reduce this "price" significantly. Salvador et al. (2004) discusses a number of strategies discovered in their research. The first is based on modular production. Although Salvador et al. (2004) claim that there is a saving in decoupling the design and logistics tasks, there will be a corresponding increase in work-in-progress (WIP) inventory costs. A method of postponement is discussed where customization takes place at the latest stage possible, close to customer delivery. Salvador et al. (2004) concludes from multiple case studies that the postponement method is seldom used. Although there is a significant use of modular products, it could not be determined if the supply chain has to be closely coupled or completely decoupled.

A windows and doors manufacturing company in Calgary, Alberta, Canada, which produces custom windows and doors, uses close coupling where custom subcomponents are supplied in order to synchronize assembly. The company has moved away from any kind of modularization of components to maintain low WIP inventory.

When researching the economics of mass customization, Piller et al. (2004) determine that there are potential savings by involving the customer in the process. They claim this potential comes from three sources: (1) from postponing some activities until the order is placed, (2) gaining more accurate information about market demands, and (3) from the increased loyalty felt by the customers. As to our knowledge, more and more companies have used these strategies to develop their production systems. In a custom windows and doors manufacturing company in Calgary, for example, customers place orders using a configuration application system over the Internet. The final specifications and delivery due date of a customer order are confirmed 5 days before the requested delivery date, and then the confirmed order is sent for production. The system would accept any design changes or delivery date change until the order is confirmed. The company's customers have developed a sense of loyalty to the company through this type of high customization and flexibility. The customers are also encouraged to participate in discussions about future market trends, product designs and future orders. The information that the customers provide through their participation is input into the company's forecasting for long-term planning of capacity and resources.

1.3.2 IT Support for Mass Customization

In order to create a configuration model for a mass customization (MC) company, MacCarthy et al. (2003) emphasize the need to correlate the MC strategy with support technology. The objective of the research carried out by van Asperen et al. (2008) is to determine whether the various operational processes of MC can be adequately supported using IT (information technology). For this purpose they discuss various MC strategies based on a known taxonomy and analyze the operational processes for each form to obtain a list of functional requirements.

A number of different models have been proposed (Pine 1993; Lampel and Mintzberg 1996; Pine and Gilmore 1997; Da Silveira et al. 2001; Duray 2002, 2004; MacCarthy et al. 2003; von Blecker and Friedrich 2006). Through a literature review, van Asperen et al. (2008) conclude that there is no generally accepted taxonomy of MC.

Their conclusion is also supported by MacCarthy et al. (2003). However, MacCarthy et al. (2003) point out that those companies in the same category should benefit from the same technologies and they identify the following three common features of MC:

1. The time between processes in MC is shorter than for standard production.
2. The use of fixed or flexible resources is determined by the extent to which processes, suppliers and materials vary.
3. The use of call-offs or design-for-orders is determined by the extent to which products will be sold repeatedly.

van Asperen et al. (2008) also define the five operational levels of MC

1. Catalogue MC (ISP, Football Club seating)
2. Call-off MC with fixed resources (Dell Computers)
3. Call-off MC with flexible resources (Motorola's Pages)
4. Design-per-order MC with fixed resources (Morningstar—investment strategies, Lightning Source—one-of-a-kind books)
5. Design-per-order MC with flexible resources (ic3D custom jean manufacturing)

They claim that different levels of MC will have different functional requirements for IT support. MacCarthy et al. (2003) also summarize the six operational processes of MC, i.e. order taking and coordination, product development and design, product validation and manufacturing engineering, order fulfilment management, order fulfilment realization and post-order processes. Both van Asperen et al. (2008) and MacCarthy et al. (2003) conclude that the post-order processing does not significantly differ from the traditional production modes and consequently they do not consider it in their research. However in OKP, the service and warranty is considered as a part of the post-order process because of the requirements for replacement of custom parts. It is obvious that from the IT support requirement point of view there is a significant difference between servicing a catalogue product and making a design-for-order product. Therefore, the research and development (R&D) of the IT system for OKP system is different from the current R&D of MC systems.

van Asperen et al. (2008) introduce a number of prototypes and applications that would provide IT support for various levels of MC. For order entry, they identify a Mass Customization Information System (MCIS), which is a Web-based object-oriented system developed by Frutos and Borenstein (2004). The system has three main components, i.e. a custom interface, a database and a database management system. It provides a real time price, stores customer information and includes links to a production system. For occasional customer orders, the parts and materials can be identified at order entry.

For resource planning and supply chain management, van Asperen et al. (2008) discuss the use of an electronic link with suppliers to support rapid sourcing. They also review Turowski's agent-based e-Commerce system (Turowski 2002), Wang and Shu's fuzzy-based SCM (supply chain management) system (Wang and Shu 2005), and Chen et al.'s fuzzy-based SCM system (Chen and Hall 2007).

Additional support is required in the areas of flexible production and product validation. Cao and Dowlatshahi (2005) propose the use of agile manufacturing that adapts processes and creates new designs quickly. Reducing setup times is important in these systems. Cheng et al. (1998) describe Agile Manufacturing based on the use of the Internet, fuzzy logic and a neural network.

For design-per-order, more flexible order realization systems are required. This increases the IT support requirements on the client side. The CAD (computer-aided design) systems are useful in these circumstances but the new designs have to be quickly translated into production schematics. This can be achieved with the CAM (computer-aided manufacturing) systems. Ito et al. (1988) and Srinivasan and Fischer (1996) describe the means to interface the CAD/CAM systems with manufacturing information management systems.

The time between design and production needs to be reduced in the design-per-order model in order to maintain delivery times. Lee et al. (2004) propose the reuse of design with a Dynamic Data Interchange Schema (DDIS) that employs the case-based reasoning method.

At the fifth level of MC, products are engineered-to-order, which requires all the IT support as mentioned above. It often means that customers and suppliers are involved in the product development process. Therefore, it requires further the adaptation of MCIS and SCM system. However according to van Asperen et al. (2008), there is very little engineer-to-order companies that meet the requirements of MC according to McCarthy's definition of MC (McCarthy 2004a).

van Asperen et al. (2008) conclude that multiple modes could co-exist within a single company, adding additional challenges and integration of various systems. They emphasize that the technologies of MCIS and DIS (Data Interchange Schema) are academic prototypes and only use for proof of concept. As a concluding statement, they write:

> There is still a long way to go before the ultimate goal of commercial available, seamlessly integrated IT support for MC across all processes will be achieved.

They further conclude that the integration of these systems presents a major challenge to the successful IT support of MC and hence it is an interesting research topic. There will not be a single solution for all models of MC. A framework that can be used as a basis for developing appropriate IT support for achieving MC is necessary.

1.3.3 Information Systems and Modelling Techniques for Mass Customization

The current state of mass customization (MC) and computer integrated manufacturing has been discussed in the previous sections. From the discussion, it is evident that the MC is common goal to be achieved by manufacturing companies, particularly for those so-called make-to-order manufacturing companies which are often SMEs (small or medium sized enterprises). To achieve MC, a company first needs to determine which MC strategy best fits its business, market and products. Then it needs to determine how the various processes will work and what IT support is required. From Sect. 1.3.2, we can see that the researchers have concluded that there is no single integrated system that will support all the processes. This conclusion leads to a challenge and yet interesting research problem. This problem may be solved by either integrating a number of systems as suggested by van Asperen et al. (2008) or adopting a framework that can be modified to the selected MC model.

In this section, the research efforts for developing an IT (information technology) framework to support an MC strategy are reviewed. Relevant information and data modelling techniques for developing such an information system framework are discussed. As a result, the background for the OKP system is described.

1.3.3.1 Central Concept in MC

Knowledge about product design and process is central to any manufacturing system. It is particularly critical in the case of MC, in order to support the high demand for information, and the need to shorten the time between product design and product production. In practice, the product design is expressed in a bill of materials (BOMs) and the manufacturing process and operation routine are recorded in a bill of operations (BOOs). These two data records are wrongly separated in current MRP II and ERP systems which limit the application of these systems in highly customized manufacturing companies, i.e. OKP companies or most of SMEs.

The concept of combining product and process data models has been around since the 1980s. Hayes and Wheelwright (1997) propose a product-process matrix. Ahmead and Schroeder (2002) determine that to maintain the benefits of flexibility and efficiency, process structure should change to meet the customer requirements. Tu and Xie (2001) propose a product production structure (PPS) that combines the design and process data to support rapidly producing OKP sheet metal parts/products.

The research by Jiao et al. (2006) about process platform planning produces an up-to-date discussion on this topic. They review the background literature and discuss the development of BOMs to GBOMs (generic bill of materials) by Hegge and Wortmann (1991) as a means of specifying product variants. They also explore the unification of BOMs and routings up to the point when Jiao et al. (2006) put forward generic bill-of-material-and-operation (GBOMO) that achieves this unification. The ultimate goal of the work of Jiao et al. (2006) is to develop a process-planning platform that supports the development of product families, and associates the process families together to produce high variety of product while maintaining economies of scale.

1.3.3.2 Generic Product Process Structure (GPPS)

The basic concept of Jiao et al.'s platform (Jiao et al. 2006) is to take a generic product structure (GPdS) and a generic process structure (GPcS) and combine them into a generic product and process structure (GPPS). Given a specific input for the variables, a GPPS could be created to produce a BOM for the product variant and a BOO for the process variant. The paper continues with a description of the class structure and its interrelationship with a GPPS.

Jiao et al. (2006) provide a good example of how the platform would work for a specific product in a product family. Following the procedure, the GPPS produces an operational description associated with each operation class in the product. Variety parameters and their instances are defined in terms of constraint types and their rules. The end-result produces a process structure and BOM for a specific example.

Jiao and Zhang (2005) confirm that as the number of product variants increased so did the use of CNC and automated equipment. They suggest that if not managed correctly, a chaotic environment would occur in which costs would escalate and production delays would increase. Consequently, it would jeopardize the achievement of MC efficiency.

1.3.3.3 Weaknesses in Current GPPS Platform

The use of GBOMO and/or GPPS to create a platform is very similar to the data modelling method that the OKP system uses, i.e. PPS. However, the PPS data modelling method in OKP system addresses more of the specific issues that relate to how to achieve mass customization (MC) in an OKP company. As a result, the class structure is more complex and a number of functional areas are expanded. The use of a work centre and its cycle time and setup time to plan and execute a MC strategy does not provide sufficient granularity in information for the appropriate support of OKP processes, e.g. where and when waiting-time occurs in the processing cycle and when to re-allocate resources needs to be clearly identified in OKP. In mass production, it is appropriate to use cycle time at the work centre level because it is producing large volumes of the same product at a consistent average speed. However, in OKP, each product is different and therefore can take a different amount of time. The way to handle this is with the following structure (Fig. 1.3).

In this case, the total setup (T_{setup}) time for the work centre is:

$$T_{\text{setup}} = \sum_{i=1}^{n} O_{\text{setup}_i} \tag{1.1}$$

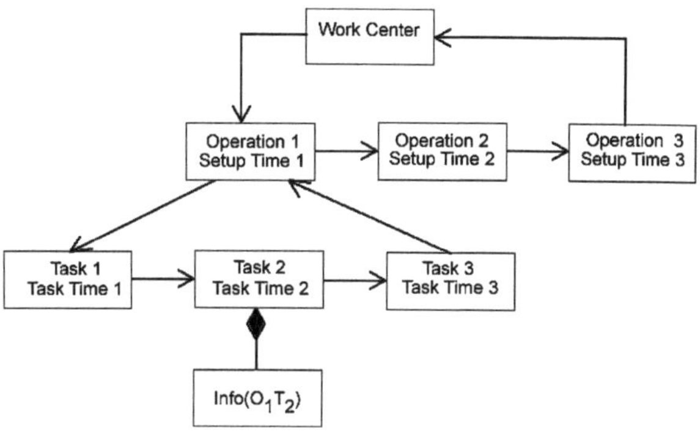

Fig. 1.3 Variable work centre cycle times

where O_{setup} is the setup time for each operation and n is the total number of operations performed to make the product variant, which may not always be the total number of operations in the work centre.

The total cycle (T_{cycle}) time is:

$$T_{cycle} = \sum_{i=1}^{n} \sum_{j=1}^{t} T_{task_{ij}} \tag{1.2}$$

where n is the number of operations, t is the number of tasks in each operation and T_{task} is the time to complete the required tasks for the specific product variant.

This structure also provides a means of defining the sequence of operations and tasks within operations. The tasks may be performed by humans or machines, but both the humans and machines will require specific information. For example, if an operation is to weld four pieces of material together to form a frame, the task sequence may be

T1—load materials
T2—weld frame
T3—unload frame from welder

The operator needs to know what pieces are to be loaded into the welder. The pieces could be the next four in sequence on a conveyer or in a numbered slot in a cart. The welding machine needs to know what size of frame to weld. The information can be input by a human, but it would be faster if it is downloaded from a computer system. Finally, the operator needs to know where to unload the frame to, e.g. a conveyer, a cart, a buffer, the next workstation, etc. All the information as mentioned here can be modelled by the PPS of a product and can be downloaded from a computer system.

Other critical aspects of GPPSs are the variables and constraints (or rules). Unless they are maintained in the GPPS, they have to be stored and maintained separately. Currently, this has been left to the role of the program in the form of code and logic. In this book, we propose that the variables and constraints be part of the PPS using the structure as shown in Fig. 1.4.

The benefit of integrating the variables and constraints into the PPS which is a combined data structure of BOM and BOO is easier maintenance, which can be performed by the knowledge worker, usual the process engineer.

There are other processes required by OKP system that are supported by this data structure. For example, it has been discussed earlier in the chapter that a rich interface is required to enable the customer to accurately and quickly specify the required product variants. This data structure can provide the information to the client program so that it can dynamically generate part of or the entire interface. Based on this interface, a customer can select and specify the specific requirements of his/her order, such as the dimension, colour, function, materials, etc. The customer's specification will be automatically recorded in the PPS of his/her ordering product as parts of the necessary product design and production data of the order. The customer interface and specification of the order formulate the

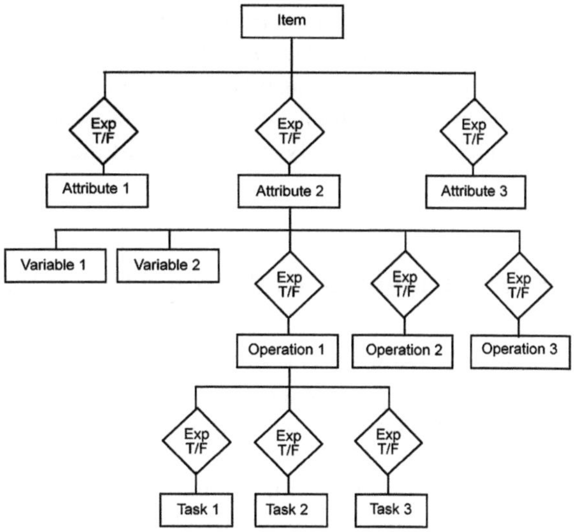

Fig. 1.4 Variable and constraint structure. *Exp* expression or condition, *Attribute* base product and/or optional feature, *Variable* run time value for expression or condition

important parts of the automatic data generation process that will be further discussed in Chaps. 5 and 6.

1.4 Risks and Disadvantages of Mass Customization

The risks and disadvantages of mass customization involve reduced quality, increased costs and delays in delivery to customers.

The two risks of mass customization identified by Svensson and Barford (2002) are reduced quality and excessive inventory.

1. Complexities and variations in the custom products often lead to mistakes in BOMs and inaccuracies in work instructions, reducing the consistencies of quality.
2. The need for custom inventory together with short lead-times often leads mass customization to over-stock.

These risks can cause high operational costs which result in failure of implementing mass customization. However, these risks may be mitigated by having accurate BOMs, BOOs, work instructions and a responsive supplier network.

The other risks for causing high operational costs may include:

1. When processes do not run smoothly, costs can increase. As a result, constantly monitoring and checking of efficiencies is required.
2. The variety of product demands a lot of information (individual BOMs and BOOs for each of the products). The information can be costly to produce, and

it would lead to production delay and increase production cost if the information generation process is not automated and it is done manually.
3. On-time delivery can also be at risk, if the supply chain of critical parts does not work in collaboration and just-in-time mode. Waiting for custom parts from suppliers can endlessly increase lead-times.

1.5 The One-of-a-Kind Production (OKP) System

In summary, the mass customization has become a common goal to be achieved by manufacturing companies today. It is also the goal to be achieved by the OKP system. However, the current research and studies on mass customization have been focused on the definition of mass customization (MC) (Pine 1993), specifications of the special requirements of MC from information and management system view (von Blecker et al. 2004), supply chain view (von Chandra and Kamrani 2005; von Holweg and Frits 2004), etc. The main methodology in these MC approaches is the Internet-based collaborations among the members in a manufacturing consortium or direct subcontracting managerial strategy (Tseng and Piller 2003). Under the direct subcontracting strategy (DSS), a master company, which deals with the customer directly, tries to decompose a customer order into a number of modules that can be cheaply made by subcontractors and then subcontract these modules to the selected subcontractors through using the e-commercial and/or Internet-based collaborative product design and production management techniques. In extreme cases, the master company can do nothing rather than subcontracting, managing subcontractors and managing customer relations since it can even subcontract the final product assembly and delivery to subcontractors. Therefore, the research on mass customization has placed its emphasis on the concepts, methods and technologies for supporting and managing DSS to achieve mass customization. The DSS-oriented MC technologies do help manufacturing companies to achieve mass customization. However, the disadvantages of these DSS-oriented technologies are:

1. The DSS requires a large and complicated supplier network or supply chain. It is often difficult to manage this supply chain to get all the subcontracted modules to deliver just in time. Especially in SMEs due to their limited buying power and discrete customer orders, they hardly control their suppliers and subcontractors to deliver the goods in time.
2. The DSS continues to lead the manufacturing business to move to low wage countries, which is common and well-known problem for manufacturing business in developed countries like Canada.
3. To adopt DSS, a company has a risk to lose its competitive strength and market share if they depend very much on the supply and subcontracting network and a supplier or subcontractor may take over their business and market share.

Therefore, in our work on the R&D of the OKP system, the emphasis has been placed on how to design and implement a production system in a manufacturing company to improve its ability to rapidly and mass-produce customized products, i.e. the OKP system.

The OKP (one-of-a-kind production) phenomenon was originated in heavy industry (e.g. ship building, power plant building, etc.), which did not follow a traditional product development cycle (i.e. sequential design, prototyping, testing, and production). Instead, OKP product development has the following features:

1. Singular successful development, i.e. no prototype is made throughout the development life cycle,
2. Product design, testing, and production are carried out concurrently, and
3. Product requirements may be changed at any time within the product development cycle due to continuous customer involvement in production development and unexpected disturbances from inside and outside of the company, e.g. operator absence or machine breakdowns, job insertion/cancelation, and unreliable supplies.

The EEC (European Economic Community) research program, "ESPRIT basic research action 3143—Factory of Future (FOF) production theory" (Rolstadås 1991), envisages that OKP would become a novel manufacturing system in twenty-first century along with a clear market trend toward customization and responsiveness. Today, more and more manufacturing firms, e.g. structural steel construction companies, custom window and door manufacturers, special equipment manufacturers, etc., are moving toward OKP to gain greater variety and rapid product development, superimposed on traditional requirements of high quality and low cost. The ability to mass-produce customized products or mass customization is critical for a manufacturing company to implement the OKP system successfully.

To improve the ability of mass customization, an OKP company normally designs a base product and tests it in its basic form. The product is tested to ensure that it meets technical specifications and the prescribed regulatory requirements according to the geographic location where it will be installed or used. The design of the base product is also based on market studies that determine appearance and functional requirements. Once the base product is designed and tested, limits of the product variations are determined and established as design constraints. These constraints include such properties as sizes (minimum and maximum), shapes, combinations, and options that are inclusive and/or exclusive. Consequently, various customized products that meet technical specifications and regulatory codes of the base product can be designed and mass-produced through variations of the base product within these pre-defined limits. By means of the OKP system to be presented in this book, the OKP companies will gain much higher production flexibility than the traditional batch production companies that normally use standard product modules to achieve limited customization with a required minimum production batch size to compensate for its production costs.

In terms of the base product, the product improvement program and process can also be defined. This can include warranty and maintenance services programs, the ability to add on new improvements after installation or complete replacement programs. This process should be a collaborative process that includes the customers and suppliers so that any improvements developed by suppliers are included as soon as they become available. Customer involvement ensures that new requirements and expectations are registered once they occur. This collaborative process leads to the design improvement process becoming shorter and more effective.

Due to the advantages of the OKP system as mentioned above, the OKP system is becoming a promising manufacturing paradigm, particularly for small or medium sized enterprises (SMEs). As summarized by Tu et al. (2006), the OKP system normally demonstrates the following characteristics:

1. "Once" successful approach on the product, i.e. no prototype or specimen will be made in OKP, and the batch size can be one;
2. Product is usually designed by modifying and combining existing products to avoid the risk of long lead-time and high cost for developing a radically innovative product;
3. Frequent changes of product design, manufacturing processes, and production schedules due to customer involvement in product development and unexpected disturbances from inside and outside of the company;
4. Mix-product production;
5. Frequent changes of production systems to adapt customized product production requirements;
6. Optimal utilization of technologies and resources to continually improve production efficiency and reduce production costs.

Dedicated manufacturing systems, e.g. mass or batch production systems, are characterized by rigid equipment designed specifically for a product or a restricted family of products. These production systems have high production efficiency but low production flexibility. They are expensive to change due to the application of dedicated machines and dedicated production lines.

During the 1980s and early 1990s, the Flexible Manufacturing System (FMS) was widely researched and developed to improve the flexibility of manufacturing companies. As mentioned in Sect. 1.2, the FMS is characterized by its use of automatic material handling systems (e.g. robots, AGVs, conveyor, etc.), CNC (computer numerical control) equipment and low volume or small batch sizes. Due to the required large investment, complicated technology support (e.g. complicated programming requirements) and low reliability, FMS has achieved very limited success in manufacturing practices today. However, the high efficiency (achieved through high automation) and flexible equipment (e.g. CNC equipment) in FMS has been widely adopted by OKP companies to achieve mass customization. From this point of view, the one-of-a-kind production (OKP) system may be viewed as an FMS with lot size of one and a lower level of automation and investment than an FMS. The OKP system rarely uses automatic material handling systems. The OKP system is able to mass-produce customized products under a base product. Traditional FMS requires a minimum batch size to compensate for its setup and programming costs during the product switch over, whereas the OKP system has a very short setup time and no programming cost.

Since an OKP system is able to produce a one-of-a-kind order, it is often confused with one-off and job-shop manufacturing companies. To clarify the confusions, we describe the major differences between OKP companies and one-off or job-shop manufacturing companies as follows:

1. OKP companies are product-oriented (i.e. the base product template as mentioned above) manufacturing companies whereas one-off or job-shop manufacturing companies are capability-oriented manufacturing companies.
2. OKP companies often adopt automated and highly efficient manufacturing processes, e.g. flow production line processes, whereas one-off or job-shop companies normally adopt high flexibility but low efficiency manufacturing process, e.g. universal equipment and job-shop or functional processes.

Due to these differences, OKP companies are able to effectively compete with large mass production manufacturing companies and gain market shares through customization, fast delivery and low cost production. On the other hand, due to lack of a major product to be offered to a market, the job-shop or one-ff manufacturing companies are difficult to gain the reputation in the market and consequently this limits their growth and competitive advantage to gain a market share. Therefore, the job-shop or one-off manufacturing companies are often small and hardly compete with large manufacturing companies.

Although the OKP system has advantages over the traditional production systems, e.g. batch or mass production systems, the characteristics of OKP result in great uncertainties and complicated product data flow, which leads to a longer product development cycle and higher production costs. As mentioned above, to increase the mass customization ability, a lot of manufacturing companies are adopting the DSS which leads to virtual manufacturing systems (VMSs). A VMS consists of manufacturing resources belonging to different physical manufacturers. Following the trend of moving toward to VMSs, the uncertainties increase and product data flow becomes even more complicated. The commercial computer-aided design (CAD) systems and the production management and control systems (i.e. MRP II and ERP systems) either overlook these problems or are not developed to meet the needs of the manufacturing companies, particular OKP companies. The DSS and VMSs require good communication channels and efficient electronic data interchange (EDI) methods to distribute manufacture of the customer order and manage the outsourcing parts or final product. The distribution of manufacturing further complicates data requirements and the whole supply chain management.

MacCarthy et al. (2003) identify the general production processes in a manufacturing company. These processes are summarized in Table 1.1.

Central to these processes is manufacturing engineering knowledge. Without accurate information about the bill of materials (BOMs), bill of operations (BOOs), engineering constraints and data requirements, it will be impossible to manage a production system to mass-produce custom products. These pieces of

Table 1.1 The general production processes

Process	Definition
Order taking and coordination	Defining the order and product specifications
Product development and design	Designing the product and its compliance with internal and external standards
Product validation and manufacturing engineering	Confirming manufacturability Producing the BOM, routing and processing instructions
Order fulfilment management	Managing the order fulfilment by determining when production can start Managing the value added supply chain Scheduling, monitoring and controlling the process
Order fulfilment process	Coordinating external supplies needed Executing the production and delivery
Post order process	Installation Warranty claims Service

information are required for all the processes from order taking to service. Order specification becomes complex and accuracy is essential.

1.6 The Requirements or Constraints for Achieving Mass Customization in an OKP Company

If an OKP company is going to achieve the mass customization through improving its internal production system efficiency flexibility rather than depending mainly on the direct subcontracting managerial strategy, it faces a number of constraints. These constraints set up the special requirements for the R&D of concepts, methods and computer and computation techniques for design and implementation of a novel production system that is able to mass-produce the customized products within a product domain, i.e. OKP system for mass-producing OKP products.

1. To achieve high production efficiency, the standardization of work is required. The standardization of work consists of three elements (Womack and Jones 2007):

   - Production cycle time (or Takt time), which is the inverse of the throughput rate at which the products must be made in a process to meet customer demand in the right amount and at the right time. Otherwise the overstock of finished goods will occur or there will be delays in delivery.
   - The precise work sequence in which an operator performs tasks within Takt time.
   - The standard inventory (including the WIP units in machines) is required to keep the process operating smoothly.

In OKP, due to high customization, it is impossible to achieve standardization of these three elements in the same manner as in batch or mass production. The standardization of work in OKP is enabled by the automatic production data generation process and accurate production management and control.

In mass production, a single design document and single set of instructions can produce large quantities of products. In an OKP company, a single design and instruction set produces one customized product. This difference has a significant impact on a manufacturing process and its performance. Mass production is based on economies of scale and this applies to the information required to operate the manufacturing process. This is demonstrated by the fact that one design and one set of instructions can be used repeatedly to produce the same product with the same process. It can take at least several hours to produce one design and instruction set, and then these designs and instruction sets are used for a long time-period to produce a large quantity of products. Therefore, the effort to produce the design and instructions is viewed to be small compared to the output.

In an OKP company, each of the customized products requires a unique set of product design and production data. Following the traditional way of manually generating the product design and production data, it would be time consuming. This is in fact one of the reasons why the cost of a customized product is normally much higher than the unit cost of mass-produced standard products. To produce a large number of customer orders in a production day, e.g. a windows and doors manufacturing company in Calgary produces about 2000 custom windows and doors daily, this traditional way of data generation would cause a big workload to its engineering department and hence the high cost of product. To solve this problem, in the OKP system, a novel concept, i.e. automatic information generation, is proposed in Chaps. 5 and 6.

Production planning and scheduling for traditional manufacturing systems is usually based on replenishment of stock as a result of forecasted demand that matches the production capacity of the plant. MRP I (materials requirement planning) is used to determine the demand for purchased parts and production orders. The factory operates on the basis of maintaining maximum utilization of resources. Under the philosophy of MRP I, the production flow is pushed in a manufacturing in a relatively large batch size. This practice results in large quantities of work-in-process (WIP) and finished goods inventories. In comparison, OKP companies often use an engineer-to-order (ETO) or make-to-order (MTO) marketing strategy and normally employ a pull production system. This practice minimizes WIP and finished goods inventories but creates an issue for production scheduling and resource planning. This issue can dramatically reduce the production efficiency and resource utilization. In order to overcome this issue it is necessary to have an accurate production data collection and control method and an adaptive production scheduling and control system. With this method and the system, the OKP company is able to schedule and adjust the production throughput so that the promised due dates of the production orders can be met, the inventory levels are properly controlled, and the resource are optimally utilized. The adaptive production scheduling and control system is described in Chap. 7.

2. To implement the OKP system in a manufacturing company, the IT support of mass customization is necessary. However, van Asperen et al. (2008) point out

   > There is still a long way to go before the ultimate goal of commercially available seamlessly integrated IT support for MC across all processes will be achieved.

   Most systems that attempt to address the issues of mass customization are modified versions of tradition manufacturing systems (TMSs) or a group of systems and/or programs that do not integrate well. Not long ago, the software companies had developed the commercial software packages (e.g. MRP II or ERP) for production management and control that only addressed part of the manufacture process, usually focused on the front part (quotations and order entry). It is only in recent years that these software companies have started to address the manufacturing process. We often see that manufacture companies use a commercial software package plus in-house developed software programs for their unique manufacturing problems or do not adopt any of these commercial computer software packages at all. The combination solution is costly since it requires in-house program development. It is also inefficient since the complicated data transfers, updates and conversions between a commercial software package and the in-house developed programs are frequently required. This solution is hindering manufacturing companies to achieve mass customization. This is particularly true in SMEs. To solve this problem, we have developed the OKP management and control software system© which is presented in Chap. 8.
3. The initial capital cost of buying automated equipment and an integrated network of computers is high. This capital cost can offset by reduced cost of inventory of finished goods and spare parts. Therefore, from a long term of point of view, it is worth for a manufacturing company to move to the OKP system.
4. The uncertainties of demand and production times in OKP can lead to chaos in the plant, resulting in high costs, low quality and long delivery times. It is difficult to forecast resource requirements for the company and suppliers. Special attention is required to address this important issue. Allocating resources to a production schedule can be difficult as the product mix can constantly change causing resource requirements to vary frequently. In Chap. 7, a computer-aided production scheduling module, named electronic Gantt chart or eGantt chart module is presented to solve this resource planning problem.
5. The forecasting of demand in OKP companies is different from the one in batch or mass production companies. Information relating to sales and resource planning in mass production tends to be both linear and cyclical in nature. Products usually have a predictable life cycle and as customer demand increases and decreases, resource requirements tend to respond in a linear manner (Krajewski and Ritzman 2002a, b). In OKP companies, these relationships are often not linear, making the prediction of demand and resource requirements more difficultly (Yan et al. 2002; Jin and Thomson 2003). A number of researchers have proposed various methods to address this problem,

including the use of aggregate planning (Feiring 1991; Vandaele and de Boeck 2003), AI (artificial intelligence) techniques (Ming et al. 1999; Kubat et al. 2003), fuzzy logic (Kuo et al. 2002) and neural networks (Priore et al. 2003). One of the ways to overcome these problems is to have a means of recording detailed history of production and analyzing this information in a way that helps with predictive models. These models can use conventional techniques or more advanced technologies like neural networks.

6. The quoting process in OKP companies is different from the one in batch or mass production companies. For mass production of standard products, a large quantity of similar standard products is duplicated with the limited number of different parts. The opposite is true for OKP, as the batch size tends to be one and the number of varieties of products and parts tends to be large. This creates a challenge for quoting an OKP product rapidly and accurately. In Chap. 4, we present a computer-aided quotation module to solve this problem.
7. To place an order to an OKP company, a customer needs to provide more detailed information for a customized product than for a standard product from a batch or mass production company. For standard products, usually a product code is used to identify the product being ordered and is followed by a batch size or quantity. For customized products, however, the information required could include options, sizes and/or features. To facilitate this function, a number of Internet-based customer interface computer systems have been developed. By using these interfaces, a customer can even virtually configure his/her product and hence to assist the customer to place a correct order. These customer interface systems can range from relatively simple option selection tools to highly sophisticated systems that apply three-dimension (3D) CAD modules and computer simulation modules.
8. The product and production data management in OKP companies is different from the one in batch or mass production companies. The batch or mass production companies usually store their finished products in large warehouses and distribute to their customers through wholesalers or retail outlets. Therefore, the products are selected by a product code and the first one on the shelf is given to the customer. If there is something wrong with the product, the customer gets the next one off the shelf. In the case of OKP companies, a specific customer product is delivered to the customer from the end of the production line. If there is something wrong with the product, it is not possible to deliver the next product in sequence because the next one maybe ordered by another customer. A new/replacement order must be processed quickly to avoid delays in delivery or backorders. When handling large quantities of custom products it is necessary to be able to track them individually. Some customers will order multiple custom products from a number of product families. It is essential to be able to produce all the products of a customer order within the same window of time. To achieve this and effectively deliver the order, the product and production data in OKP companies must be managed in a way somehow similar to the practice in those engineering consulting companies or construction companies to manage their projects. The OKP companies must manage their product and production data by individual customer

orders rather than managing by product families that is often adopted by batch or mass production companies. In this way, each of the customer orders can be tracked in an OKP company, the production of the products in a customer order can be synchronized, and eventually the customer order can be delivered just in time. From the data management point of view, we refer the OKP system as project-based production system. By the integrated product design and production data structure, named product production structure as described in Chap. 3, and the OKP management and control software system© as presented in Chap. 8, we demonstrate the principle and feasibility of the product and production data management technique.

This data management technique also helps the customer service of the OKP companies. For customer service, batch or mass production companies can mass produce replacement parts and store these parts for warranty and service. Based on statistical analysis and life-cycle management of products, volumes of replacement parts can be predicted with a fair amount of accuracy. For OKP companies, it is not possible to stock replacement parts as each can be a custom size, colour or features. It is necessary to be able to recall the details of the original custom order to make the required replacement of customized parts.

This data management technique further helps the knowledge management in an OKP company since the detailed information on production design and production of each of individual customer orders is stored in the company database. These individual data sets in the database can be used to analyze the best practice or worse practices in the company.

9. Manufacturing companies back-flush the records of inventory levels and resource usage. From the analysis of inventory levels and resource usage, the production efficiency and the average scrap rate can be calculated for a period. In case of batch or mass production, this can be performed at the summary level. However, in case of OKP, it is necessary to retain the historical data at a detailed level since each product is managed and produced differently in the company. As mentioned in the previous point, the product and production data are even managed individually according to each of the customer orders. Consequently, the OKP system will result in a very large amount of data records along the time. For large quantities of information, some data mining techniques can be used to identify the usage of options and patterns in custom products by various categories such as region, market, and type of customers. Jiao and Zhang (2006) identify patterns of customer requirements using a data mining approach. In their research, patterns of customer requirements are associated with the functional requirements of products for modelling portfolios of products. Shao et al. (2006) classify customer requirements into groups using data mining methods and associate the configurations of requirements with configurations of products.
10. Due to the high customization and customer involvement or influence in product development in OKP, OKP companies need an effective and efficient method and computer aided-engineering system to change product design, process planning and production schedule concurrently. Otherwise, the

lead-time and the cost of a customer order can be increased too much to be acceptable by the customer. This concurrent engineering approach in batch or mass production companies may be optional but in OKP companies it is inevitable. By the product production structure (PPS) and the prototype-based incremental and concurrent product development method (see Chap. 3); we present a solution to realize the concurrent engineering approach of product design, process planning and production schedule in the OKP system.

From operational point of view, the major issues for implementing the OKP system are identified as follows, which include but are not limited to the following 11 issues.

1. The product variations become too many to be handled manually or with conventional computer-aided production management and control systems.
2. The customization in OKP requires that the inventory management, production schedule, resource planning, and product design and manufacturing data (i.e. BOM and BOO) have to be flexible and changeable.
3. The company has to make to order rather than make to stock.
4. The product flows in the company piece by piece (i.e. single piece flow technique normally used in lean production) according to a pre-defined and adaptively changeable sequence.
5. There is a requirement for a large variety of components, many of which need to be made by the company rather than through the outsourcing according to the DSS.
6. The OKP system can lead to excessive amounts of custom inventory or delayed deliveries.
7. The OKP system can lead to excessive scrap rates because of specific customer requirements, e.g. special product sizes, colours, processes, etc.
8. The special custom processes can lead to requirements for unique skills that may not be fully utilized.
9. The workloads must be balanced to meet customer delivery sequences while maximizing utilization of resources.
10. Replacing defective products during manufacturing causes disruption to production schedules and delivery schedules.
11. When service requires a replacement part, it has to be customized.

The issues as mentioned above draw specific requirements for design and implementation of the OKP system in a company. These issues motivate us to carry out the numerous R&D projects that eventually result in the OKP system as presented in this book.

1.7 The Novelty of the OKP System

The OKP system as presented in this book can be viewed as an extension of the research in the field of Computer Integrated Manufacturing (CIM). It makes contributions by addressing the problems in mass customization from an alternate approach than the ones in existing literature. The existing approaches to mass customization have placed its emphases on manufacturing consortiums and the managerial strategies, such as the DSS (direct subcontracting strategy) and virtual manufacturing systems (VMSs). The IT solutions, such as MRP I, MRPII and ERP systems, are recommended by these approaches. These IT solutions for computer integrated manufacturing are developed based on the production management and control practices in batch or mass production companies rather than SMEs that often produce custom products. van Asperen et al. (2008) propose the use of a mixture of six systems to support different operational issues in mass customization. Other researchers propose the similar combinations and integration techniques.

The mass production processes, like a flow line, have been developed to make the standard products efficiently so that a company can occupy a large market share or even dominate the market. Consequently, the IT systems have been designed to help manage these repeated processes. Figure 1.5 shows the common feature of those traditional approaches, i.e. a bottom-up view of a production process. As the competition is increasing, it is more and more difficult for a manufacturing company to dominate a market and customers are also demanding individualized products. Therefore, manufacturing companies have looked for ways to gain a competitive advantage. Custom options are introduced to make products more appealing and to satisfy diversified customer requirements. The traditional manufacturing companies, that follow the bottom-up view of a production process, have been trying to accommodate the customization by modifying manufacturing processes and/or extending the computer-aided production management and systems.

The OKP system as presented in this book is based on a reverse principle of looking at the problem as an information management problem rather than a product production problem.

In the OKP system, the product information is modelled and the relevant constraints are captured that determine allowable variations. A large number of

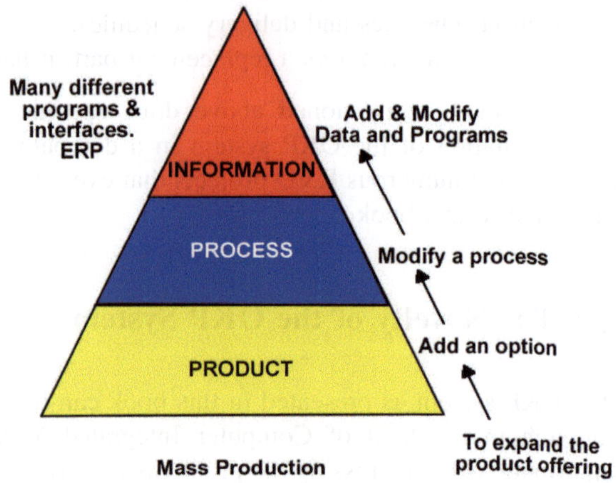

Fig. 1.5 The view of production process in traditional production systems

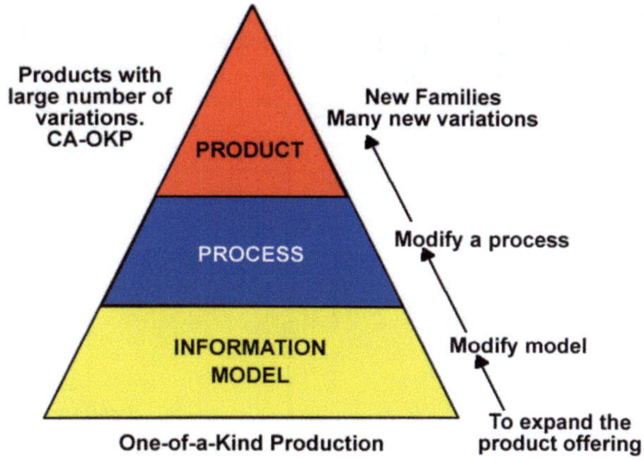

Fig. 1.6 The view of production process in the OKP system

variations on a product are available to satisfy a wide range of diversified customer requirements. When a new option or custom product is introduced, the model is modified along with processes and it leads to many new variations in products. This novel idea provides a much greater flexibility for the OKP systems than the traditional production approaches as illustrated in Fig. 1.5.

The OKP companies can maintain their competitive advantage much more effectively and at a much lower cost than a one-off or job-shop producer. This reversed idea of viewing production process in the OKP system is shown in Fig. 1.6.

To address the problems of variety and uncertainty as specified by McCarthy (2004a, b) and the problems of integration across all operations as described by van Asperen et al. (2008), the OKP management and control software system© has been developed. The framework of this software system is depicted by the diagram in Fig. 1.7.

In comparison with the existing computer aided production management and control systems, e.g. commercial MRP II or ERP systems, the OKP management and control software system© has the following novel features, in addition to the novel concepts, methods, data modelling structure and computation technique as mentioned before in this chapter:

1. The bill of information (BOI) has been added to the product design and production data model. This requires that the information requirements should be defined in a similar manner as parts and operations for products. For example, when a customer specifies a product, the configurator (or the information generator) will generate the BOM, BOO and the information required to make the product.
2. The OKP management and control software system uses its own meta-language to describe custom specifications of products. The customer interface stores the selected values in a configuration string using the meta-language. Only those values required to specify the customer's variations are stored for use in later functions of the system.

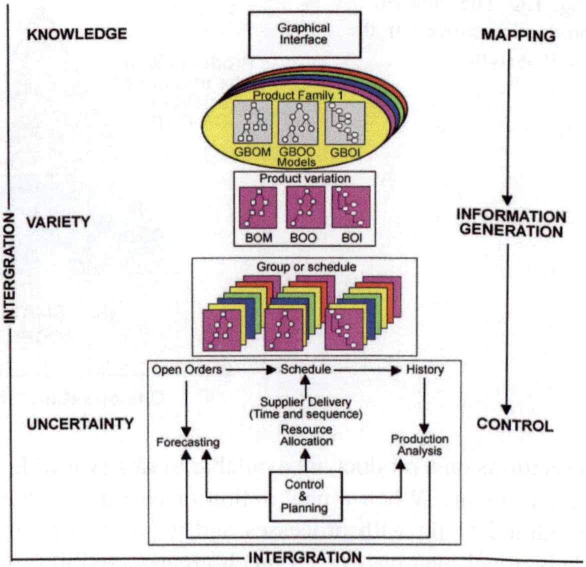

Fig. 1.7 The framework of the OKP management and control software system©

3. An engineering constraints evaluator has been added to the OKP management and control software system©. Xie et al. (2005) recognize that the current configurators are not adequate to solve an engineered product configuration problem. The current configurators use the semantics of "pre-defined rules". The method used by the OKP management and control software system© to solve the problems that are expressed by mathematical formulas and computational procedures is an expression class. This expression class receives values for all the variables from the "configuration string" and uses a mathematical expression evaluator to calculate the results.
4. The OKP management and control software system© has added a "library" of parameterized primitives that specify the characteristics of components and the relationships between the components. A component can be modelled using parameters and subsequently used for all component sizes by modifying its parameter set. If a customized window has two pieces of glass, for example, the glass size can be determined by the evaluation mathematical formula, (half the size of the window) − (the thickness of the frame) + (the depth of the cavity). Here the size of the window is specified by the customer. The thickness of the frame and depth of cavity are primitives.
5. The data structure in the OKP management and control software system has been designed to include the rules, constraints and parameters with the evaluator class that can evaluate both Boolean and mathematical formulas. Previous models maintain the rules in the logic of the program or in a separate database.
6. The current configurators in MRP II or ERP systems provide information (BOM and BOO) of one product at a time, and leave the processing of a group or schedule of products to other parts of the system which are not integrated. The OKP management and control software system© addresses the issue of

integration by including the BOI in the product and production data modelling structure. The manufacturing processes and production schedule are part of the data structure and become available to other parts of the system.

1.8 Chapter Summary

Findings from literature and experimental practice indicate that there is an urgent need to develop the ability for mass customization in manufacturing companies. This is particularly true for SMEs which often produce one-of-a-kind products. The one-of-a-kind production (OKP) system is a novel and promising manufacturing paradigm which adopts some achievements from research areas of computer integrated manufacturing (CIM), flexible manufacturing systems (FMSs) and mass customization but it is different from these approaches for obtaining economies of scale and customization. This chapter defines the OKP system and identifies the differences between the OKP system and these production systems through the review and analyses of relevant production systems, particular those production systems which are often confused with the OKP system, such as one-off or job shop production systems and flexible manufacturing systems. The discussion has been made in this chapter on the differences between the current approaches for mass customization and our R&D efforts for the OKP system, which is also an often confused question. In this chapter, we also define the special requirements of design and implementation of the OKP system in a manufacturing company. To meet these requirements is a challenge and the solutions presented in this book to meet these requirements are novel.

References

Agrawal M, Kumaresh TV, Mercer GA (2001) The false promise of mass customization. McKinsey Q 38(3):62–71

Ahlstrom P, Westbrook R (1999) Implications of mass customization for operations management: an exploratory survey. Int J Oper Prod Manage 19(3):262–274

Ahmead S, Schroeder RG (2002) Refining the product-process matrix. Int J Oper Prod Manage 22(1):103–124

Berman B (2002) Should your firm adopt a mass customization strategy? Bus Horiz Kelley School Bus 45(4):51–60

Cao Q, Dowlatshahi S (2005) The impact of alignment between virtual enterprise and information technology on business performance in an agile manufacturing environment. J Oper Manage 23:531–550

Chen ZL, Hall NG (2007) Supply chain scheduling: conflict and cooperation in assembly systems. Oper Res 55(6):1072–1089

Cheng K, Harrison DK, Pan PY (1998) Implementation of agile manufacturing—an AI and internet-based approach. J Mater Process Technol 76:96–101

Comstock M, Johansen K, Winroth M (2004) From mass production to mass customization: enabling perspectives from the Swedish mobile telephone industry. Prod Plan Control 15(4):362–372

Da Silveira G, Borenstein D, Fogliatto FS (2001) Mass customization: literature review and research directions. Int J Prod Econ 72:1–13

Duray R (2002) Mass customization origins: mass or custom manufacturing? Int J Oper Prod Manage 22(3):314–329

Duray R (2004) Mass customizers' use of inventory, planning techniques and channel management. Prod Plan Control 15(4):412–421

Feiring B (1991) Production planning in stochastic demand environments. Math Comput Model 15(10):91–95

Feitzinger E, Lee HL (1997) Mass customization at Hewlett-Packard: the power of postponement. Harv Bus Rev 75(1):116–121

Frutos JD, Borenstein D (2004) A framework to support customer–company interaction in mass customization environments. Comput Ind 54:115–135

Fujimoto H, Ahmed A (2003) Assembly process design for managing manufacturing complexities because of product varieties. Int J Flex Manuf Syst 15:283–307

Hayes RH, Wheelwright SC (1997) Link manufacturing process and product life cycles. Harv Bus Rev 57:133–140

Hegge HMH, Wortmann JC (1991) Generic bill-of-material: a new product model. Int J Prod Econ 23:17–128

Heilala J, Voho P (2000) Modular reconfigurable flexible final assembly systems in electronic industry. In: Proceeding of the delft assembly automation workshop, pp 11–12, May

Ito Y, Shinno H, Saito H (1988) A proposal for CAD/CAM interface with expert systems. Robot Comput Integr Manuf 4(3–4):491–497

Jiao JX, Zhang YY (2005) Modelling of modularity and scaling for integration of customer in design of engineer-to-order products. Comput Aided Design 37(2):149–172

Jiao J, Zhang L, Pokharel S (2006) Process platform planning for mass customization production. Int J Mass Cust 1(2–3):237–259

Jin G, Thomson V (2003) A new framework for MRP systems to be effective in engineered-to-order environments. Robot Comput Integr Manuf 19:533–541

Kotha S (1996) From mass production to mass customization: the case of the National Industrial Bicycle Company of Japan. Eur Manage J 14:442–450

Kotler P (1989) From mass marketing to mass customization. Plan Rev 17:11–47

Krajewski LJ, Ritzman LP (2002a) Forecasting: operations management, 6th edn. Prentice Hall, Upper Saddle River, p 539

Krajewski LJ, Ritzman LP (2002b) Resource planning: operations management, 6th edn. Prentice Hall, Upper Saddle River, pp 731–780

Kubat C, Taskin H, Topal B (2003) Comparison of OR and AI methods in discrete manufacturing using fuzzy logic. J Intell Manuf 15:517–526

Kuo RJ, Wu P, Wang CP (2002) An intelligent sales forecasting system through integration of artificial neural networks and fuzzy neural networks with fuzzy weight elimination. Neural Netw 15:909–925

Lampel J, Mintzberg H (1996) Customizing customization. Sloan Manage Rev 38(1):21–30

Lee CKM, Lau HCW, Yu KM, Fung RYK (2004) Development of a dynamic data interchange scheme to support product design in agile manufacturing. Int J Prod Econ 87:295–308

MacCarthy B, Brabazon PG, Bramham J (2003) Fundamental modes of operation for mass customization. Int J Prod Econ 85:289–304

Matta A, Semeraro Q (2005) A framework for long term capacity decisions in AMSS. Design of advanced manufacturing systems. Springer, Dordrecht, pp 1–35, Chap. 1

McCarthy IP (2004a) Special issue editorial: the what, why and how of mass customization. Prod Plan Control 15(4):347–351

McCarthy IP (2004b) Manufacturing strategy—understanding the fitness landscape. Int J Oper Prod Manage 24(2):124–150

Ming XG, Mak KL, Yan JQ (1999) A hybrid intelligent inference model for computer-aided process planning. Integr Manuf Syst 10(6):343–353

Piller FT, Moeslein K, Stotko CM (2004) Does mass customization pay? An economic approach to evaluate customer integration. Prod Plan Control 15(4):435–444

Pine BJ (1993) Mass customization. Harvard Business School Press, Boston

Pine BJ (1999) Mass customization: the new frontier in business competition. Harvard Business School Press, Boston

Pine BJ, Gilmore JH (1997) The four faces of mass customization. Harv Bus Rev 75(1):91–101

Priore P, Fuente D, Pino R, Puente J (2003) Dynamic scheduling of flexible manufacturing systems using neural networks and inductive learning. Integr Manuf Syst 14(2):160–168

Rampersad HK (1993) Integrated and simultaneous design for robotic assembly. Wiley, New York

Rolstadås A (1991) ESPRIT basic research action no. 3143—FOF production theory. Comput Ind 16:129–139

Salvador F, Forza C, Rungtusanatham M (2001) Operations configurations for mass customization. In: Proceedings of the European Operations Management Association, 8th international annual conference, Bath, UK, pp 754–764

Salvador F, Rungtusnathan M, Forza C (2004) Supply-chain configurations for mass customization. Prod Plan Control 15(4):381–397

Selladurai RS (2003) Mass customization in operations management: oxymoron or reality? OMEGA. Int J Manage Sci 32(4):295–300

Shao XY, Wang ZH, Li PG, Feng CXJ (2006) Integrating data mining and rough set for customer group-based discovery of product configuration rules. Int J Prod Res 44(14):2789–2811

Srinivasan V, Fischer GW (1996) Direct interface integration of CAD and CAM software—a milling application. J Mater Process Technol 61:93–98

Svensson C, Barford A (2002) Limits and opportunities in mass customization for 'build to order' SMEs. Comput Ind 49(1):77–89

Tseng MM, Piller FT (eds) (2003) The customer centric enterprise: advances in mass customization and personalization. Springer, New York

Tu YL (1996a) A framework for production planning and control in a virtual OKP company. Technical papers of North American Manufacturing Research Institution of SME, pp 121–126

Tu YL, Xie SQ (2001) An information modelling framework to support sheet metal parts intelligent concurrent design and manufacturing. Int J Adv Manuf Technol 18:873–883

Tu Q, Vonderembse MA, Ragu-Nathan TS (2004) Manufacturing practices: antecedents to mass customization. Prod Plan Control 15(4):373–380

Tu YL, Dean PR, Xue D, Li W (2006) One-of-a-kind product design and manufacture. In: Proceedings of the 7th international conference on frontiers of design and manufacturing, Guangzhou, China, vol 3, pp 185–190

Turowski K (2002) Agent-based e-commerce in case of mass customization. Int J Prod Econ 75:69–81

van Asperen E, Pinchetti T, van Wezel M (2008) IT support for mass customization. Int J Mass Cust 2(3–4):358–374

Vandaele N, de Boeck L (2003) Advanced resource planning. Robot Comput Integr Manuf 19:211–218

von Blecker T, Friedrich G (eds) (2006) Mass customization: challenges and solutions. Springer, New York

von Blecker T, Friedrich G, Kaluza B, Abdelkafi N, Kreutler G (2004) Information and management systems for product customization. Springer, New York

von Chandra C, Kamrani AK (2005) Mass customization: a supply chain approach. Kluwer Academic Publishers, New York

von Holweg M, Frits KP (2004) The second century: reconnecting customer and value chain through build-to-order. Moving beyond mass and lean production in the auto industry. MIT Press, Cambridge

Wang J, Shu Y (2005) Fuzzy decision modelling for supply chain management. Fuzzy Sets Syst 150:107–127

Womack JP, Jones DT (2007) Lean thinking. Free Press, New York
Wortmann JC, Muntslag DR, Timmermans PJM (1997) Customer-driven manufacturing. Chapman & Hall, London
Xie H, Henderson P, Kernahan M (2005) Modelling and solving engineering product configuration problems by constraint satisfaction. Int J Prod Res 43(20):4455–4469
Yan HS, Zhang XD, Ma XD (2002) Karmarkar's and interaction/prediction algorithms for hierarchical production planning for the highest business benefit. Comput Ind 49:141–155

Chapter 2
Framework of Computer Aided One-of-a-Kind Production System

The first step in creating the framework of a computer aided production management and control system is to define the manufacturing strategy that the framework will support. After this, the next step is to determine the requirements of the system that supports the operation of the chosen strategy. It is recognized that a formal method of specification is needed in order to provide a rigorous definition that can be re-used by clients when adopting and modifying the specifications and configuration of the system for particular applications. In this book the RTPA (real-time process algebra) methodology (Wang 2002) will be adopted as the formal method of specification. In Sect. 2.1 the methodology for specifying the framework is discussed. In Sect. 2.2, the mass production (MP) and the one-of-a-kind production (OKP) systems are compared at a conceptual level. This comparison demonstrates the different requirements that OKP has from tradition manufacturing systems (TMS). In Sect. 2.3, the OKP strategy is defined and its adaptation to other known strategies mentioned in Chap. 1 is discussed. Section 2.4 presents the results of the first analysis phase, which determines key requirements of an IT (information technology) support system for this strategy. This is followed by a conceptual view of the framework and a definition of the three main components in Sects. 2.5, 2.6 and 2.7. This includes the creation of framework classes as they pertain to the data modelling function. The summary of OKP management and control solutions is provided in Sect. 2.8. The chapter is then summarized in Sect. 2.9.

2.1 Framework of the OKP Management and Control Software System

In this section the use, specification and development of the framework of computer aided OKP management and control system is discussed. According to Booch et al. (2007)

Y. Tu and P. Dean, *One-of-a-Kind Production*,
DOI: 10.1007/978-1-84996-531-6_2, © Springer-Verlag London Limited 2011

A framework is a collection of classes that provides a set of services for a particular domain; a framework thus exports a number of individual classes and mechanisms that clients can use or adapt.

Classes and mechanisms are described as follows:

- A key abstraction is a class that forms part of the vocabulary of the problem domain. A class defines how an object behaves.
- A mechanism is a structure whereby classes collaborate to provide some behaviour that satisfies a requirement of a problem. It is a design decision about how collections of objects cooperate, which represents a pattern of behaviour.

It is necessary to identify the key abstractions in the domain of information technology (IT) in order to support for computer-aided one-of-of-kind production management and control. This involves two processes according to Booch et al. (2007), which are discovery and invention. Through discovery, abstractions are recognized, and through invention, new classes and objects are created that are useful in the design and implementation of the system. For example, in this domain, products, parts, operations, machines and resources are mentioned. In developing a system, these terms will be incorporated but new ones need to be included too, such as databases, lists, queues, messages, event managers and so on.

Another objective is to identify mechanisms or patterns of behaviour. This is done by reviewing the requirements of the system and then determining how various classes and objects can work together to satisfy these requirements. For example, in this domain, there is a requirement to determine how long it will take to make a group of customized products. Note the difference between mass production and one-of-a-kind production for this requirement.

For mass production (MP), the total time of making a batch of n products is

$$T_{\mathrm{MP}} = nt \tag{2.1}$$

In formula (2.1), n is the total number of products and t is the cycle time of a production line. Note, in mass production, the products are often made on a flow line with a constant cycle time t.

For one-of-a-kind production (OKP), the total time for making n customized products can be calculated below:

$$T_{\mathrm{OKP}} = \sum_{i=1}^{n} t_i \tag{2.2}$$

where n is the total number of products and t_i is the cycle time when the ith product is making at the last work centre of a production line. Note that in OKP the times to make customized products can vary significantly and the cycle time of the OKP line for making each of the products could be different.

The discovery of key abstractions leads to the creation of classes including products, parts and operations and recognizes the need for a class called "production list" which is the creation of a set of products for a production run. The requirement for the total time results in a class list having behaviour called "total-time" which performs the summation of the individual cycle times.

The difference between MP and OKP is noted—cycle time t for a product model or family in MP is a constant and it comes from a database or class property when the product model or family is created in the system. In OKP, a mechanism is required to calculate this property for each and every customized product during the production run.

During the final analysis phase, the levels to which the mechanisms are abstracted and included in the framework are determined, while leaving other parts to be customized later by the client. For example, in the mechanism of calculating the lead-time to produce a custom product, the total time is

$$T_P = \sum_{k=1}^{m} tc_k \tag{2.3}$$

where m is the total number of components of the product and tc_k is the time of making kth component.

$$tc_k = \sum_{h=1}^{o} \left(\sum_{j=1}^{p} ta_{khj} + ts_{kh} \right) \tag{2.4}$$

In formula (2.4), o is the number of operations required to make the kth component, p is the number of tasks in the hth operation, ta_{khj} is the time to perform the jth task in the ith operation for making component k, and ts_{kh} is the setup time for the ith operation.

In an OKP strategy using automated equipment, one of the objectives is to reduce the setup times (ts_k) of operations to zero, so that the system can change over between each variation without incurring a setup cost. The use of setup time (ts_k) may or may not be required initially in the framework. It would be up to the client to add this functionality when adapting the framework.

2.1.1 *Information Technology Patterns*

When talking about buildings, Alexander (1977), who is an architect, wrote:

> Each pattern describes a problem which occurs over and over again in our environment, and then describes the core of the solution to that problem, in such a way that you can use this solution a million times over without doing it the same way twice.

The notion of patterns in systems engineering (SE) was started by Beck and Cunningham (1987) for user interface design patterns. This was followed by others like Gamma and the Gang of Four (GOF) who produced a book on Design Patterns (Gamma et al. 1995). The pattern concept was further expanded upon in a series of books on Pattern-Oriented Software Architecture (POSA) (Buschmann et al. 2007a, b). According to Buschmann et al. (2007a) patterns provide us with a vocabulary to express architectural visions as well as examples of representative designs and detailed implementations that are clear and to the point. A pattern is a set of interacting classes (Wang 2007). To date, there is no standard for pattern definition although there are a growing number of patterns being developed and written about.

Wang (2007) presents a number of definitions of patterns and pattern modelling. Wang's use of real-time process algebra (RTPA) provides a formal method for modelling patterns. He observes that conventional pattern notation using UML class diagrams are inadequate to denote what the behaviours and functions are. Wang (2007) explains this by stating that UML class diagrams only provide a rough conceptual model. In contrast Wang (2007) presents the RTPA notation system and methodology to provide a rigorous means and generic formula for modelling patterns.

2.1.2 *Real-Time Process Algebra (RTPA) Methodology*

The RPTA methodology is a formal method used for software system design, specification and modelling (Wang 2002, 2007). The method is both mathematical and rigorous. Wang (2002, 2007) presents this methodology in response to the notion that the current empirical approach to programming is inadequate and therefore leads to confusion and uncertainty in its accurate representation of requirements. The RTPA has both structure and notation, which is strongly typed and language independent, so the resulting specification is accurate and usable in any software environment.

A software system modelled in the RTPA has three subsystems: architecture, static behaviours and dynamic behaviours. Each of the three subsystems can be systematically extended by a three-level refinement process at the system, class and object levels. In the RTPA specification there are two key modelling methods, the component logical model (CLM) and the process model. Wang's method of specification involves using the RTPA meta-processes, algebraic relations and system architectural notations along with a set of primary and abstract data types and the refinement process. During the specification of the framework in our research project a number of additional abstract data types (ADT) are defined to support the specification. For example the Relational Database and Tables are specified below:

$$\begin{aligned} \text{ADT} = {} & \text{RelDB(Relational Database)} \\ & \text{RelDB : ST} \\ & \text{RelDB}, \{\text{Connect, Disconnect, Execute...}\} \end{aligned}$$

ADT = RelTbl(Table in Relational Database)

ADT = RelTbl : ST

RelTbl, {Create, Drop, Read, Insert, Update, Delete, GetNext, GetPrev, GetFirst, GetLast, ...}

A simplified RTPA specification of a relational database in the OKP Management and Control Software (OKPMCS) system© is provided in Appendix B.

2.1.3 UML Class Diagrams and RTPA

The UML (unified modelling language) has been developed by the Objective Management Group (OMG). The OMG is a consortium that aims at setting standards for distributed object-oriented systems, and is now focused on modelling (programs, systems and business processes) and model-based standards. The OGM was initially founded in 1989 by eleven companies, including Hewlett-Packard, IBM, Sun Microsystems, Apple Computer, American Airlines and Dada General. The UML is a general purpose and standard modelling language in software engineering. It is commonly used to specify, visualize, modify, construct and document the artefacts of an object-oriented software system. It combines techniques from data modelling (entity relationship diagram), business modelling (workflows), object modelling, and component modelling.

To help with the specification of the framework, a combination of UML class diagrams and a formal methodology will be used to define classes and behaviours in a systematic and rigorous manner. The UML can be used to sketch basic concepts during the initial phase of specification. It can be used by a designer to create the "blueprint" for a programmer (Fowler 2004). A model driven architecture, according to Kleppe et al. (2003), is a standard approach to use the UML as a programming language. The intent here is to develop a framework for a domain specific application in OKP. This framework will contain some key functions and the UML is used to define the basic concepts. For simplicity and to be easily understood by a wider range of readers, this book intends to avoid any language dependant definitions. Consequently, a model driven architecture is not used.

In order to define the framework rigorously and at the meanwhile to avoid being language dependant, a formal method is used to specify major components of the framework of the OKPMCS system. A formal specification uses mathematical notations to describe, in a precise way, the properties of a system. This formal specification describes what the system must do without saying how to do it (Ambler 2002). This abstraction provides a clear definition of what a system does without getting involved in the programming code of the system. It also removes the ambiguity often found in natural human languages.

The way mathematical notation helps is with use of mathematical data types or abstract data types (Wang 2002, 2007). These data types obey a collection of mathematical laws that make it possible to reason effectively about the way that a specified system behaves. The RTPA provides a notation that supports both primitive and abstract data types, which have a predefined structure and allowable behaviours (Wang 2002). The RTPA is a set of formal notations and rules for describing algebraic and real-time relations of software processes. It can describe a system in a formal manner and eliminate any ambiguities while remaining language independent. It is a formal methodology that is similar to current programming languages and practices. For these reasons, the framework and its main components have been described using the RTPA notation. For example, the Product (or Item) Class can be expressed in the RTPA as:

Abstract Data Type: Item:ST

ItemST ≙ Architecture // Class structure
|| Static Behaviors
|| Dynamic Behaviors

1. Item.Architecture ≙ CLM:**ST** // Component Logic Module
|| AccessModel**ST**
|| Events:**S**
|| Status:**BL**

Item.Architecture.CLM ≙ ItemIDS: // Item class properties
(< Itemdesc:**S** >,
< ItemProduct:**S** >;
< ItemConfig: **S** >;
< ItemLine:**S** >,
< Attributes:**ST** >
)

Where Attribures**ST** = $\overset{n}{\underset{i=1}{R}}$ AttributeST

Item.Architecture.Events ≙ @AddAttribute // Item class events
|| @Calcstd
|| @CopyConfig

Item.Architecture.Status ≙ ⓈItemFound // Item class status
|| ⓈConfigNotCopied
|| Ⓢ

2. Item.StaticBehaviors // Item class static behaviors
{
Item.addAttribute (<**I::** Attribute**ST**, AttKeyS>; <**O::**( )>)
|| Item.CalcStd (<**I::**( )>; <**O::**StdHrsN>
|| Item.CopyConfig ( ) // copies template of operations to item
|| Item.toDisplay ( ) // prints full operations and tasks list for Item
|| Item.toCfgString ( ) // converts attributes and variables to output string
|| Item.toCfgString ( ) // sets all attributes isSelected and variables
}

Item. addAttribute (<**I::** AttributeST, AttKeyS>; <**O::**( )>) ≙ // add Attribute
{
↳ Attributes.add(attribute**ST**, AttKey) // adds a new attribute object to Attributes collection
}

Item.CalcStd (<**I::**( )>; <**O::**StdHrsN>) ≙ //calculate item standard
{
→ StdHrs := 0

```
      NofAtrributes
→          R        ( ? AttributeST(i).isSelectedBL = T
          i=1

                 NofOperations
          →           R        StdHrsN:= StdHrsN + AttributeST(i).OperationST(j)
                     j=1

                          (<I::AttributeST(i).VariablesST>;<O::StdHrsN>)
          → ⓢItemFoundBL := T           // Standard calculated

          | ? ~
          → ⓢItemFoundBL := F           // Standard not calculated
          )
}
3. Item.DynamicBehaviors          // Item class dynamic behaviors
{
(@AddAttribute   ↳  Item.addAttribute (<I:: AttributeST, AttKeyS>; <O::( )>)
|| @CalcStd   ↳  Item.CalcStd (<I::( )>; <O::StdHrsN>
|| @CopyConfig
) → ⊗
}
```

For the detailed notation system of the RTPA, please refer to Appendix C.

2.1.4 Software Development Methodology

Software development processes include requirements, architecture, design, implementation, testing and deployment. The method used in our work is based on the agile unified process (AUP) developed by Ambler (2002). The AUP has four phases: inception, elaboration, construction and transition. During inception, an approximate vision for the system is developed, along with defining the business case, scope and some vague estimates of the work involved. The elaboration phase refines the vision and implements the core architecture on an iterative basis. High-risk elements of the system need to be resolved while identifying most of the requirements. A more realistic estimate of the time to complete the system can then be made. In the construction phase, the remaining lower-risk and easier elements are implemented on an iterative basis, and the preparation for deployment starts. Finally, in the transition phase, the beta testing is completed and the deployment is finished.

The AUP is made up of a number of disciplines from which a development case can be constructed based on the circumstances of the project. In short, the methodology for developing a computer aided production management and control software can be summarized in Table 2.1. This methodology is also applied for the research and development of software systems in other areas.

Iterations can involve all or most of the disciplines but not in the same proportions. For example, in the elaboration phase, the majority of time could be spent

Table 2.1 Software development methodology

Phase	Discipline	Artifact	Comment
Research	Business modelling	#1 Concept model	Visualizes noteworthy concepts in the application domain
	Requirements	#2 Use case #3 Supplementary specification	Captures functional and non-functional requirements
	Design	#4 RTPA specification	Designs the software object
Case study	Implementation	Code	Programs the software objects
	Test	Use case data	Tests that the behavior meets the specification
	Deployment	Live running of system	User acceptance

on requirements and design, and much less time spent on business model refining and implementation considerations.

2.2 Modelling of Mass Production and One-of-a-Kind Production

A traditional mass production process can be modelled by the chart in Fig. 2.1. Products are initially designed and prototyped. Following this, manufacturing and purchasing instructions are developed. Products are identified by a finished goods inventory code and its BOMs (bill of materials) and BOOs (bill of operations) are stored in a database. A product line is created and the suppliers of raw materials and components are organized as part of the supply chain. The system is based on a flow process with standardized work centres which have balanced workloads and the cycle time is determined according to a planned throughput. The raw material is ordered from suppliers on a regular basis according to forecast demand, economic ordering quantity (EOQ) and safety stock levels. The finished products are stored in warehouses from which they are delivered to wholesalers, retailers or direct to customers. The demand for products is based on sales forecasts and replenishment of warehouse stock.

In developing a system for managing and controlling one-of-a-kind production (OKP) operation of customized products, it is first necessary to consider mass producing the customized products using OKP strategies at a high level, in comparison with the mass production process as illustrated in Fig. 2.1. Consequently, the OKP process is modelled in Fig. 2.2.

In OKP, a base product is designed and produced (once) (Tu et al. 2006). During the design process, possible variations of the product are defined and established as constraints in a rule-based system to be applied later when the customer selects among these variations (Sanoff and Poilevey 1991; Kumar et al. 2004). Manufacturing and purchasing instructions are captured in templates that have the rules and constraints attached and stored in a knowledge database. When

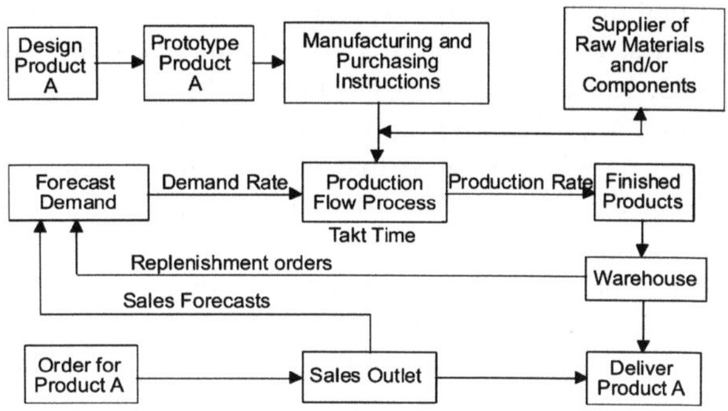

Fig. 2.1 Mass production process

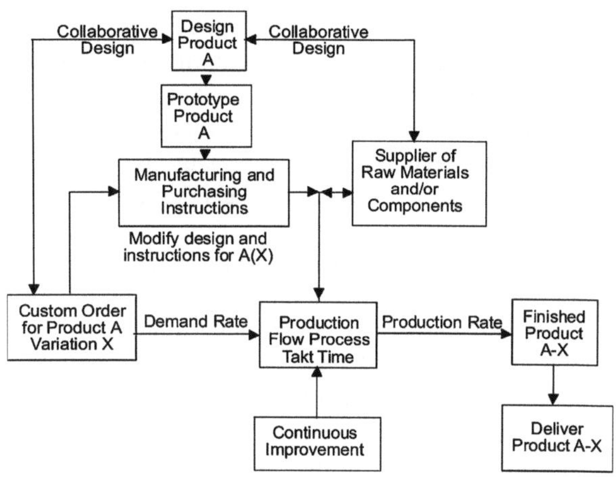

Fig. 2.2 One-of-a-kind (OKP) process

the customer makes an order, a graphical user interface is used to capture detailed specifications of the variations. The customized order is then automatically processed through a product production specification generator that creates all the information required to manufacture the product in an OKP system. The material is ordered from suppliers and the delivery is synchronized with the manufacturing schedule to minimize waiting. Manufacturing instructions are available for each customized product as it flows through the production process. The finished customized product is then delivered to the customer.

2.3 One-of-a-Kind Production (OKP) Strategy

The OKP strategy enables a manufacturing company to mass-produce one-of-a-kind products in a manufacturing company instead of mass customization in a manufacturing consortium through direct subcontracting strategy (DSS) or virtual manufacturing systems (VMSs). This type of manufacturing consortium is called virtual manufacturing companies. Most research papers have placed the emphasis on mass customization in a virtual manufacturing company (Cao and Dowlatshahi 2005; Rolstadas 2000; Onosato and Iwata 1993). These authors believe that the virtual manufacturing is the way to realize mass customization. As discussed in Chap. 1, this type of virtual manufacturing consortium has the disadvantages to hinder the growth of an OKP company. In fact, the OKP strategy can be used or realized by either a single manufacturing company or a manufacturing consortium. The basic principle of the OKP strategy is that a company makes a group or domain (or kind) of products that can be produced on a production line. Each product can have nearly infinitive number of variations based on the addition of options, combinations and any number of variable parameters. Products are grouped based on the ability of the production lines to quickly switch from one product to another or between variations with a small or zero setup time. One of the basic assumptions of this strategy is that the past product orders will not be re-ordered except for service requirements. This is because of the high number of possible variations. The necessity is to create a computer aided production management and control system to be very effective in handling large amounts of data required by a manufacturing company using the OKP strategy.

In short, the OKP strategy means that each product is unique and is ordered by a specific customer at a particular time. It is commonly assumed that there are no products held in finished goods inventory that will satisfy the customer's requirements. Customization takes place at the time of entering the order, normally through an Internet-based computer-aided customer interface. The OKP strategy allows customers to change the specifications of any product on the order until the time the order is released to the shop floor.

The manufacturing process is performed with batch-of-one or one-piece flow. The one-piece flow production technique is often adopted in lean production. Therefore, the OKP strategy supports Lean production. As mentioned in Chap. 1, the OKP is referred as project-based production since each customer order is managed in an OKP company as a project. By adopting one-piece flow technology, the project-based production also reflects the true nature OKP, i.e. an OKP company handles every customer's order as a specific project not only for data management but also for production flow in the company. Since each product is unique, potentially each part or sub-component from internal and external sources is also unique. For one-piece flow production to work effectively, the special production scheduling and control methods are required, which will be described in Chap. 7.

Orders are delivered from work-in-progress (WIP) and therefore no finished goods (FG) inventory is required in an OKP company. This means that production schedule and distribution schedule need to be synchronized so that the excess WIP inventory does not build up at the end of the production line, and customers receive their products just-in-time.

The main advantage of this strategy is that a customer receives a specifically ordered product that matches their requirements at a closer unit cost as a mass-produced product but with a much shorter lead-time. This creates competitive advantages in the market place and loyalty in customers.

Besides the OKP strategy, there are other strategies that can work for improving customization in manufacturing companies. These include delaying customization as late as possible in the production cycle, using a modular design, and making small batches when customers order multiple items of the same variation. These strategies can also applied in OKP and the OKPMCS system supports not only the OKP strategy but also these strategies used in manufacturing to improve the customization and production efficiency. The OKP strategy is presented in this book because it has been developed to meet the specific requirements of OKP as identified in Chap. 1. When a batch of products of the same variation are made on the OKP system, the system can look at these as individual items and produces the products at the same production efficiency as these products are made on a mass production flow line since the cycle time for making the same products on an OKP line is a constant and will not fluctuated. In short, the OKP system is more easily to handle a batch of similar orders than a tradition manufacturing system (TMS) handles unique items.

Current manufacturing philosophies favour Lean production, which encourages small batch sizes, flow line production, and using a pull system where parts, components and products are ordered and made according to customer orders. This practice moves away from making-to-stock, which is considered widely as a waste of resources. As mentioned above, the OKP strategy supports Lean production. All these facts imply that the OKP strategy would be a promising, if it is not the best, strategy for small or medium sized enterprises (SMEs) to effectively compete with large manufacturing companies in the world.

2.4 Analysis of Requirements

The method used for the analysis of requirements in OKP employs the "cases" in comparisons between the operations of OKP and the methods to manage operations in tradition manufacturing systems (TMSs).

Case 1 An order needs to be processed from order entry to invoicing without having to create a part number.

Case 2 An order needs a detailed BOM and routing to be processed.

Case 3 An order needs purchased and/or manufactured parts without creating an item master record.

Most tradition manufacturing systems (TMSs) require that a part number, BOM and routing are created before the order process starts. This is effective if the order contains standard products. If the order contains custom or one-of-a-kind products, maintaining detailed BOM and routings in the system is very time consuming and can dramatically affect lead-time.

For example, a standard part may have a part number MW-1000-1500-BW-S with properties of metal window (MW), 1000 mm wide, 1500 mm high, black paint on the outside, white paint on the inside and silver coloured hardware.

If there were ten models of windows, sizes ranging from 300 mm to 3000 mm wide and 300 mm to 3000 mm height (in increments of 1 mm), 20 colours inside and outside and ten different kinds of hardware, $10 \times 2700 \times 2700 \times 20 \times 20 \times 10$ part numbers would be required, i.e. 2.9×10^{11}. Another method is required that uniquely identifies the product during the order fulfilment process but also provides the detailed specification of the custom product while being able to automatically price the product at the quotation stage, thus requiring access to BOM and routings.

Case 4 The company wants to ship items from WIP (work-in-process) without using FG (finished good) inventory.

Case 5 Produced items need to be quoted using actual costs of material and labour.

To ship products, a TMS requires that an order be closed and items transferred to finished goods inventory. This is effective if standard products are being built to stock. Also, items are quoted at standard quotations rather than actual costs. For OKP, the creation of finished goods (FG) inventory is not needed, and products are unique items that can be shipped immediately from WIP (work-in-process). OKP items should also be quoted using actual costs because of variable material and labour involved.

Case 6 The salesperson wants to create quotes or estimates without having to create item master records, BOMs or routings.

Items in the TMSs require that a part number, BOM and routing are created before an estimate or quote can be made. This is time consuming and creates redundant or useless data that requires extra storage and maintenance. Typically, this is resolved by having a standalone estimating system and the acceptable data is re-input into the TMS.

Case 7 The system needs to associate purchase materials or parts with a custom product order.

Case 8 The system needs to assign the cost of custom parts to an ordered product.

Case 9 The planning process for purchasing materials and parts needs to consider current and forecasted orders.

TMSs are designed to purchase material for stock using material requirement planning (MRP I) technique. The MPR I technique is the fundamental inventory management method used in MRP II (manufacturing resource planning) systems and ERP (enterprise resource planning) systems. The MRP I plans the purchasing orders and inventory replenishment using aggregate demand of production items. It uses time horizons, order size and stocking policies, then makes replenishment order suggestions. In OKP, the demand for common materials, e.g., adhesives, fasteners, etc. in a customized window and door manufacturing company, can be planned using an MRP I based system, such as an MRP II system or an ERP system. However, the majority of the main parts and materials which are required in variable quantities for each customized product as opposed to standard products need to be managed and planned by the OKPMCS system. Furthermore, the custom parts need to be purchased in association with a specific product being made, so that these parts match the specification of the customer orders and actual costs as quoted or planned.

Case 10 The customer needs to be able to provide enough information so that design and production specifications can be generated.

Case 11 The order needs to generate the BOM and BOO (bill of operation).

Case 12 The order needs to generate the data for purchase orders for custom parts.

Case 13 The order needs to generate specific information to manufacture the product. The information may be required in text or graphical form.

Because OKP companies design and build products according to customer specifications, a significant amount of time and cost goes into the design stage of an order. To reduce this amount of design time, an OKP product is often designed through modification of a base product (or template) in a group or kind of products. In an engineer-to-order (ETO) company, the computer-aided design (CAD) packages like AutoCAD and Solid Works have helped with speed and productivity of product design. Engineers create the BOM of a product in a CAD system and then this piece of information is automatically imported or re-keyed into an MRP II or ERP system for production management and control. This product design practice of using a CAD package is not efficient or even inapplicable in OKP companies because of product uniqueness (or high customization) and consequent design cost and the learning curve involved. A more intuitive and user friendly computer aided customer interface is required that will enable customers to specify options and properties of custom products in terms of a prototype which is a product made in the past and has the most similar functions and properties to the customer's order. Therefore, the OKP product design is not a design from scratch, but a specification of design options that produces a one-of-a-kind variation. From this specification, the OKPMCS system generates a BOM and BOO for product variants and allows the order to be processed without creating an item master record.

Case 14 Products with their parts and subcomponents need to be uniquely identified.

Case 15 Parts and subcomponents need to be delivered to work centers in sequence and when required.

Case 16 Resource planning for the shop floor needs to know expected production times for each item.

Case 17 Material process optimization needs to know the exact specification for each part and subcomponent.

Case 18 Performance matrices need to know the expected and actual production time of each item manufactured.

Most TMSs have shop floor execution support for batch processes, not for production of individual products. Resource allocation, load balancing and monitoring routines are based on planning and controlling the production of standard products in batches. For OKP, because of the individual item control required, shop floor execution becomes more complex and has different requirements. Resource allocation is complex because of variable demands, unlike the consistent demand of TMS items. The workload balancing and sequencing become major problems because of the unique nature of parts and subcomponents and variable labour requirements. Tradition manufacturing systems (TMSs) base their performance and utilization matrices on production batches. For OKP, this has to be done on an individual basis in order to get an accurate picture.

Case 19 Each manufactured product needs to have its own true actual cost.

Most TMSs are designed around standard costing. Standard cost for an item is defined by a detailed BOM and routing with its costs rolled up using standard material costs, labour hours and overheads. In OKP, it is important to show the actual cost of an item by creating a detailed BOM and routings and accumulating actual costs for materials and labour, while adding an overhead factor. Each variant of a product will have its own different cost.

Case 20 Specifications for manufactured products needs to be available for replacement or spare parts.

For traditional manufacturing companies that deliver products from stock, completing the manufacturing process is usually the end task. However, OKP companies provide maintenance, service and warranty. In this situation it is essential to track warranty and service information for each specific product. An OKP company cannot create an inventory of spare parts as each product has unique parts and the spare part or replacement product can be manufactured only when service or warranty is required.

At a conceptual level, the framework of the OKPMCS system, that meets these requirements, is shown in Fig. 2.3.

From initial analysis, three main components of the framework are identified. The data modelling module (DMM) is the first component. In order to generate information it is necessary to capture all design and production knowledge about product families, and constraint knowledge with regard to how variations are created and which can be manufactured. Second, and central to the framework, is the information generation module (IGM) that is essential to the process, because it reduces the time between design and production and also the cost. This facilitates the objective of keeping the cost and delivery time of customized products in the same range as standard products. The third component is the manufacturing

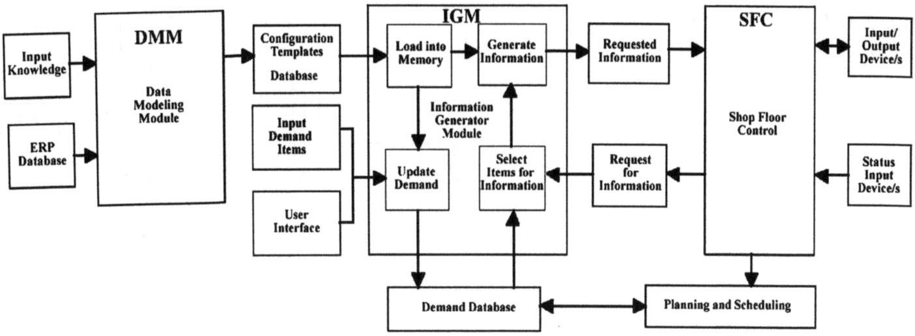

Fig. 2.3 The conceptual diagram of the framework

process itself, using the information, planning and control required on the shop floor to create custom products efficiently and on time, i.e. shop floor control (SFC) module.

2.5 Data Modelling Module

The basis of information production lies with the utilization of templates or models. These are sometimes referred to as product configurations. During the order entry phase, a customer will select a product from a set of product families where each product family may have one or more sets of variations that fall into specific categories. A single configuration or template can represent its BOM and BOO. A customer can select specific configurations either from a catalogue or set of graphics of sample products (or product templates).

2.5.1 BOM Template

The use of a BOM template is similar to a conventional structure, but includes constraints or conditions and parametric functions. An example is shown in Fig. 2.4, in which EP, end product; BP, base product; A, attribute; SC, subcomponent; IP, intermediate part; PP, purchase part; RM, raw material; IS, is selected.

The example structure shows that an end product is made up of a base product and one or more attributes which can be selected by the customer. It is possible to construct a BOM template that can contain optional parts. As shown in the example, it is possible to construct a tree for a customized product that does not contain sub-component SC3. The BOM tree generator knows that the customer's specification does not require an SC3 based on the value of IS3 and therefore the constraint "If IS3 = TRUE" is false and that part of the BOM tree is not included in the output PPS.

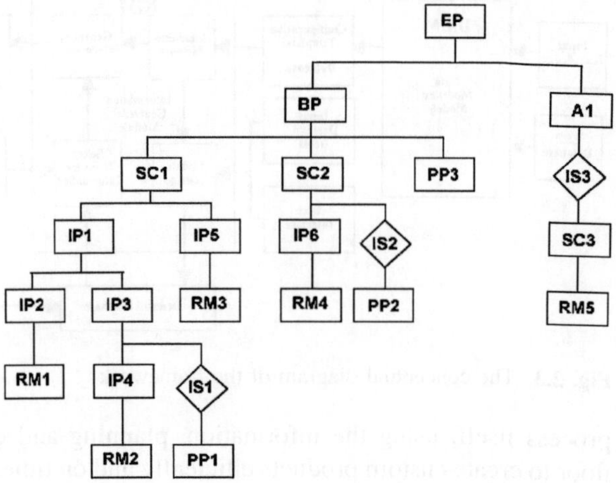

Fig. 2.4 Bill-of-materials (BOM) template

Parametric functions can be resolved from the customer's specifications when input variables are given values. Resolution of the parametric is determined by the design of the product and can be dependent on properties of multiple components or parts.

Based on these input parameters, other engineering constraint rules can be resolved. These constraints can involve the calculation of dimension for intermediate parts, sub-components and purchased parts. They can also determine when parts need to be changed or added based on the condition of specified constraints. These constraints lead to constraint variables.

2.5.2 BOO Template

Resource requirements are also mapped with a similar template structure based on process graphs. Figure 2.5 illustrates an example. The graph is based on two different types of operations, assembly (A) and machining (M). Each operation can have one or more tasks (T). For example, a cutting operation can have several tasks including picking up material, setting up the machine to cut, loading the material, operating the cutting machine and then tearing down the cut material. Each of these tasks will take a certain amount of time which depends on some of the input parameters. As the process is improved or changed, some of these tasks may be automated, e.g., the setting of the machine to cut can be performed by a CNC machine connected to a network to get instructions. A method of determining the sequence of tasks for an operation is important, as sometimes these may be optimized or handled by a CAD/CAM (computer aided design/computer aided manufacture) system.

The cycle time of task *T* is a function of some of the inputs and/or constraint variables. The requirement for an operation is sometimes dependent on the

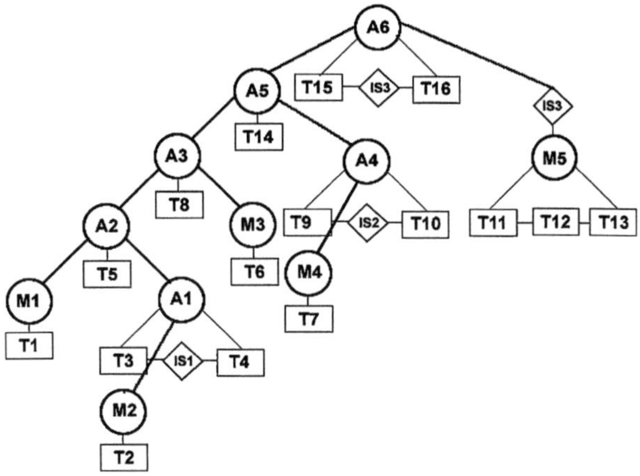

Fig. 2.5 Bill-of-operations template

selection of an attribute as in the case of M5 (see Fig. 2.5). In Fig. 2.5, A, assembly operation; M, machine operation; T, task; IS, is selected. An assembly or machining operation is defined as one or more tasks performed in sequence by one or more resources. When an operation is set up, it is defined with minimum and maximum numbers of resources that can perform the operation at the same time. Consequently, if the total time of an operation is t minutes, then, if one person is assigned to the operation, the cycle time of the operation is t. If n (where $n \leq$ the maximum) resources are allocated to the operation then the cycle time of the operation is t/n minutes. Within an operation some tasks are dependent on the selection of attributes as shown in the example of A1, A4 and A6 of Fig. 2.5.

One or more operations are assigned to a workstation. A group of related workstations constitutes a production line. A production line is defined as a group of workstations that makes a range of variations of one or more products belonging to a product family.

There is a correlation between these two mappings (Figs. 2.4 and 2.5) that can be seen in the example shown in Fig. 2.6, in which EP, end product; BP, base product; A, attribute in a box; SC, subcomponent; IP, intermediate part; PP, purchased part; RM, raw material; IS, is selected. A1 through A6 in circles represent assembly operations and M1 through M5 in circles represent machine operations.

The resource template is "pruned" at the same time as the BOM template to produce the BOO. For example, if attribute A1 is not selected, operation M5 will not be required and will be excluded from the BOO. The generated PPS (product production structure including BOM and BOO, see Chap. 3) for each custom product will contain only those resources required to make the custom product and as a result it complies with the variable structure of the assembly line.

In addition to defining the relationship between materials and operations, it is necessary to identify input variables, engineering constraints, material properties and task variables.

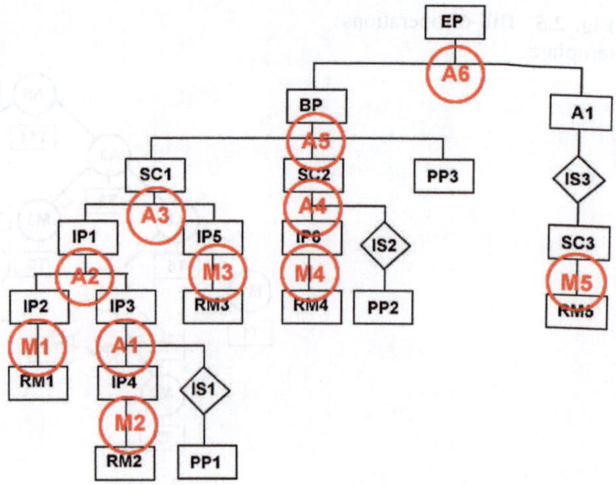

Fig. 2.6 Relationship between materials and operations

Input variables are contained in item specifications and defined by the customer. They include key variables that define the version of the product required, such as attributes, materials, dimensions, materials and descriptive properties. These variables are associated with the base product and attributes.

Engineering constraints are defined by constraint rules using input variables and producing values for constraint variables. Rules are associated with the end product and used throughout the tree pruning process.

Material properties are defined by input and constraint variables to specify details, such as material type, dimensions of materials, colour and other properties. Values are the result of constraint calculations.

Task instructions are defined by input and constraint variables to specify details such as dimension, location, direction and other types of variables in manufacturing instructions. Values are the result of constraint calculations.

During process design, operations are grouped together in defined work centres. Production flows from the first work centre to the last, producing custom products. Figure 2.7 shows an example of the distribution of operations in work centres. The example shows 5 work centres and the use of "shared resources" in work centre W1. Although one of the objectives is to minimize the amount of travelling time for material, sometimes it is more efficient to have a common area for some shared resources like raw material processing.

In order to facilitate one-piece flow production, work centres and inventory storage should all be located together. This can either be done in a straight line or in a work cell layout. An example of a U-shaped cell layout is shown in Fig. 2.8.

This example shows the flow of the end product as it is being made, the flow of raw materials after initial processing and storage of raw materials and purchased parts. For some companies, these work cells can be several hundreds or thousands of square feet in size. It is necessary to minimize the distance that operators and materials have to travel. It is also necessary to have information about the

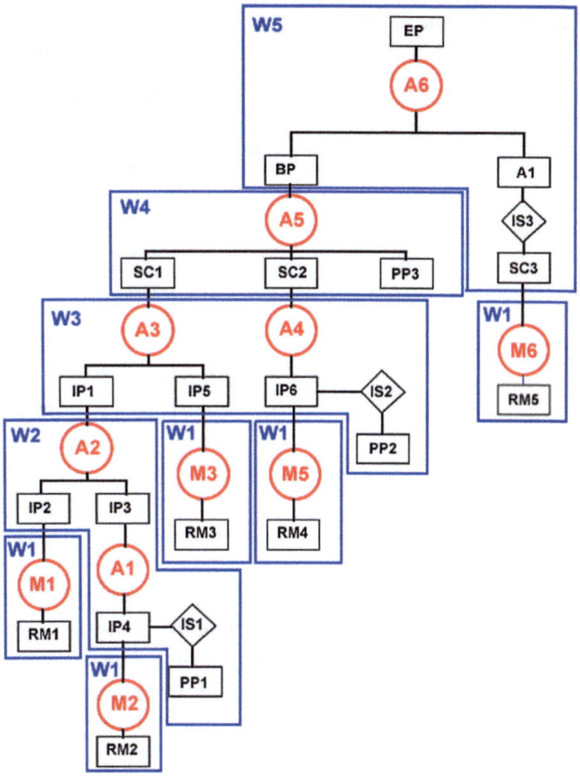

Fig. 2.7 Operations and work centres

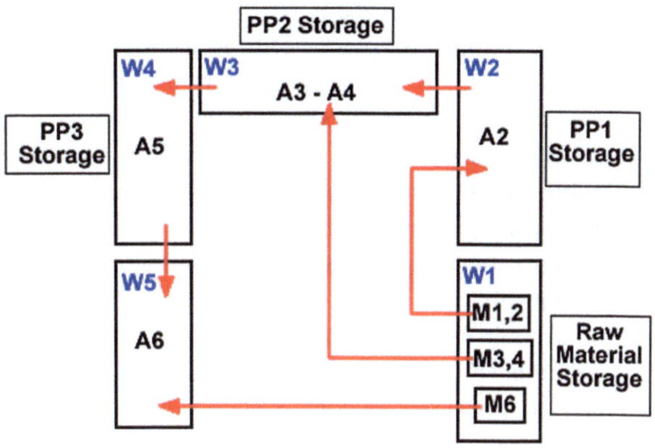

Fig. 2.8 An U-shaped work cell

movement of material and resources at appropriate times. Not all customized products require the use of all materials and operations. To avoid confusion when large quantities of custom parts are being made accurate directions for the movement of materials and sub-components are essential.

Fig. 2.9 Information flow

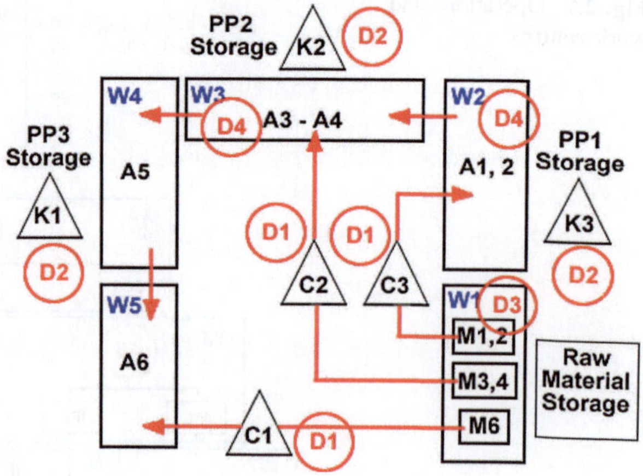

2.5.3 BOI Template

The third template maps information required for processes, including materials and operations and the movement of materials. Manufacturing instructions are required in a number of different forms. These can include: pick lists for materials, processing instructions for single or batched parts, download files for CNC equipment, assembly instructions and delivery instructions. In a flow process, parts and components move from work centre to work centre, or from machine to machine. Information needs to flow through the process and be available when required in these different formats. Figure 2.9 shows a simple example of mapping information to the process. It also illustrates four types of reports, D1, cart (moveable pallet) delivery; D2, non-moveable bin parts storage; D3, cutting machine download; D4, optional part installation and purchase order report.

In a one-piece flow production process, parts and sub-components are usually passed from work centre to work centre. There are times when they are transferred on moveable pallets or travelling carts (C1, C2, and C3 in the example). This creates the situation where intermediate parts and sub-components are moved from one work centre to another in quantities larger than one. Because some customized products do not require all parts and sub-components, it is necessary to identify these moving parts and their location in the cart or pallet. The same idea applies to non-movable storage units (K1, K2, and K3 in the example). Customized parts need to be kept in storage in the same sequence as products are being made. This generates the requirement for information about the storage and movement of materials.

In the example as shown in Fig. 2.9, work centres W2 and W3 receive intermediates parts in carts C3 and C2 respectively. When required by IS3, work centre W5 receives parts in cart C1. The cart identification is determined by the report type D1. Purchased parts are delivered by various suppliers and stored in non-movable storage units K1, K2, and K3. They are stored in the sequence of production and are based on a first-in/first-out (FIFO) system. Report type D2 defines

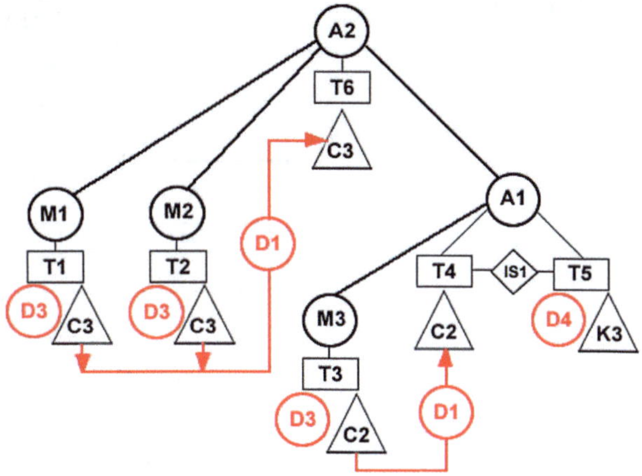

Fig. 2.10 Information template

the purchase and storage requirements for these parts. If machine cutting is performed by CNC machines, the download files of work instructions are required. These instructions are defined in the D3 output format. Optional instructions for the installation of parts determined by IS1 and IS2 are defined in report type D4.

The mapping process shown in Fig. 2.10 determines what information is required. In Fig. 2.10, A1 and A2 are assembly operations; M1, M2 and M3 are machine operations; T1–T5 are tasks; C2 and C3 are moveable carts or pallets; K3 is a non-moveable bin parts storage; D1 is a cart report; D3 is a machine instruction download report; D4 is an optional parts installation and purchase order report. Information comes from values of various input and constraint variables. Each set of values will be different for each customized product. Since the main sequence of production is maintained in all reports, one-piece production flow will match the flow of information.

2.5.4 *Classes*

From the case analysis in Sect. 2.4 and discussion on the templates as mentioned above, the initial classes and class relationships are identified. These classes are divided into three groups: BOM, BOO and information classes.

2.5.4.1 BOM Classes

BOM classes include: Item, Attribute, Component, Raw Material, Property, Constraint and Variable. An item or end product is made up of one or more product attributes. Since the specification is a variation of a base product, the first attribute is the base product. Additional attributes specify added options to the base product. Each attribute has a set of input variables which defines the order entry

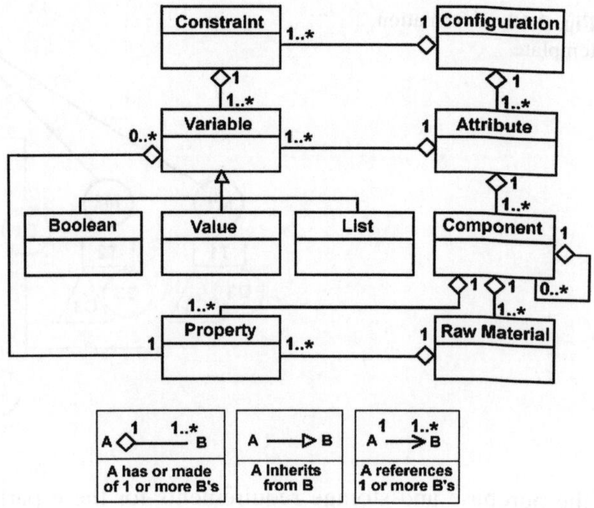

Fig. 2.11 BOM classes

interface. Variables are the main parameters that specify a unique custom product fully. An item will also have one or more engineering constraints that determine how individual components are made.

Each attribute is made up of a number of components which are made from other components or raw materials. This produces the hierarchical structure of a BOM. Each raw material has a set of properties that are defined by input and/or constraint variables. The Variable class is an abstract class from which three inherited classes are formed as Boolean, List and Value Variable classes. Figure 2.11 shows the relationship between these classes.

2.5.4.2 BOO Classes

BOO classes contain: Work Centre, Resource, Machine, Operation, and Task. Each attribute is made by one or more operations. These operations are assigned to a work centre which is made up of resources, i.e. workers and/or tools/machines. Each operation is performed by completing one or more tasks which transform raw materials into components. All the components of the product are made and assembled together to produce the end product.

Figure 2.12 shows the relationship between these BOO classes and the BOM classes.

2.5.4.3 Information Classes

Information classes include: Information and Reports.

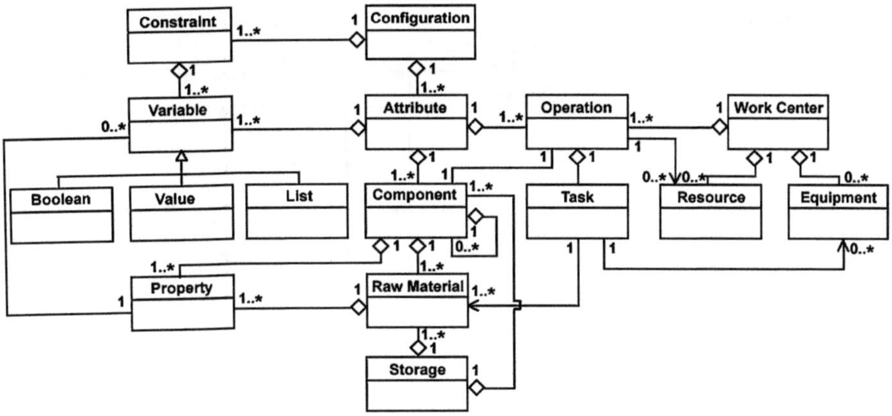

Fig. 2.12 BOM and BOO classes

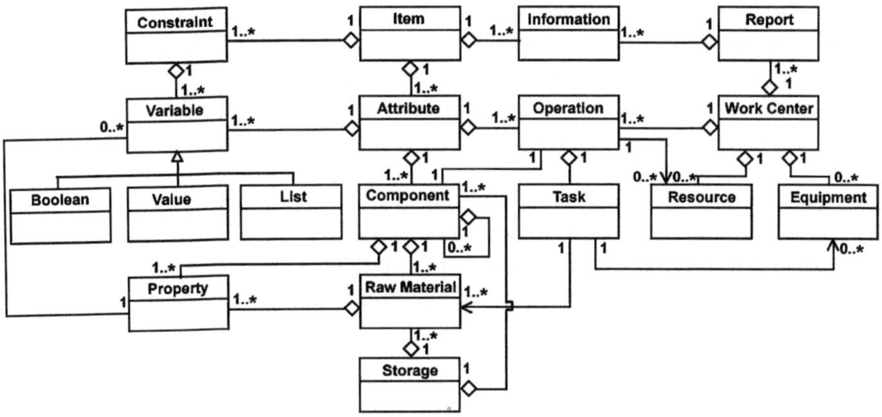

Fig. 2.13 BOM, BOO and information classes

Each item has its own set of information. As reports and outputs are required by the users of the system, they are defined to contain specific pieces of information. When information is generated for an item or group of items, various report objects are created and are available on request.

Figure 2.13 shows the relationships among all these classes.

This group of classes constitutes what is called the template or configuration. When a customer specifies the variation of a product, a configuration is used, and input variables and engineering constraints are applied to produce the BOM, BOO and BOI (bill-of-information) for that variation.

Fig. 2.14 Information generation module

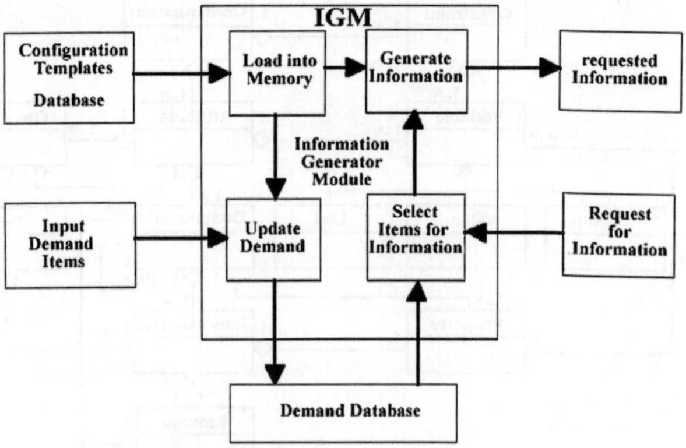

2.6 Information Generation Module

The information generation module takes a customer's specifications and selects the appropriate template from the database. Figure 2.14 shows the relationships among various parts of this module. The information generation module applies specific values and selects options for the template. A BOM and operation routings are generated for customized products along with the manufacturing instructions in a product production structure (PPS) which will be further described in Sect. 3.4.1. This process makes it feasible to have unique information about each custom product and supports a one-piece flow production system. Information is then available to all the functions of the information system including resource planning, scheduling, shop floor control, and distribution and service data. The whole process is managed by the control system, which monitors manufacturing processes and provides various feedback mechanisms.

2.7 Shop Floor Control Module

Each production line or production cell in a manufacturing company is configured differently for producing various products. An example of a production shop floor is shown in Fig. 2.15. The layout shown in Fig. 2.15 is fairly common for a flow shop, and under the following conditions it can support mass production of customized products. The first condition is flexibility, which comes with the use of computer controlled equipment (e.g., CNC machines) that can reset itself for the next customized product (Matta and Semeraro 2005). Production system flexibility effectively reduces the setup time to nearly zero. This means that the production line can change from one product variation to the next without wasting time for setup. The second condition is availability of individual

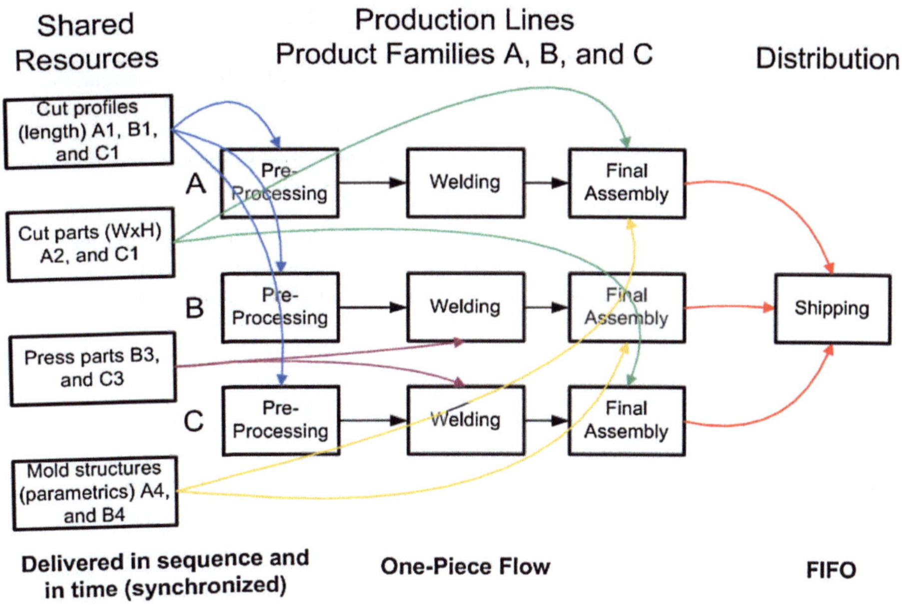

Fig. 2.15 An example of a flow shop

instructions, one-piece at a time, enabling the operators or machines to perform specific operations for making customized products efficiently, without too much delay or waiting time.

Planning and control of this kind of process becomes complex because of the customization of each product. Customized products require different amounts of materials and different amounts of time to process. The biggest challenge is getting the right material to the right place at the right time and balancing workloads on each production line so that the wasted time (or waiting time) is minimized. A feedback system is required to assist planning of the next group of products depending on the performance of the lines (Li et al. 2006). This is important, as an order for a customer may contain products from multiple lines and the total order should be delivered on time without back orders. To satisfy all these conditions, it is necessary to synchronize production of each line as well as the parts and components from shared resources.

2.8 Summary of OKP Management and Control Solutions

From the initial data structure and overview of the main three modules, the solutions for management and control of OKP can be summarized in Table 2.2 by using the cases as discussed in Sect. 2.4.

Table 2.2 Summary of OKP management and control requirements

Case	Description	Solutions
1	An order needs to be processed from order entry to invoicing without having to create a part number	An order is created from a configuration and specific parameter values and not from a part number. Each item is given unique identification to track it from order entry to invoicing
2	An order needs a detailed BOM and routing to be processed	Each item generates a complete BOM and BOO along with information for manufacturing instructions
3	An order needs purchased and/or manufactured parts without creating an item master record	When a product requires a purchased or intermediate part, information is generated by the system without the use of an item (product) master record
4	The company wants to ship items from WIP without using FGI	No two products are the same so the use of FGI is not required. The scheduling system arranges for all products of an order to be completed in the same time frame so they can be shipped together directly from the shop floor
5	Produced items need to be quoted using actual cost of material and labour	The detailed BOM and BOO of a product enable the system to quote each product independently
6	The salesperson wants to create quotes or estimates without having to create item master records, BOMs or routings	Products on a quote use the same generation function as orders to determine the exact cost of making a product. Based on pricing rules the system can generate accurate prices for quotes and estimates. Note: The retention of BOMs and BOOs generated by quotes is a user decision
7	The system needs to associate purchase materials or parts with a custom product order	From Case 3, the generated purchase and intermediate parts are uniquely identified and associated with the corresponding unique product
8	The system needs to assign the cost of custom parts to an ordered product	When the system generates the custom part for a product it uses the actual cost from the purchase order, not the standard cost
9	The planning process for purchasing materials and parts needs to consider current and forecasted orders	Current orders in the system will have their associated BOMs and BOOs from which materials can be ordered. Forecasts will use aggregate BOMs to add their material requirements to the planning process
10	The customer needs to be able to provide enough information so that design and production specification can be generated	The order entry interface is dynamically generated from the product's configuration. All necessary specification elements will be presented to the customer
11	The order needs to generate the BOM and BOO	The central concept of the system is to generate BOMs and BOOs as well as manufacturing information

(continued)

Table 2.2 (continued)

Case	Description	Solutions
12	The order needs to generate the data for purchase orders for custom parts	The system does this as mentioned in Cases 3 and 9
13	The order needs to generate specific information to manufacture the product. The information may be required in text or graphical form	The system does this as mentioned in Case 11. This knowledge is contained in the BOM, BOO and bill-of-information
14	Products with their parts and subcomponents need to be uniquely identified	The system does this as mentioned in Cases 3 and 7
15	Parts and subcomponents need to be delivered to work centres in sequence and on time when they are required	The master schedule determines when an order has to be made and the sequence of order production. The information generator knows details of each unique BOM and BOO and it can arrange the "flow" of information in the same sequence as the product
16	Resource planning for the shop floor needs to know expected production time for each item	From the detailed BOO of each unique item the information generator can calculate the expected production time for each item
17	Material process optimization needs to know the exact specification for each part and subcomponent	From the detailed BOM the system can provide the exact specification (properties) of each part and sub-component for optimized processing
18	Performance matrices needs to know the expected and actual production time of each item manufactured	By establishing a feedback system from the shop floor when parts and products are completed, the system can generate performance matrices from the expected and actual production times
19	Each manufactured product needs to have its own true actual cost	The system will generate the true cost of each product as it has the unique BOM and BOO from which costs can be generated (Case 8)
20	Specification for manufactured products needs to be available for replacement or spare parts	Information that generates the BOM and BOO for each product can be retained in history files for this purpose. A decision has to be made about whether the configuration at the time of making the original product should be used or the current version of the configuration

2.9 Chapter Summary

In this chapter the OKP strategy for mass-producing customized products has been defined. The software development methodology is presented and the framework the OKP Management and Control Software (OKPMCS) system© is described at the conceptual level. The analysis process produced 20 cases which identify the main functional requirements of the OKPMCS system. The three main modules of the OKPMCS system are identified along with the modelling templates and the expert knowledge that are required to make the function of the system correctly meet the OKP needs as discussed in Chap. 1. A group of classes is created to form the basis of the templates or configurations for product variations in OKP. A summary of the OKP management and control solutions are also provided.

References

Alexander C (1977) Towns building construction. Oxford University Press, Oxford

Ambler S (2002) Agile modeling: effective practices for extreme programming and the unified process. Wiley, Chichester

Beck K, Cunningham W (1987) Using pattern languages for O-O programs. In: Proceedings of OOPSLA 1987 workshop on specification and design for O-O programming, Orlando, Florida

Booch G, Maksimchuk RA, Engle MW, Young BJ, Conallen J, Houston K (2007) Object-oriented analysis and design with applications, 3rd edn. Addison Wesley, Reading

Buschmann F, Henney K, Schmidt DC (2007a) On patterns and pattern languages, pattern-oriented software architecture, vol 5. Wiley, Chichester

Buschmann F, Henney K, Schmidt DC (2007b) A pattern language for distributed computing, pattern-oriented software architecture, vol 4. Wiley, Chichester

Cao Q, Dowlatshahi S (2005) The impact of alignment between virtual enterprise and information technology on business performance in an agile manufacturing environment. J Oper Manage 23:531–550

Fowler M (2004) A brief guide to the standard object modelling language, UML distilled, 3rd edn. Pearson Education, Inc, Upper Saddle River

Gamma E, Helm R, Johnson R, Vlissides J (1995) Design patterns: elements of reusable object-oriented software. Addison Wesley, Reading

Kleppe A, Warmer J, Bast W (2003) MDA explained: The Model Driven Architecture™: practice and promise. The Addison-Wesley object technology series. Addison-Wesley, Reading

Kumar KD, Karunamoorthy L, Roth H, Mirnalinee TT (2004) An infrastructure for integrated automation system implementation. Int J Flex Manuf Syst 16:183–199

Li W, Tu YL, Xue D (2006) Adaptive production scheduling and control for one-of-a-kind production shop floor. Transactions of NAMRI/SME, Milwaukee, Wisconsin, USA, pp 159–166

Matta A, Semeraro Q (2005) A framework for long term capacity decisions in AMSS. Design of advanced manufacturing systems. Springer, Dordrecht, pp 1–35 Chap. 1

Onosato M, Iwata K (1993) Development of a virtual manufacturing system by integrating product models and factory models. Ann CIRP 42(1):475–478

Rolstadas A (2000) Production management in a virtual company. Comput Ind 16:129–139

Sanoff SP, Poilevey D (1991) Integrated information processing for production scheduling and control. Comput Integr Manuf 4:164–175

Tu YL, Dean PR, Xue D, Li W (2006) One-of-a-kind product design and manufacture. In: Proceedings of the 7th international conference on frontiers of design and manufacturing, Guangzhou, China, vol 3, pp 185–190

Wang Y (2002) The real-time process algebra (RTPA), annals of software engineering, vol 14. Kluwer Academic Publishers, Dordrecht, pp 235–274

Wang Y (2007) Software engineering foundations. Auerback Publications, New York

Chapter 3
Rapid Product Development in One-of-a-Kind Production

The keen global competition has caused many manufacturing companies to cut "fats" (or wastes/costs) through strategically moving their manufacturing bases and facilities from developed areas, e.g. Hong Kong, Europe, USA, Canada, etc., to developing countries and areas which provide favourable conditions, such as low labour costs and other cheap resources. This movement is leading to job lose and economic pressure in those developed countries. As mentioned in previous chapters, the OKP strategy could help keeping or re-establishing manufacturing business in these developed countries.

In terms of the discussion and definition of OKP in Chap. 1, an OKP company could be loosely understood as an advanced factory which can quickly and mass-produce customised products in a certain manufacturing domain (i.e. "Kind" in OKP), e.g. a ship building manufacturer, a structural steel construction company, a special or custom equipment and machine tool manufacturing company, a sheet metal product production plant, a customized window and door manufacturer, etc. In short, an OKP company is required to produce one-of-a-kind products of high quality at a competitive price within a short lead-time, i.e. rapidly and mass-producing customized products.

To realize the goal of OKP, which maybe simply called rapid mass customization, there are a number of issues raised from the special requirements and/or problems in rapid mass customization. These requirements and/or problems include highly customised products, the concurrent approach of product development and production, incomplete product and production data, the need for quickly capturing marketing opportunities and early response to the customer's demands, continuous customer influence upon production, and the optimal selection of suppliers and subcontractors.

Furthermore, in rapid mass customization, concurrent product development and the synthesis of suppliers and subcontractors for product realisation are two important activities. The concurrent product design, process planning, and shop floor production scheduling and control will play a key role in linking these two activities. The current computer aided process planning (CAPP) systems have been developed based on conventional process planning theory which is a linear process

Y. Tu and P. Dean, *One-of-a-Kind Production*,

DOI: 10.1007/978-1-84996-531-6_3, © Springer-Verlag London Limited 2011

model. It can only run in series, not concurrently. In OKP, continuous customer influence on production runs concurrently with product development and production. The conventional process planning theory and the existing CAPP systems become very difficult in this situation.

In this chapter, a framework for concurrent product design, process planning and production scheduling and control is presented. This framework consists of a stepwise concurrent product design and process planning management module, a novel incremental design and process planning module, an integrated data modelling technique which is named as product production structure (PPS), and a reference control structure for production planning and control in OKP. It also includes three data models for data modelling and structuring of the product, the process and manufacturing plants in OKP. The stepwise concurrent product design and process planning management module and the incremental design and process planning module constitute the prototype-based incremental and concurrent product development method that is facilitated by the PPS data modelling technique.

To avoid confusing readers, we would like to define the computer aided process planning in the following according to the conventional definitions of process planning (Link 1976; Chang and Wysk 1985; Alting and Zhang 1989) and the particular needs of rapid mass customization in OKP.

> CAPP in OKP is a computer aided decision making support and planning tool which aims to generate the methods and cost estimates to economically and competitively convert the technical specifications of customer's requirements into a wanted product. The technical specifications of customer's requirements are understood as the design of a product which is modelled by a proper product modelling tool.

Before the framework of concurrent OKP product development is introduced, a brief historical review of the relevant technologies and theories are given in the following section.

3.1 Historical Review

As described by Kidd (1994), the characteristics of agile manufacturing (AM) can be summarised as follows:

1. AM is not synonymous with lean production (Womack et al. 1990; Jones 1992), flexible manufacturing, computer integrated manufacturing or other paradigms.
2. The concept of AM is based on the synthesis of a number of enterprises, each bringing core skills or competencies to the joint venture, and uses each partner's facilities and resources. These joint venture enterprises are called virtual corporations because they do not own significant capital resources.
3. The deployment of advanced information technologies and the development of nimble organisational structures to support highly skilled, knowledgeable and empowered people in those joint ventures, are critical to the success of AM.

4. AM is capable of responding rapidly to changes in customer demands. It is able to take advantage of the windows of opportunity that occasionally appear in the market place.
5. The advanced technologies developed through lean production, flexible manufacturing, computer integrated manufacturing, and global manufacturing can be used in developing AM. However, AM places particular emphasis on the participation of skilled, knowledgeable, flexible, motivated and responsive people.

Recently, agile and virtual are alternatively used to describe a type of manufacturing company that is commonly referred as virtual manufacturing company today. From the summary points of Kidd (1994), the virtual corporation is a basic feature of a so-called agile manufacturing company. In short, the "agile" description focuses on the technical functions of virtual manufacturing companies, whereas the "virtual" description focuses on the managerial and strategic formats of the companies.

One-of-a-Kind production (OKP), as predicated by scholars like Rolstadås (1991), Wortmann (1991) and Hirsch and Thoben (1992), could be a promising manufacturing model for the factory of the future. At the same time OKP poses some challenges to the factory of the future. Tu (1996a) characterised the OKP philosophy as: (1) high customisation, (2) successful product development and production in one go, (3) optimal or rational utilisation of technologies and resources, (4) adaptive production planning and control, (5) continuous customer influence throughout the production, (6) incremental process planning, (7) distributed control and inter-organisational autonomy, and (8) virtual company structure and global manufacturing.

Process planning has been defined as "the sub-system responsible for the conversion of design data to work instruction (Link 1976)." A more specific definition of process planning is given by Chang and Wysk (1985) as,

> Process planning is the function within a manufacturing facility that establishes the processes and process parameters to be used (as well as those machines capable of performing these processes) in order to convert a piece-part from its initial form to a final form which is predetermined on a detailed engineering drawing.

According to the surveys conducted by Alting and Zhang (1989), Sood et al. (1993), Maropoulos (1995), and Kiritsis (1995), the approaches to developing computer aided process planning systems for various part types are based on one or more of three basic methods: variant, semi-generative, and generative. The variant approach is sometimes referred to as a data retrieving method. By a coding method called GT (group technology) code, a standard process plan can be stored in a data base. A process plan for a new part is created by recalling, identifying, and retrieving an existing plan for a similar part which is called a master part (Alting and Zhang 1989). The generative approach can be defined as a system that synthesises process information in order to create a process plan for a new part automatically (Chang and Wysk 1985). The semi-generative approach is intermediate between two other approaches. The variant method has advantages such as being close to the human expert's way of process planning. Hence it is easily implemented. However, its application has been limited by its lower automation level and flexibility. The generative method has a higher automation level and flexibility. Due to some unresolved technical and theoretical problems, however, the CAPP system developed by the generative method cannot efficiently create high quality process plans. According to a survey conducted by Maropoulos (1995) during a travel study project to several university research centres in USA, UK, The Netherlands, Germany, Sweden and Denmark, it seems that most research in CAPP is based on the generative approach. The problems which are currently solved using the generative approach may be seen from four aspects. They are tooling technology and management (e.g. wear balancing, intelligent tool selection), process modelling (modelling tools and methods, integrated modelling of products and processes, automatic feature recognition), process planning (e.g. interface between design and scheduling, distributed process planning, feature-based design and process planning, geometric reasoning), and the integration of CAD/CAPP/CAM.

All the problems exist in the research and development (R&D) of CAPP systems would be the problems of developing a CAPP system in OKP. In addition to these common problems in the development of CAPP systems, however the development of a CAPP system in OKP will be further faced with the special problems such as supply chain synthesis, manufacturing process planning and testing in different plants which include the ones in suppliers and/or subcontractors, special needs of data modelling in OKP, selecting suppliers/subcontractors and building a manufacturing network for producing an OKP product through an integrated and concurrently approach of product design, process planning, shop floor scheduling and cost analysis. These special problems of developing a CAPP system in OKP have not been particularly attacked and meaningfully covered with the past and existing researches and developments in the area of so-called computer aided process planning. Hence this chapter does not particularly place its emphases on the common problems in the area of CAPP. Instead it focuses on the special problems in the development of a CAPP system in OKP. In the following, we would like to further summarise several interesting developments in the relevant areas.

A number of ESPRIT projects have been carried out with regard to CAD/CAPP/CAM integration. IMPPACT (ESPRIT 2165, Integrated Manufacturing of Products and Processes using Advanced Computer Technologies) has proposed several reference models, such as product model, process model and factory model, for the integration of CAD/CAPP/CAM (Bjorke and Myklebust 1992). In the process model, manufacturing activities are placed in four layers including process group, process, operation, and pass. The factory model specified the components of production systems or production mechanisms and the physical and logical relations between the mechanisms. The production mechanisms include shop floors, work centres, workstations, tools, tool shops, tool adaptors, workers, means of transport, and carrying supports.

Khoshnevis et al. (1994) and Park and Khoshnevis (1993) propose a real time computer aided process planning (RTCAPP) system as a support tool for economic product design. The RTCAPP system is designed as a manufacturing cost estimator to support the design of a part or a product. After a new feature has been added, the RTCAPP system will suggest a process plan and estimate the consequent manufacturing cost. The RTCAPP system is assumed to interactively work with a design system in real time.

Bernardi (1993) presents a skeletal plan based methodology for emulating the human expert's process planning procedure. In this method, a skeletal plan (abstracted plan or fragments of a plan) is associated with manufacturing features. The generation of a process plan is consequently subject to this sequence: abstraction, selection and refinement. The geometrical and/or technological representation of a part in Bernardi's method allows the process planning system to recognise the relevant features. Then the skeletal plans that are related to these features will be selected, merged and refined by the process planning system until a complete process plan is created.

Cho et al. (1994) presents an integrated system of CAPP and shop floor control. In their presentation, the authors consider that the CIM architecture generally consists of three hierarchical levels: shop floor, workstation and equipment. The CAPP consists of machining feature identification, definition, classification, representation, and reasoning, provided through a CAD model of a product. The design of a part is converted into a feature graph. The feature graph will be converted by the CAPP into a task graph that is a process plan with alternative feasible processes. Finally, the task graph will be hierarchically decomposed into operation schedules at the three CIM architecture hierarchical levels. To find an optimal shop floor schedule with a process plan consisting of alternative processes, Chen and Khoshnevis (1993) have proposed a heuristic method consisting of a search method and a concurrent assignment algorithm.

The STEP project (ISO 1993) has defined an international standard for the representation and exchange of computerised product information throughout the life cycle of a product. This standard is independent of particular computer systems, and enables consistent implementations across multiple applications and systems.

Candadai (1994) analyses the information necessary in agile manufacturing and defines three types of information for product design and production. These are product data, process data and manufacturing plant data.

Chu and Holm (1994) develop a control system for product manufacturability control in concurrent engineering environments. A hierarchy of manufacturability criteria is established for measuring the manufacturability of a product. Methods for calculating and setting manufacturability criteria are also developed.

Tu (1996a) proposes a reference architecture for an agile OKP company. An agile OKP company is viewed as a flexible firm which consists of marketing networks, manufacturing networks, and several centralised intelligent agents. The networks are seen to formulate the flexible extensions of the firm. The intelligent agents constitute the core of the firm. These intelligent agents include an information processing agent, an engineering agent, a managerial agent, and manufacturing agents.

Zhang (1996) proposes several methods of product data modelling and partner company data modelling for the development of a conceptual database model to support partner synthesis in virtual manufacturing. This research was carried out through a case study conducted in an electronics company in Hong Kong which produces customised products in its distributed manufacturing network. The contribution of this research is that it combines the variant of products and the variant of production resources and capacities in the partner companies through the data modelling of products and partner companies.

3.2 Stepwise Concurrent Product Design and Process Planning Management Module

According to our studies and observations on various manufacturing companies, the OKP strategy has been widely applied, particularly in small or medium sized enterprises (SMEs) (Tu 1996a; Zhang 1996). However, most companies have no agility to cope with market changes. To solve this problem, a stepwise concurrent product design and process planning management module is proposed in Fig. 3.1.

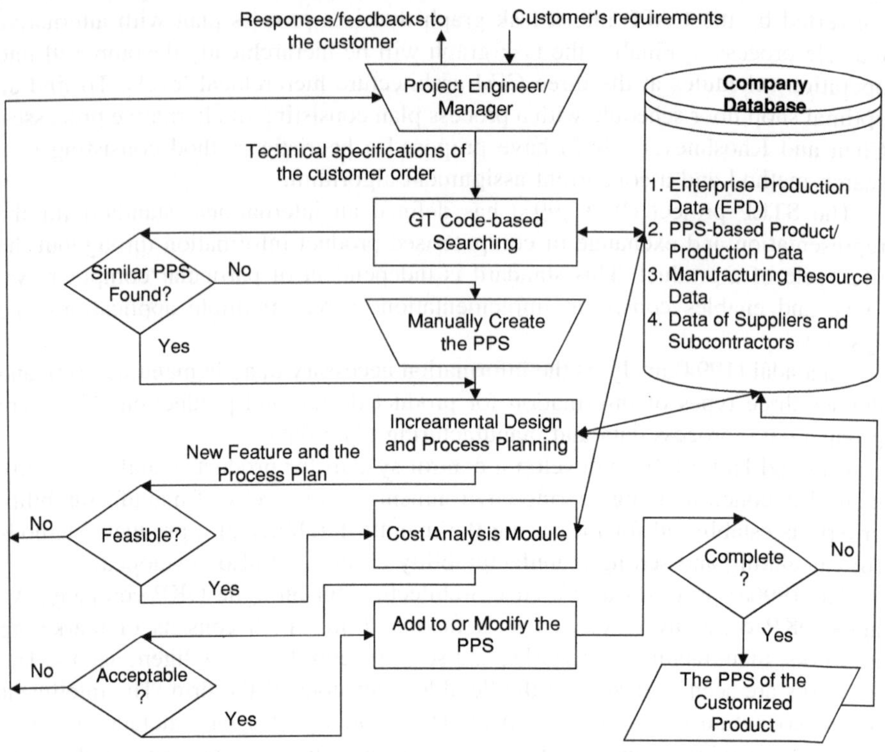

Fig. 3.1 The stepwise concurrent product design and process planning management module

As shown in Fig. 3.1, the *Project Engineer/Manager* (*PEM*) communicates with customers directly or uses a customer interface module to get the customer's requirements and convert the requirements into the technical specifications. Then by GT (group technology) code, the *PEM* or other engineer(s) can use the *GT Code-based Searching* model to search from the Company Database for a similar PPS (product production structure, see Sect. 3.4.1) that is close to the customer's ordered product. This similar PPS is called a *prototype*. If a prototype or similar PPS cannot be found, a PPS for the customer ordered product needs to be created manually. After the PPS of a customized product has been found or created, the *Incremental Design and Process Planning* module and the *Cost Analysis Module* will be used to concurrently design the product and plan the relevant manufacturing processes feature by feature until the whole product has been designed, and a relevant and feasible process plan is hereby generated. The merits of this type of stepwise concurrent product design and process planning management method will be further discussed in the following section. The Cost Analysis Module is described in Chap. 4 and the Enterprise Production Data (EPD) is described in Chap. 8.

Since the *GT Code-based Searching* model can easily be found from several existing variant or generative CAPP systems (Alting and Zhang 1989; Sood et al. 1993; Kiritsis 1995), our focuses mainly on the development of the Incremental Design and Process Planning module and the Cost Analysis module, and the particular problems for data modelling and structures in OKP.

3.3 Incremental Design and Process Planning (IDPP) Module

As summarised by many authors, including Hirsch and Thoben (1992), Rolstadås (1991), Kuhlmann (1991) and Tu (1996a), the production of an OKP product is started with incomplete product data and is carried out under continuous customer influence. The development and manufacture of an OKP product is concurrently approached in an OKP company. The product data become clear through production and a consistent collaboration between the manufacturer and the customer. This evolutionary and concurrent product development and production makes it difficult for an OKP company to gain the agility to change the process plan and respond to the customer's input. To cope with this problem, a novel concurrent product design and process planning method, called incremental design and process planning (IDPP) method, and a cost analysis module have been developed to solve the problems as mentioned here and to expedite the OKP product development cycle.

A prototype or similar PPS, as mentioned in Sect. 3.2, is an integrated data model of product design or bill of material (BOM) and the operation routine or bill of operation (BOO) for making a product that is close to the customer's requirements or customized product. However, this prototype or PPS is not the final and even may not be feasible PPS for making the customized product. It needs to be further designed and modified in an iterative loop as shown in Fig. 3.1.

In this iterative loop, the incremental design and process planning module and the cost analysis module will be called to design the product and plan the processes in a stepwise manner, i.e. feature by feature. However, this PPS or prototype does consist of an initial design and a primitive process plan. According to our observations from OKP companies, although the OKP products are different from each other, the differences between two OKP products are often less than 20% of the overall product designs. Therefore, to develop a new product through modifying the prototype or PPS can significantly reduce the lead-time for product design and process planning. Therefore, this stepwise concurrent product development method is also called prototype-based incremental and concurrent product development method. Using this method not only reduces the lead-time of a product development but also increases the manufacturability of a customized product since the major portion (i.e. about 80%) of the product design and process plan is come from the previous successful experiences that are stored in the prototype or the PPS. Therefore, this method can effectively reduce the variable part (see Preface of this book) of the cost of an OKP product, which is the major part of the cost.

The IDPP module and the cost analysis module are used to extend or modify the prototype (or PPS) according to the differences or new features (including new functionalities, new geometric features and new packaging features) that are identified through comparing and contrasting the prototype and the technical specifications of the customized product. The iterative loop for incremental product design and process planning as shown in Fig. 3.1 will be carried out until no more new feature or difference can be found. The IDPP module is illustrated in Fig. 3.2.

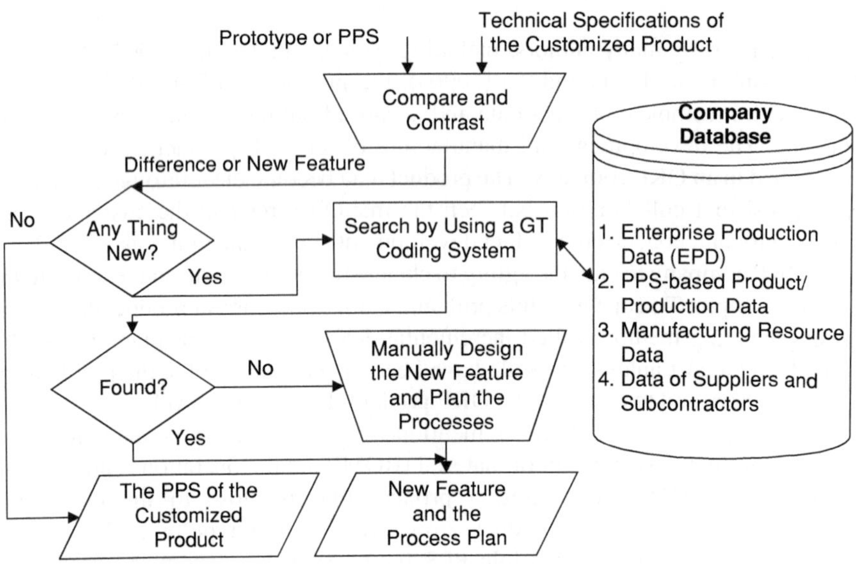

Fig. 3.2 Incremental design and process planning module

As shown in Fig. 3.2, the incremental design and process planning (IDPP) module starts with a manual process to compare and contrast the prototype and the technical specifications of a customized product according to the shape type, size accuracy, surface roughness and overall structure of an identified product design feature, as well as material, weight and production volume of a part. If a difference cannot be found or the current PPS meets the customer requirements perfectly, the IDPP module will jump to the final step in the stepwise product design and process planning module as shown in Fig. 3.1, i.e. outputting "The PPS of the Customized Product". Otherwise, the IDPP module finds a difference or new feature and it will use a GT coding system to search for the solution of design and process plan from the PPSs of the products which were made in the past. If the solution is found, the IDPP module outputs the solution, i.e. "New Feature and the Process Plan" to the cost analysis module for checking the manufacturability according to economic criteria. If the solution cannot be found, the engineer in an OKP company needs to design the new feature and plan the necessary processes for making this feature according to his/her experiences or industrial tests, e.g. to cut an uncertain sculptured surface of a mould, or to make a mock-up for a structural steel frame.

In order to evaluate the manufacturability of a part, a hierarchy of manufacturability criteria has been developed. The manufacturability criteria are divided into two categories (Chu and Holm 1994), i.e. economic criteria and technical criteria. The economic criteria are used and checked by the cost analysis module (see Chap. 4) as shown in Fig. 3.1. The technical criteria include standardisation of the part shape, manufacturing processes to be used, material and tolerance of the part, rationality of the machining dimension pattern, machinability and quality of a supplier or subcontractor. These technical criteria are recorded in the company database as the rules for manufacturability assessment.

Finally, it should be pointed out that a new feature and its relevant processes can often be designed and planned in a number of alternative ways. This means that a part can be processed by alternative machines in alternative sequences. The part can also be produced by or subcontracted to a subcontractor. This needs to be further determined by the cost analysis module under considerations of cost and lead-time (or economic criteria).

3.4 Basic Concepts, Methods and Control Structure for Production Planning and Control in OKP

Presently, the market available computer aided production control and management systems are commonly called Enterprise Resource Planning (ERP) systems. The leading ones are SAP, Oracle, Microsoft Dynamics, Consona ERP, Epicor Enterprise, SYSPRO ERP, Plex Online, Infor ERP, IFS Manufacturing, Global Shop One-System ERP Solutions, and Sage Software Manufacturing Solutions

(Business-Software 2010). These software systems have some difference in functionality, but the most functions are common. These common functions include: (1) Customer Relationship Management; (2) Supply Chain Management (including: procurement, supplier management); (3) Product Configuration; (4) Estimating and Quoting; (5) Inventory, Production Control, and Manufacturing; (6) Costing, Accounting and Finance; (7) E-commerce; (8) Human Resource, Time-and-Attendance, Payroll (Business-Software 2010; Aberdeen Group 2006). These market available ERP systems have been developed from a concept called materials requirement planning (MRP I), which assumes the companies to control their production inventory based on a hierarchy of data structure, i.e. Bill of Materials (BOM). All the production data (inventory, lead-time, and due date) are stored and updated according to the BOM. In other words, the BOM is used in these software systems as a key index, which is referred by professionals as data structure. Lately, these software vendors have realized that the production routine could be another key index in addition to BOM. Therefore, they simply add the routine into their data structure. However, these two data indexes are managed and retrieved separately (Jiao et al. 2007). These ERP systems work well in batch or mass production companies in which the BOM and production routines are relatively stable. However, in the OKP companies, both BOM and production routines change frequently. This has frustrated OKP companies in their use of ERP systems. Managers in Canadian SMEs have stated that: "there are a lot of holes in these ERP systems".

3.4.1 Product Production Structure (PPS)

Yao et al. (2007) discuss the need to integrate the relationship between product designs and manufacturing processes to facilitate computer-aided manufacturing. They propose an object-oriented system model that would be adaptive to changes in product design. Current research has primarily focused on the modelling of generic product families and creation of customized products from the generic product families based on requirements of individual customers (Siddique and Ninan 2006). The modular design approach has also been used in mass-customization research by separating functions of a design into modules and creating customized products by combining these modules (Zou et al. 2004; Mukhopadhyay and Setoputro 2005). The basis of producing information for OKP systems lies with the utilization of parametrics and templates (or models). Rolstadås (1991) proposes a concept for modelling information for one-of-a-kind products using material and resource graphs. Other researchers confirm that applications based on generic bill of material (GBOM) have been developed. The addition of operations in GBOM has been proposed by Hegge and Wortmann (1991), Du et al. (2002), and Huang et al. (2005). Tu et al. (2000) and Tu (1996b) propose a PPS (product production structure) for scheduling and controlling ship-web assembly operations. Different methods have been proposed to address the requirements of mass customization including product configuration (Zhang et al. 2004) and constraint-based knowledge systems (Li et al. 2005) in OKP or make-to-order companies. Zeng and Jin (2006) do a study on product configuration based on product models and conclude that product configuration and modelling are key technologies for realizing mass customization.

The concept of Product Production Structure (PPS) was initially proposed by Nielsen (1992) and Nielsen and Holm (1992). The PPS of a product is a joint logic illustration of the product design and process planning. In Nielsen (1992) and Nielsen and Holm (1992), a PPS consists of three types of legends, circles/ovals, boxes and arrow lines. A *box* of a PPS represents a final or an intermediate product state (e.g. a component or a work-in-process part). A *circle/oval* on a PPS represents an operation or a manufacturing process. The arrow lines indicate the sequence of the processes by which the product is produced.

To model the evolutionary and concurrent product development and production in OKP, the concept of PPS needs to be modified. The modified PPS still consists of three legends, i.e. circles/ovals, boxes and arrow lines. The box is re-defined as a transformation state by which a desired product state (e.g. a part) can be achieved from the starting product states (e.g. some components or raw materials). By this definition, a transformation state (i.e. a box) may consist of sub-PPS which is defined or panned to create this transformation product state. For the integrated production control in a global manufacturing environment, the three generic operations (i.e. circles/ovals) in the PPS are defined as follows:

1. "OS" means outsourcing operation.
2. "SC" means subcontracting operation.
3. "OP" means shop floor operation to be carried by an OKP company in house. It could be a manufacturing operation (e.g. machining, welding, cutting, assembly, surface processing, etc.), a material handling operation, or a storage operation.

By the concept of modified PPS, OKP product design and production decomposition is illustrated in Fig. 3.3. As shown in Fig. 3.3, the design and production decomposition of an OKP product is modelled in a three-dimension (3D) space, i.e. time dimension (T axis), product design dimension (Pd axis) and production specification dimension (Ps axis). Along the T axis, an OKP product is produced step by step and the production data are getting clearer and clearer. Along the Pd axis, an OKP product is designed from a general specification into detailed parts, intermediate assemblies and components. Along Ps axis, the product production is specified or decomposed from a general requirement or guideline into the detailed processes.

Therefore, using the modified PPS dada structure, a complicated OKP product and its production can be modelled or specified through the modification of a prototype or PPS according to the prototype-based incremental and concurrent product development method as presented in Sects. 3.2 and 3.3. As mentioned before, a prototype is the PPS of a product which was made in the history and has the most similarities to the customer's requirements or customized product. For the simplicity, in the following description of this book, the modified PPS is referred

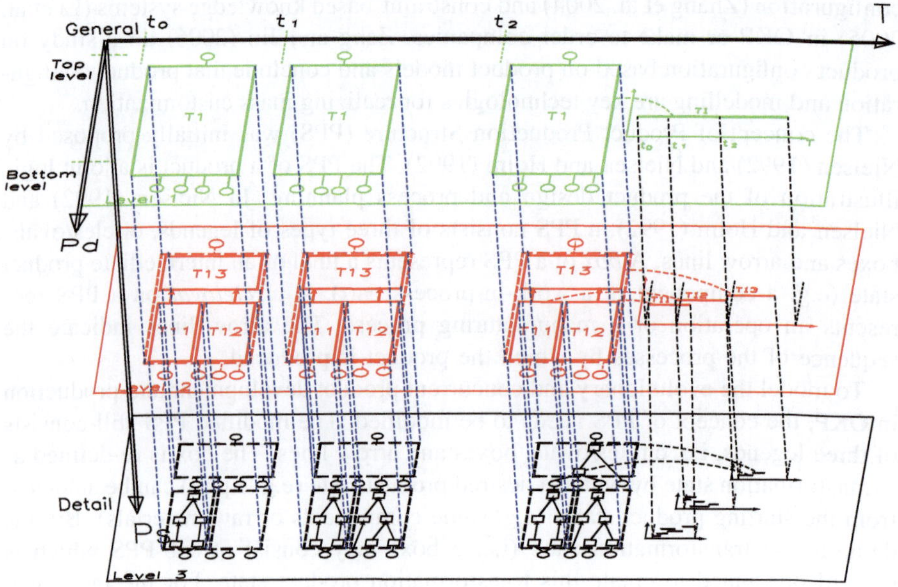

Fig. 3.3 Illustration of concurrent design and production specification of an OKP product

as PPS. With the PPS integrated data structure and the product design and process planning methods, an OKP product can be quickly specified through using a computer aided customer interface.

To illustrate the principles of the PPS, a case product of "Electrical Generator ACSG500" as shown in Fig. 3.4 is modelled or specified. The BOM of ACSG500 is listed in Table 3.1.

ACSG500 will be produced by the following operations:

OS31: Purchase EG500 (Electrical Generator Engine, See Table 3.1) from "Custom Part Supplier#2".
OP31: Build OC-18 (Outer Casing 18 cm) casing for the engine.
OS32: Purchase CP100 (Control Panel).
OP32: Build H24-18 (CP Housing 24 cm × 18 cm) for CP (Control Panel).
OP21: Install the engine (EG500) in the casing (OC18). It formulates the intermediate assembly part P2 (EG500 and OC18).
OP22: Install the control panel (CP100) in the housing (H24-18). It formulates the intermediate assembly part P3 (CP100 and H24-18).
OP11: Build SS24 (Steel Stand 24 cm) for ACEG500.
SC11: Install and wire the control panel (CP100) to the engine (EG500). It formulates the intermediate assembly part P1 (P2 and P3).
OP01: Assemble SS24 (Steel Stand 24 cm) to P1. It formulates the final product ACSG500 (electrical generator).
OP02: Quality control and testing.

Fig. 3.4 ACSG500 electrical generator

Table 3.1 Bill-of-material (BOM) of ACSG500

Part ID	Description	Qty	Unit cost	Supplier
CP100	Control Panel	1	2000.00	Custom Part Supplier#1
EG500	Electrical Generator Engine	1	1500.00	Custom Part Supplier#2
OC-18	Outer Casing 18 cm	1	300.00	Material in stock
SS24	Steel Stand 24 cm	1	120.00	Material in stock
H24-18	CP Housing 24 cm × 18 cm	1	200.00	Material in stock

Based on the BOM and planned processes or operations, the PPS of ACSG500 is drawn in Fig. 3.5.

The PPS as shown in Fig. 3.5 looks very similar to a project network diagram. However, the PPS is a combined flow chart of BOM and production routines or Bill of Operation (BOO), whereas the project network diagram can only be viewed as a BOO or activity flow chart. The OKP management and control software (OKPMCS) system©, which is described in Chap. 8, provides a graphic interface for OKP companies to easily create and modify the PPS of a customized product. Therefore, using the PPS data modelling technique and the OKPMCS system, an OKP company can easily change the BOM and the BOO of a customized product. After these changes have been made, the OKPMCS system will automatically update the relevant data in the central database. This provides a much easier way for OKP companies to cope with the frequent changes of BOM and BOO (or production routines), which overcomes the logic error of separating BOM and BOO in existing MRP II and ERP systems. From a cost or budget control point of view, the PPS model of a product provides a vehicle to concurrently record and control the material cost and operation cost of a project, whereas in project

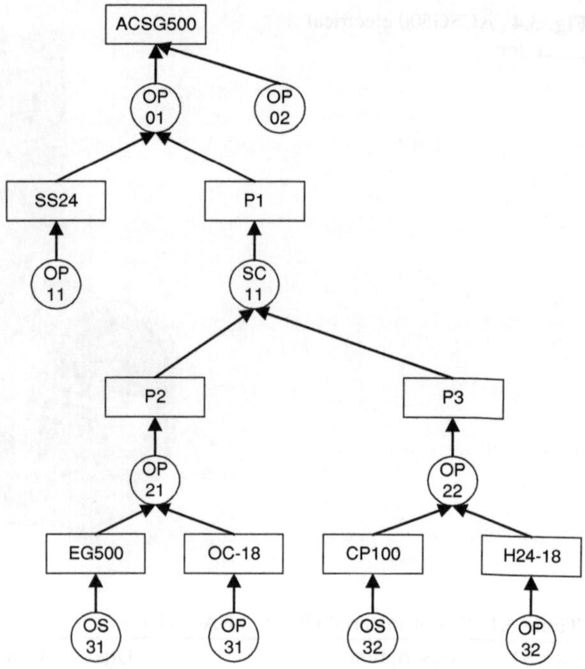

Fig. 3.5 Product production structure (PPS) of ACSG500

management software system (e.g. Microsoft Project 2007), these two types of cost are recorded separately. As mentioned in Sects. 3.2 and 3.3, the prototype-based incremental and concurrent product development method needs to be facilitated by the PPS data modelling technique.

3.4.2 Production Planning and Control

Onosato and Iwata (1993) divide a manufacturing system into two main subsystems. One is named Real and Informational System (RIS) which refers the engineering and management parts in a manufacturing enterprise. The other is named Real and Physical System (RPS) which refers the manufacturing part in the company. Corresponding to the RIS and RPS respectively, a computer simulation system can be found or built. The computer simulation systems are named Virtual and Informational System (VIS), and Virtual and Physical System (VPS). By the definitions of RIS, RPS, VIS and VPS, Onosato and Iwata (1993) define a virtual manufacturing system as RIS + VPS or VIS + VPS.

Following the ideas presented in (Onosato and Iwata 1993), two simulation systems are expected to be developed to support the production planning and control in an OKP company. One is used to model and simulate the managerial and engineering processes in an OKP company. This simulation system is named *Virtual Engineering & Management System.* A virtual engineering and management system can be developed by computer simulation technologies or well-defined "games". A game is an organisational simulation which is designed and played according to the former experience and knowledge as well as the estimations of the new production. Through playing the game, some conflicts and problems for a coming production project can be identified on beforehand. Then the production can be planned based on the experience achieved through playing the game (Riis 1994).

The other simulation system is used to simulate the OKP shop floors. This simulation system together with the virtual engineering and management system composes a virtual manufacturing system for an OKP company. A joint illustration of real and virtual systems of an OKP company is shown in Fig. 3.6. According to the concepts of virtual engineering and management system and virtual manufacturing system, the concept of virtual production control is described by two phases. Phase 1 is named *current production control* phase and phase 2 is named *production estimation and plan* phase.

In *current production control phase*, the production in the real manufacturing system is directly under control of the real engineering and management system. The productions in an OKP company are modelled or simulated by the simulation system. Virtually, the real engineering and management system views this simulation system as a complementary part of the real manufacturing system. It controls the production simulated by the simulation system just as it controls the production of a shop floor or shop floors inside the company. The production progresses or states in OKP company will be fed back to the simulation system and then the simulation results will be updated correspondingly. In this manner, the production planning and control tasks in the OKP company will be simplified and the productions in the company will be managed or controlled more precisely.

In *production estimation and plan phase*, the virtual engineering and management system will link the virtual manufacturing system to estimate and further plan the production in the next production period. By connecting the virtual engineering and management system to the computer simulation system, a total

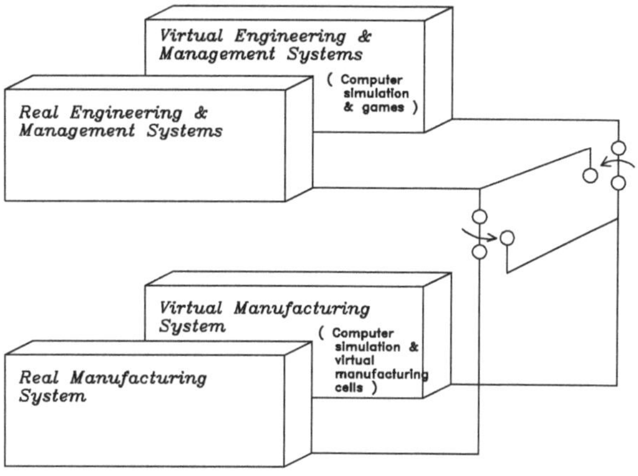

Fig. 3.6 Illustration of a virtual and real combined production system

production simulation system can be formulated. By this production simulation system, the production of an OKP product or OKP products can be simulated. By the simulation results achieved from this production simulation system, the detailed negotiations between the OKP company and its suppliers and subcontractors can be conducted and the further detailed production plan for the productions will be approached. By alternatively or even concurrently using the two phases of virtual production control principle, an OKP company can evolutionarily and concurrently develop and produce an OKP product. The objectives for OKP shop floor planning and control are high customisation, critical due date, low price and high quality.

3.4.3 Production Planning and Control Structure

In Fig. 3.7, a production planning and control structure in OKP is presented, which aims to cope with the marketing uncertainties in OKP. This control structure has been used to build the production control system in an OKP company. It can be flexibly extended or contracted according to the different customers' orders.

On the left side of Fig. 3.7, the shop floor control system of an OKP company is shown. It consists of four hierarchical floors. Except floor A, on the other floors, there may be several control systems. Each of the systems from floor A through floor D can be hierarchically built up by the sub-control systems on the different

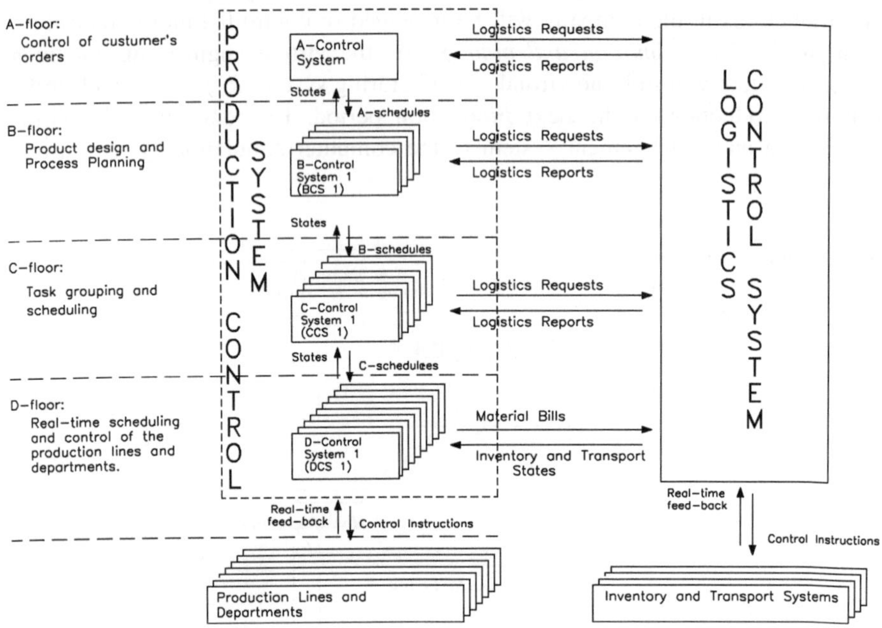

Fig. 3.7 A reference structure for OKP control systems

hierarchical levels if the control task is complicated. On the right side of Fig. 3.7, it shows the Logistics Control System in the company. The concepts and a reference control structure for structuring the logistics control system have been presented in Tu (1993), Tu and Holm (1993), and Tu et al. (1992a, b). Here, only the production planning and control system is briefly described in the following.

A-floor Control of customer's orders. The main task for the control system on this floor is to control the lead time of an OKP product. The decisions on this floor are normally made by the top managers in an OKP company according to a fixed due date for delivering a product to the customer and the knowledge about the production states and capacity of the company.

B-floor Product design and process planning. On this floor, each of the OKP products is decomposed individually and the necessary production processes are planned. An OKP product is decomposed into the standard pre-fabricated blocks or parts as many as possible and the customer's order driven fabricated blocks or parts as few as possible to reduce the cost and to improve production efficiency. On the B-floor, the control systems will set the rough time schedules for all different kinds of fabricated blocks of the product in coincidence with the schedules made on the A-floor. If several different OKP products are produced concurrently in an OKP company, every one of these products will be decomposed and scheduled by one of the control systems on the B-floor.

C-floor Task grouping and scheduling. On the C-floor, first of all, the standard fabricated blocks decomposed from different OKP products will be grouped so that the batch productions can be organised in shop floors inside the company as well as in subcontractor's or supplier's companies. Secondly, the production schedules will be made by the control systems on the C-floor in accordance with the production schedules made on the B-floor.

D-floor Real-time scheduling and control of the shop floors or subcontractors and suppliers. Each of the control systems on the D-floor controls one and only one shop floor or a subcontractor/supplier. The main control object of a control system on the D-floor is to manage and schedule the production activities on a shop floor or in a subcontractor's or supplier's shop floor according to the production schedules made on the C-floor. The control tasks of a control system on the D-floor include production activities scheduling, and planning and real-time control of the critical resources. The critical resources mean all kinds of critical equipment (e.g. machines, robots and local transport systems), and the workers. An adaptive production scheduling and control system to be described in Chap. 7 is adopted for the control systems on this floor.

Finally, it shall be noted that a reference control structure presented here is understood as a reference model for developing a production planning and control system in an OKP company. It does not mean that all kinds of production planning and control systems in OKP shall include all four floors completely. Sometimes, if the control task is simple, the control system may consist of only one or two floors.

3.4.4 Data Modelling for Managing OKP

To create the database as shown in Fig. 3.1, three data models, i.e. product model, process model and manufacturing plant model, are developed. We briefly describe these three data models in the following.

Product Model. As shown in Fig. 3.1, a PPS based product reference model has been developed. The primary use of this model is to describe all attributes of a product from its design to its manufacture. This model consists of four sub-models: geometric feature model, shape model, tolerance model and administration model. The geometric feature model is used to describe portions of the skin of a part that conforms to some machining methods and the relationships among these manufacturing features. The shape model is used to represent the part shape. The tolerance model describes the geometric and shape position dimension tolerance of a part. The administration model includes product material, overall shape dimension, batch size, predicted number, and GT code.

Process Model. It is a reference model of manufacturing processes. In this model, manufacturing processes are classified as primary processes, primary/secondary processes and tertiary processes (Candadai 1994). Primary processes are main-shape generating processes such as casting and forging. Primary/secondary processes generate both the main-shape of the part and the manufacturing features. Tertiary processes do not affect the geometry and comprise mainly grinding or finishing operations. We use a set of data to describe the capabilities of manufacturing processes, which include process name, type, economic production quantity, machining accuracy and surface roughness, and suitable material shape.

Manufacturing Plant Model. As shown in Fig. 3.1, a diverse amount of data about the suppliers and subcontractors is needed for the IDPP and the cost analysis modules. Hence a manufacturing plant model is developed. The data in this model include production resources, production states of the subcontracting parts or outsourcing components, relationships with the company, operational data, quality standards and plant capabilities.

3.5 Chapter Summary

In this chapter, the concepts, methods and computer modules for rapid product development in one-of-a-kind production (OKP) are presented. The emphasis is placed on the special problems of product design, process planning and production control in OKP. These problems are summarized in Sect. 3.1. The OKP product is often designed and manufactured concurrently and the production sometimes starts with ambiguous customer requirements and uncertain product design and process plan. Therefore, the OKP product development involves high risk and uncertainties, and it requires concurrent engineering approach. These problems have caused a long product development cycle and a high product development cost in OKP. To solve these problems, we have developed a number of novel and effective methods and computer modules. These include:

1. Prototype-based incremental and concurrent product development method that consists of the stepwise product design and process planning management module and the incremental design and process planning (IDPP) module;
2. Product production structure (PPS) that is an integrated data modelling technique;
3. Virtual production control technique which is an application of computer simulation technology for OKP shop floor control (Phase 1), and OKP scheduling and control (Phase 2);
4. A four-floor system hierarchy for adaptive OKP scheduling and control.

These novel methods and computer modules are developed to particularly address the special problems of product development in OKP. Some of these have been implemented in a heavy engineering company in Hong Kong, a customized sheet metal product manufacturing company in New Zealand, a customized window and door manufacturer in Canada, and special equipment manufacturers in Canada and China.

References

Aberdeen Group (2006) Benchmarking ERP in SMB. http://www.aberdeen.com/summary/report/. Accessed 1 June 2010

Alting L, Zhang H (1989) Computer aided process planning: the state-of-the art survey. Int J Prod Res 27(4):553–585

Bernardi A (1993) PIM-skeletal plan-based CAPP. Comput Ind 23:87–97

Bjorke O, Myklebust O (1992) IMPPACT: integrated modelling of products and processes using advanced computer technologies. Tapir, Trondheim

Business-Software (2010) Top 10 manufacturing software vendors REVEALED, profiles of the leading manufacturing software vendors, 2010 edn. http://www.business-software.com/erp/manufacturing.php. Accessed June 2010

Candadai A (1994) Information needs in agile manufacturing. Engineering data management: integrating the engineering enterprise, pp 101–110

Chang TC, Wysk RA (1985) An introduction to computer-aided process planning systems. Prentice Hall, Englewood Cliffs

Chen Q, Khoshnevis B (1993) Scheduling with flexible process plans. Prod Plan Control 4(4):333–343

Cho H, Derebail A, Hale T, Wysk RA (1994) A formal approach to integrating computer-aided process planning and shop floor control. J Eng Ind 116:108–116

Chu X, Holm H (1994) Product manufacturability control for concurrent engineering. Comput Ind 24(1):29–39

Du X, Jiao J, Tseng M (2002) Graph grammar based product family modelling. Concurr Eng Res Appl 10(2):113–128

Hegge HMH, Wortmann JC (1991) Generic bill-of-material: a new product model. Int J Prod Econ 23:17–128
Hirsch BE, Thoben KD (eds) (1992) Future research in one-of-a-kind production, one-of-a-kind production: new approaches. Elsevier Science Publishers B.V., North-Holland, pp 87–94
Huang GQ, Zhang XY, Liang L (2005) Towards integrated optimal configuration of platform products, manufacturing processes, and supply chains. J Oper Manage 23:267–290
ISO (1993) TC 184/sc4 n81 ISO DIS 10303-1 product data representation and exchange—part 1: overview and fundamental principles
Jiao J, Zhang L, Pokharel S (2007) Process platform planning for variety coordination from design to production in mass customization manufacturing. IEEE Trans Eng Manage 54(1):112–129
Jones DT (1992) Beyond the Toyota production system: the ear of lean production. In: Voss CA (ed) Manufacturing strategy: process and content. Chapman & Hall, London, pp 189–210
Khoshnevis B, Park JY, Sormaz D (1994) A cost based system for concurrent part and process design. Eng Econ 40(1):101–124
Kidd PL (1994) Agile manufacturing, forging new frontiers. Addison-Wesley, New York
Kiritsis D (1995) A review of knowledge-based expert systems for process planning, methods and problems. Adv Manuf Technol 10:240–262
Kuhlmann T (1991) A revolving planning and control system. In: Proceedings of new approaches towards 'one-of-a-kind production', Bremen, pp 95–112
Li B, Chen L, Huang Z, Zhong Y (2005) Product configuration optimization using a multi-objective genetic algorithm. Int J Adv Manuf Technol 30:20–29
Link CH (1976) CAPP-CAM-I automated process planning system. In: Proceedings of the 1976 NC conference
Maropoulos PG (1995) Review of research in tooling technology, process modelling and process planning, part I: tooling and process modelling, part II: process planning. Comput Integr Manuf Syst 8(1):5–20
Mukhopadhyay SK, Setoputro R (2005) Optimal return policy and modular design for build-to-order products. J Oper Manage 23(5):496–506
Nielsen JA (1992) Concept for real-time control of automatic shop floor systems in one-of-a-kind production. Ph.D. thesis, Aalborg University, Denmark
Nielsen JA, Holm H (1992) Real-time control system for automatic one-of-a-kind production: specification and implementation of a test bed. In: Proceedings of 1992 IEEE international symposium on intelligent control, Glasgow, Scotland, pp 430–435
Onosato M, Iwata K (1993) Development of a virtual manufacturing system by integrating product models and factory models. Ann CIRP 42(1):475–478
Park JY, Khoshnevis B (1993) A real-time computer-aided process planning system as a support tool for economic product design. J Manuf Syst 12(2):181–193
Riis JO (1994) Games and organizational development, production management methods. In: Proceedings of the IFIP WG5.7 working conference on evaluation of production management methods, Gramado, Brazil, 21–24 Mar 1994. IFIP Transactions B-19 North-Holland 1994, ISBN 0-444-81910-X, 11–18
Rolstadås A (1991) ESPRIT basic research action no. 3143—FOF production theory. Comput Ind 16:129–139
Siddique Z, Ninan JA (2006) Product portfolio identification based on association rule mining. Integr Comput Aided Eng 13(2):133–148
Sood S, Wright PK, MacFarlane J (1993) Process planning: a review. DE-Vol. 66, intelligent concurrent design: fundamentals, methodology, modelling and practice, ASME, pp 45–54
Tu YL (1993) Real-time scheduling and control of material handling and equipment set-up in automatic shop floor for one-of-a-kind production. Ph.D. thesis, Aalborg University, Denmark, 300 pages, ISBN: 87-89867-30-0
Tu YL (1996a) A framework for production planning and control in a virtual OKP company. Technical papers of North American Manufacturing Research Institution of SME, pp 121–126

Tu YL (1996b) Automatic scheduling and control of a ship welding assembly line. Comput Ind 29(3):169–177

Tu YL, Holm H (1993) Control system for material handling and equipment set-up in one-of-a-kind production. In: Proceedings of the IASTED international conference on robotics and manufacturing, 23–25 September 1993, Oxford, England, pp 61–64

Tu YL, Holm H, Nielsen JA (1992a) The structure of service task control system in an automatic shop floor for one of a kind production. In: Proceedings of the IMACS/SICE international symposium on robotics, mechatronics and manufacturing systems, Kobe, Japan, 16–20 September 1992, II, pp 1203–1208

Tu YL, Holm H, Nielsen JA (1992b) The service task control system of automatic shop floor for one of a kind production. In: Proceedings of 6th IPS (integrated production system) research seminar, 23–25 March 1992, Fuglsø, Denmark, pp 49–65

Tu YL, Chu XL, Yang WY (2000) Computer aided process planning in virtual one-of-a-kind production. Comput Ind 41:99–110

Womack JP, Jones DT, Roos D (1990) The machine that changed the world. Rawson Associates, New York

Wortmann JC (1991) Towards one-of-a-kind production: the future of European industry. In: Eloranta E (ed) Advances in production management systems. Elsevier Science Publishers BV, North-Holland, pp 41–49

Yao AN, Yang X, Rong Y (2007) Computer-aided manufacturing planning for mass customization: part 3, information modelling. Int J Adv Manuf Technol 32(1–2):218–228

Zeng FS, Jin Y (2006) Study on product configuration based on product model. Int J Adv Manuf Technol 33:766–771

Zhang WJ (1996) On methodology of developing a conceptual database model to support partner synthesis in virtual enterprises. In: Proceedings of the 3rd CIRP workshop on design and implementation of intelligent manufacturing systems, 1996, Tokyo, Japan, pp 137–144

Zhang J, Wang Q, Wan L, Zhong Y (2004) Configuration-oriented product modelling management for made-to-order manufacturing enterprises. Int J Adv Manuf Technol 25:41–52

Zou G, Hu Y, Wang X, Li C (2004) The representation of conceptual product based on component-connector design feature with P/T net approach. Int J Adv Manuf Technol 26:1193–1201

Chapter 4
Cost Estimate and Control in One-of-a-Kind Production

As mentioned in previous chapters, one of the goals of One-of-a-Kind Production (OKP) is to achieve mass customization that aims to achieve economies of scope (or customization) and scale (mass production) (Tseng and Jiao 1996). In the traditional manufacturing paradigm, customization is in conflict with mass production with respect to product development lead-time and cost. Normally, a trade-off must be made between customization and low cost through mass production. However, the rapid developments of computer and Internet communication technologies, concurrent engineering and computer aided design methodology are starting to allow for a greater involvement of customers and suppliers in the development of a product, which formulates a value chain and is called the product development (PD) chain (Poirier and Reiter 1996; O'Grady and Chuang 2001) or product-oriented supply chain (Huang and Mak 2000). The involvement of customers and suppliers in the whole product development life cycle through e-commerce technologies is a promising and possible approach of mass customization that has the potential of reaping substantial benefits (O'Grady and Chuang 2001; Huang and Mak 2000; Wasti and Liker 1999); yet, it is not well studied as indicated by O'Grady and Chuang (2001). In a PD chain, the cost control through a proper or optimal plan and a selection of various PD processes or suppliers are very critical to the success of one-of-a-kind production.

The PD cost can be controlled in two stages, i.e. product specification and design stage and product production stage, through a PD chain. At the first stage, it is, in turn, a problem on accurately estimating the possible costs associated with a PD cycle, whereas at the second stage the problem is on how to control the production to stick to the planned budget or cost. Following the concept of two-stage PD cost control, we discuss cost estimate in stage one and cost control in stage two respectively in this chapter.

Estimate of PD cost is closely related to the selection of a mix of PD processes and their relevant suppliers, the selection of which can meet the customers' requirements with the lowest sum of the costs. This refers to several issues or problems. First, product development in a global manufacturing environment has

Y. Tu and P. Dean, *One-of-a-Kind Production*,
DOI: 10.1007/978-1-84996-531-6_4, © Springer-Verlag London Limited 2011

been proved to have advantages, such as agility, and quick and cheap solutions for parts and technologies from suppliers, which are critical to the success of the whole product development, customer satisfaction, etc. These advantages are obviously the goals that need to be achieved through mass customization (Pine 1993). However, a well-established PD management system (e.g. cost analysis module) in mass customization is still not available in the commercial market. Although a relatively large number of papers and reports have been found to address the cost estimate and control problems in various PD processes, these methodologies were normally developed based on cases in traditional manufacturing companies rather than OKP companies which are often running in a global manufacturing environment to gain agility and mass customization. In a global manufacturing environment, a manufacturing company (normally called a master company in a PD chain) may carry out a PD cycle through collaboration with its customers, suppliers and subcontractors. These customers, suppliers and subcontractors may use different PD management systems and computer-aided engineering software systems. To communicate effectively, share the necessary information, integrate the efforts and optimally control the PD cost between the master company and its customers and suppliers/subcontractors are problems that need to be addressed through the development of a computer-aided PD cost estimation and control module for one-of-a-kind production.

In generally speaking, PD cost estimate and control is an interdependent and correlated problem. It is influenced and dynamically determined/changed by a number of preconditions. The cost of a product design, for instance, will be strongly influenced by the correctness and accuracy of the product definition. A correct understanding of customers' requirements and hence a clear product definition will obviously result in a quick and cheap product development. Likewise, the cost of manufacturing a part is normally affected by the design of the part. The final form of the part design, which includes the geometry, surface, tolerance and property specifications of the part, either directly or indirectly influences the selection of manufacturing processes and hereafter the cost of manufacturing the part. The bad experiences gained from manufacturing industries (e.g. sheet metal manufacturing companies, compressor manufacturers, customized window and door manufacturers, etc.) show that decisions made at the earlier design stages without PD costing considerations normally result in an incorrect product design and development process. An incorrect PD cycle will cause the company to suffer a long PD cycle time due to many reworks, and poor product design and manufacture, which often results in low customer satisfaction and consequently a high PD cost.

The PD cost estimate and control are, no doubt, effective tools to monitor a PD cycle in OKP. The cost estimate results can be used as important reference parameters or indicators to check the feasibility and profitability of a product to be developed through a PD chain. Reworks in the downstream PD process, e.g. reworks in manufacturing, are more costly than corrections made in earlier PD stages, e.g. in the design stage. PD cost estimate is obviously a planning and control activity that takes place at earlier PD stages. Hence, the accurate estimation and optimal control of the cost of PD are crucial to the reduction in PD cycle time and PD cost that is necessary for mass customization. The development of useful methods, algorithms and a system framework for building a computer-aided semi or fully automatic cost estimate and control module, to support this planning and control activity at earlier PD stages, is urgently needed for the successful implementation of the OKP strategy and technology to achieve mass customization in manufacturing companies.

Before we discuss the cost estimate and optimization methods and algorithms, we would like to define a PD cost in OKP as follows:

$$PD_{cost} = C_{dd} + C_p \tag{4.1}$$

where C_{dd}, product definition and design cost; C_p, product production cost.

Product definition and design cost is mainly determined by the time consumed by the product design engineers and by the communications among engineers, customers, suppliers and subcontractors. These costs and times can be planned using a mathematic programming model. We have developed this mathematic programming model to maximize customer satisfaction under the constraints of customer requirements, technical requirements, resource availability and a limited or fixed design budget (measured in dollars) (Tu et al. 2002a). This means the product definition and design cost can be planned by using this mathematic programming model and does not need to be estimated and controlled by the computer-aided cost estimate and control module. On the other hand, the product design in OKP companies is in fact the modifications or revisions of a base product or prototype (see Chap. 3) according to customer requirements. The cost of product design can be easily calculated or quoted by:

$$\text{Cost of Product Design} = (\text{The total design hours per engineer}) \times (\text{hourly rate of engineering})$$

This part of the cost is relatively clear and it will not affect the PD process. The ambiguous and troublesome part of the PD cost estimate and control is the production cost. Therefore, the discussion of this chapter focuses on the estimate and control of the total product production cost.

The product production cost is the sum of the costs of all the manufacturing processes or operations and consumption of resources and materials. The cost of a manufacturing process or operation includes:

Cost 1 Test and possible rework fees due to the difficulty of the process/operation.

Cost 2 The total consumption of the necessary manufacturing/production resources to carry out the process/operation. It includes the consumption of labour, machines, and tools.

Cost 3 The total logistics cost for purchasing and/or transporting a raw material, component or subcontracted manufacturing/production process, and relevant logistics management cost (e.g. communication cost and paper/data management cost) through a PD chain. If a manufacturing process or project operation is outsourced or subcontracted, it normally includes the logistics cost. If a raw material or component is within the company, the logistics cost is equal to the price of the material or component plus inventory and material handling costs within the company. In short, to simplify the cost estimate and control in a global manufacturing environment, we consider material handling and purchasing cost as a part of the logistics cost of a manufacturing process or project operation.

4.1 Review of PD Cost Estimate and Control Techniques

The rapid and economical development of a product has attracted a lot of attention from practitioners and scholars. Quite a few research papers have been found that address the problems of rapid and economic product development from the point of view of PD cost estimate and control such as: design for cost (DFC) and economical manufacturing (Ou-Yang and Lin 1997), neural network based costing and control (Bode 2000), cost estimation based on information management (Kals et al. 1999), concurrent costing (Ou-Yang and Lin 1997), activity-based costing (Koltai et al. 2000), etc. Besides our past work (Tu et al. 2007), to our best knowledge, no paper has been found that meaningfully addresses a computer-aided cost estimate and control module in OKP. The papers only addressed the problems in a traditional manufacturing company and did not take into consideration of the dynamic relationships among manufacturing partners and the complexity of cost estimation and control in OKP.

As indicated by Becker and Prischmann (1993), the accuracy or reliability of a PD cost estimate is increased along with the PD progress. A cost estimate at earlier PD stages is normally very rough and not reliable. Sometimes, an over-estimated PD cost of a product at an early PD stage may mislead a company into giving up a product that may have had a great potential market. On the other hand, an under-estimated PD cost may trap the company into a nearly endless investment and may result in a product that has a low or even no market share. Hence, the correct estimation of the PD cost at earlier PD stages is very important for rapid and economic PD, particularly in mass customization.

To support decision-making at early PD stages, quite a few research efforts have been made for cost estimation at earlier PD stages. Typical approaches include parametric-oriented costing, similarity-based costing, activity-based costing, detailed costing, life cycle costing, learning curve costing and neural network based costing (Bode 2000; Smunt 1999; Koltai et al. 2000; Bode 1998). Monden and Lee (1993) report a cost estimate system called Kaizen Costing. This system was used for budgetary control and cost reduction of product development. Niemand and Wolbold (1994) propose a standard costing system to meet cost limits (or standards). Kawada and Johnson (1993) present an accurate cost estimation system to support strategic management accounting. Cawthorne-Nugent et al. (1989) develop a computer-aided costing system that employs knowledgebase technology and the cost models associated with part features. Marx et al. (1998) further report another application of knowledgebase technology for PD cost estimation. To solve the problem that, at earlier PD stages, there is not enough information and clear functional relationships between the product attributes and costs, as well as the ambiguity of the prices or costs from suppliers, some artificial intelligence and fuzzy logic techniques, such as learning curves (Becker and Prischmann 1993), neural networks (Tu and Jiang 1997) and self-learning methodologies (Zhang and Fuh 1998) were applied to PD cost estimation.

Manufacturing cost in OKP is always an important factor that influences early decision-making on the design, process planning, manufacture and final market share of the product. It is a major part of the overall PD cost. Usually, to determine the manufacturing cost, a lot of detailed data is needed such as: selected manufacturing processes and the sequence of the processes, the processing times of the processes, the hourly rates of labour forces and machines, raw material cost, subcontract cost, transportation cost, logistic chain (or supply chain) management cost, overhead cost, and equipment depreciation and maintenance cost. Traditionally, manufacturing cost is estimated through the addition of the aforementioned costs. A computer is often used as a tool to speed up these additive calculations. Typical examples for applying this traditional manufacturing cost estimate method can be found in (Casey 1987; Goldberg 1987; Ostwald 1984). Winbourne and Toolsie (1991) propose a manufacturing cost estimate system that can be used at earlier product design stages to estimate the manufacturing cost of a product under the uncertainty and ambiguity of the manufacturing process plan. A major drawback for this approach is that the proposed system requires some initial estimates of the cost in terms of product design and manufacturing parameters, e.g. surface treatment, machine set-up or tear-down times, etc. These parameters are difficult to determine at earlier PD stages through a PD chain in mass customization. All the approaches, as reviewed above, did not explain how to estimate accurately the cost values of the relevant product design and manufacturing parameters. Wang and Bourne (1995) present a manufacturing cost estimating system, called the totally integrated manufacturing cost estimating system (TIMCES), which integrates the computer-aided design (CAD) system, computer-aided process planning (CAPP) system and computer-aided cost estimate (CACE) system into one system. Inputs to this system include the selected manufacturing processes, the sequences of the processes, and cost data for standard components and manufacturing processes. However, the TIMCES is just another application of the traditional manufacturing cost estimate method. It needs a large amount of complicated input data that is not normally available or is uncertain in the earlier PD stages due to high customization in OKP. Kamarthi et al. (1993) propose a manufacturing cost estimate system for sheet metal parts/products manufacturing. In fact, this system employed cost as an important parameter in the selection of the manufacturing processes and available machine tools, in order to have lower costs at every process planning stage. Their work implies that the cost estimate and product design as well as process planning need to be integrated to consider in a customized sheet metal manufacture company. However, their system only considers the standard process sequence and a two-dimensional (2D) shape of sheet metal parts/products. It is lack of flexibility to cope with high customization in OKP.

The research efforts, as reviewed above, have made great progress in developing and implementing a computer-aided semi or fully automatic PD cost estimate system. However, from the viewpoint of OKP, two common drawbacks of these approaches can be identified:

1. They all focus on the problem in a traditional manufacturing company which normally produce standard products in large batches rather than in an OKP company which mass-produces customized products and is often working in a global manufacturing environment. Hence, the methods and systems developed from these research projects cannot meet some special requirements in a PD process in OKP such as: a relatively frequent change of a product design or a manufacturing process due to changes of the suppliers or customer requirements; a quick price or cost change in the global market; relatively high uncertainty and ambiguity of product design and manufacturing processes in earlier PD stages in OKP; and, complicated cost estimation and optimal control when searching globally for an available manufacturing resource.
2. All the previous and existing research work on computer-aided cost estimate systems assume that the processes through a PD cycle are carried out sequentially rather than concurrently. However, an OKP company often develops its products concurrently in order to achieve responsiveness to customer needs and, hence, to seize a market opportunity rapidly.

These two major drawbacks limit the application of these cost estimation methods and systems in OKP. Therefore, novel concepts, methods and a framework are presented in the following sections of this chapter, which are the fundamentals for the development of the computer-aided cost analysis module. To validate the feasibility of this computer aided cost analysis module, the concepts and the methods, case studies are presented in a later part of this chapter.

4.2 Cost Index Structure

To estimate and control the production cost in OKP, a cost index structure has been developed, as illustrated in Fig. 4.1. It has a hierarchy similar to the product production structure (PPS) as presented in Sect. 3.4.1.

As shown in Fig. 4.1, every operation in the PPS of a customized product will be associated with a cost index. A cost index includes three generic data classes, viz. adding, resources and logistics, which are associated with the possible three costs of a manufacturing process. "Adding" means costs added to a manufacturing process. It is used to pay for possible tests and reworks due to the difficulty or

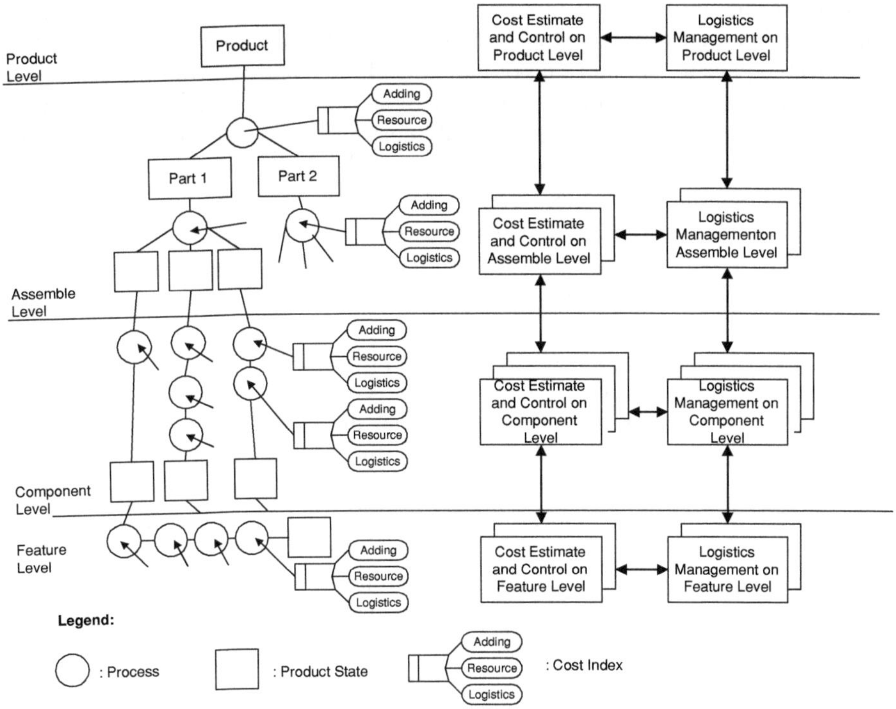

Fig. 4.1 Cost index structure

uncertainty of the process. The difficulty or uncertainty of a process is determined by the design of a product, part, component or feature (e.g. geometry, surface finishing or treatment requirements, special tolerance and machining requirements, etc.). This is the first part of product production cost (i.e. Cost 1) as defined at the outset of this chapter. "Resources" mean the cost of consumption of manufacturing resources as the second part of the product production cost, which was defined as Cost 2 at the outset of this chapter. "Logistics" means logistics cost, as Cost 3 discussed at the outset of this chapter. These cost data classes are coded as data entrances in a central database so that the computer aided cost analysis module can retrieve and update these data as well as the other computer software systems in a PD chain can access and retrieve these cost data. This is particularly important for OKP, in which possible subcontractors may use different computer systems. The data, which are in a relational database, are text format and can be easily transferred via the Internet using XML (extensible mark-up language) formatting.

The idea to build the cost index structure in a hierarchy, on one hand, is to keep in line with the hierarchical nature of a product production structure (PPS). On the other hand, this is also a natural or logical way to estimate, control and record the cost of product production from the feature level on the bottom up to the product level on the top in OKP. It is obvious that the cost of a feature that is added to a component is a part of the cost of the component. In other words, the cost of the component equals to the total sum of the costs of the features. After the costs of all the components are determined, the cost of a part, if it is assembled in these components, will be equal to the assembly cost plus the total cost of the components. Likewise the cost of the product, if it is assembled in several parts, will be equal to the assembly cost plus the sum of the costs of all the parts. Hence, it is natural and logical for the computer-aided cost analysis module to estimate and control the cost of a product production from the bottom level up to the top level of the hierarchy, as shown in Fig. 4.1. The models, e.g. cost estimate and control models on the component level, will estimate or search for the cost data (i.e. adding, resources and logistics costs) from suppliers or the company database and, thereafter, will assign values to the generic cost index associated with each manufacturing process. It should be noted that the cost data from the cost estimation models on a lower level (e.g. cost estimate and control models on the feature level), can be used by the cost analysis module to derive or infer cost data in the cost index on a higher level (e.g. component level). The logistics management model, as shown in Fig. 4.1, includes a number of logistics management sub-models that are organized hierarchically from the feature level on the bottom and up to the product level on the top. These logistics management sub-models are Internet- and agent-based searching models, which are able to communicate with and search for cost data from suppliers and subcontractors (Tu and Kam 2004).

With the cost index hierarchy shown in Fig. 4.1, a PD cost can be decomposed from level to level into detailed expenditures or indexes. In fact, most of these expenditures on different levels of the cost index structure can be directly estimated by using the price of a standard component and part, or the cost of a common manufacturing process. This is in line with the theme of the mass customization, which encourages the concepts of modular design and common product family architecture (Tseng and Jiao 1996; Jiao et al. 2003). In practice, most non-standard manufacturing processes can normally be viewed as combined manufacturing processes, which may be derived by the linear combination of these standard prices or costs of common processes. To facilitate this cost estimate process and method, a costing knowledgebase needs to be developed as an important part of the computer-aided cost analysis module to record structurally the standard prices and costs of common processes. However, it is difficult to propose a general data structure to build such a costing knowledgebase for all kinds of OKP companies. The data structure of a costing knowledgebase for a particular OKP company, e.g. a sheet metal manufacturing company (Tu et al. 2007), can be built according to the company's experiences, special manufacturing requirements and practices.

Although a costing knowledgebase is business-oriented or case-based, the cost index structure, as proposed in Fig. 4.1, is generic as it is defined by three generic data classes, i.e. adding, resources and logistics. It can be applied to all kinds of OKP companies. To be in line with this cost index hierarchy, the costing knowledgebase needs to be developed to store the cost data in a similarly hierarchical structure, i.e. the data tables in the database are linked and indexed from a table that records the product data information on the top, and then hierarchically from level to level down to the data tables that record the feature data information on the bottom.

4.3 Cost Estimate

With the cost index structure described in previous section, two cost estimate methods, viz. the generative cost estimate method and the variant cost estimate method, can be described in the following two subsections. These two cost estimate methods have been developed through our research for the computer-aided cost analysis module in OKP to search or derive semi or fully automatically the cost values or instances for these data classes in a cost index.

4.3.1 Generative Cost Estimate Method

The generative cost estimate method is a pipeline method. With the cost index structure shown in Fig. 4.1, through searching price or cost data in the costing knowledgebase, the computer-aided cost analysis module should be able to estimate or calculate the data or instances of the cost indexes associated with manufacturing processes from the bottom level up to the top level of the cost index structure. The cost index of an operation, e.g. "Install engine in casing" for making the electrical generator as shown in Fig. 3.4, can be determined by calculating the adding, resources and logistics costs of the operation. As mentioned before, the adding cost is decided according to the difficulty of the process, i.e. installing engine in casing in this example, and it is the sum of the costs of the possible tests and reworks. Since this operation is not so difficult, the adding cost is set as zero. The resources cost for this example will be calculated by multiplying the Hourly Rates of the manufacturing resources used (e.g. assembly tools and the workers) to the Work Efforts (i.e. the total time used by these manufacturing resources). The Hourly Rates are recorded in the costing knowledgebase and the time used can be obtained from the manufacturing process planning or dedicated computer simulation packages (e.g. numerical control (NC) simulator in commercial CAD systems). For this example, the Hourly Rate is $30 and the time used is 40 h. Consequently, the resource cost is $1200. The logistics cost is determined according to if this operation will be carried out on a new (or added) material or component, and how and from where this material or component can be provided. As mentioned before, the logistics cost is the sum of the costs of the purchased item, transportation (material handling plus inventory cost if the item is in the company), and communication management. For the example of "Installing engine in casing", its logistics cost is set as zero since the parts for assemble

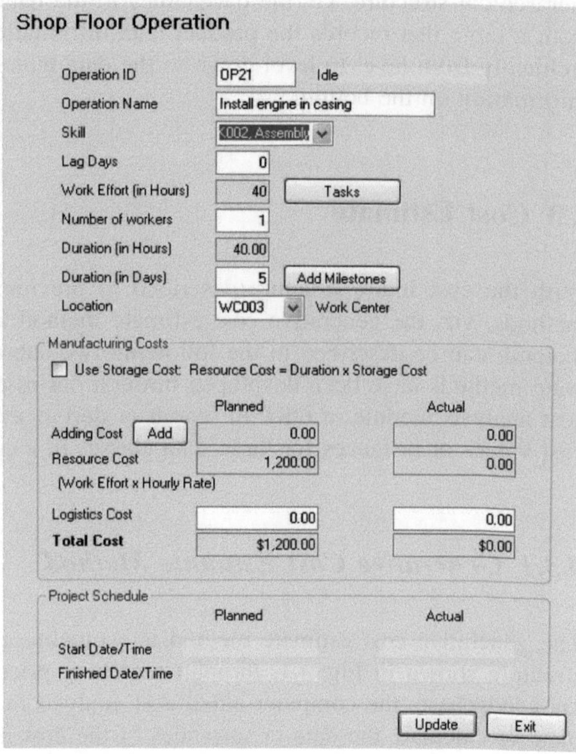

Fig. 4.2 Cost estimation of an operation making an electrical generator

engine and casing have the logistics costs quoted or estimated on lower levels of the PPS of the case product (see Fig. 3.5). The screen snapshot for this example from the cost analysis module of the OKP management and control software (OKPMCS) system©, which will be described in Chap. 8, is illustrated in Fig. 4.2.

Normally, a supplier will quote the transportation cost as a part of the quotation for providing an item. Hence, the logistics management model can provide the costs of the purchased item and relevant transportation cost by inquiring and selecting a quotation from suppliers. The communication management cost for sourcing an item depends on the difficulty of sourcing this item. In practice, the manager of a company can set up some guidelines or controlled budgets for sourcing various items. These controlled budgets can be recorded in the costing knowledgebase and be used by the computer-aided cost estimation and control system to quote or estimate the communication management cost as a part of the logistics cost of getting a required item. Refer to the three types of operation as defined in Sect. 3.4.1, a shop floor operation (OP) may not have the logistics cost, an outsourcing operation (OS) may not have the adding cost, and a subcontracting operation (SC) may have all three costs.

When all the cost indexes have been assigned data or instances, the production cost of a component, part or product can be calculated by the computer-aided cost analysis module by adding up all the cost data in the cost indexes associated with the processes that are used for making the component, part or product in the cost index structure.

4.3.2 *Variant Cost Estimate Method*

The generative cost estimate method, as described in Sect. 4.3.1, can only be used to estimate the costs of standard processes. For those non-standard processes (e.g. new parts) that are required by the product design to address some particular customer requirements, the variant cost estimate method needs to be employed. The variant cost estimate method estimates the cost of a process by searching and comparing the similarities between the developing product and products that were made in the past.

Using the PPS data structure (see Sect. 3.4.1), products made in past can be recorded in the company database (see Fig. 3.1). In fact, the cost indexes can be recorded as part of the PPS of a customized product. By searching the company database, the computer-aided cost analysis module may find the cost index of an operation that is part of the PPS of a past-made product and is similar to the operation required for making the new product. The computer-aided cost analysis module searches for a similar process by following a GT (group technology) coding system (Tu et al. 2000a, 2002b). The data in PPSs are stored and retrieved in the company database according to similarities in or similar digits of GT codes, from left to right along a GT code. Similar processes or operations of the PPSs in the database have the similar GT codes. If a process or operation of the new product is a combination of several primary processes or operations, the computer-aided cost analysis module will search for the cost indexes of these primary processes/operations and then determine the cost index for this combined process/operation by adding up the cost indexes of these primary processes/operations.

4.4 Cost Estimation and Optimization Procedure

With the cost estimate methods as described in the previous subsections, the cost estimation and optimization procedure of the computer-aided cost analysis module in the OKPMCS system can be described in the following two stages.

4.4.1 *Stage 1: Cost Estimate*

In this stage, the module needs to estimate various costs (or values) for the cost indexes associated with the processes/operations (see Fig. 4.1) from the feature level on the bottom up to the product level on the top. The inputs to the system in this stage include manufacturing processes/operations (including all the possible alternatives), manufacturing resources (including all the possible alternatives) and their cost rates (dollars per hour), quotations or prices from all suppliers and subcontractors that are provided by the suppliers or subcontractors or stored in the company database, and the costs or prices of standard components and manufacturing processes from the company database. With these input data, the computer-aided cost analysis module will estimate cost values or instances for the generic data classes (i.e. adding, resources, logistics) in the cost indexes associated with the planned processes/operations, which include alternative processes/operations, by using the two cost estimate methods (i.e. generative cost estimate method and variant cost estimate method as described in Sect. 4.3).

The cost estimate is automatically carried out by the computer module through an incremental or evolutional process that estimates the cost indexes in a feature-by-feature manner. This incremental process follows the prototype-based incremental and concurrent product development as discussed in Sects. 3.2 and 3.3. Traditionally, the cost of a product is estimated after a product has been completely designed and planned for production. However, this traditional cost estimate method is not suitable for the concurrent approach of a PD in OKP. This is because:

1. The ambiguity and uncertainty of customer requirements interrelated with suppliers' capability and the master company's capacity need to be identified in a feature-by-feature incremental way through an iterative cycle: customer requirements, supplier or subcontractor selection and solution identification, design, feedback to the customer.
2. The product has to be made through a concurrent engineering approach in mass customization (Tseng and Jiao 1996; Pine 1993; Jiao et al. 2003).

Therefore, the incremental way of designing a product, planning the manufacturing processes/operations, scheduling manufacturing resources, controlling inventory, managing logistics and estimating cost feature-by-feature from the component manufacture on the bottom level up to the final product assembly on the top level in a production hierarchy can dramatically simplify the complexity of simultaneously considering all these processes/operations or parameters. This simplified concurrent engineering method can equip the OKP system with a strong flexibility to cope with the changes and uncertainties of customer requirements and, hence, enhance the master company's and its suppliers' abilities and synergy to quickly identify the right or true needs of customers and design the right product to meet the customer needs. In particular, this incremental way of developing a product and estimating the cost can provide in-time feedback or reference to support the engineers' decision-making on the selection of alternative designs, manufacturing processes, manufacturing resources and suppliers/subcontractors in OKP. The shortages of this incremental method are:

1. a lack of global optimization
2. frequent and complicated communications among the master company, the customers, suppliers and subcontractors

Computer and Internet communication technologies as well as e-commerce technologies have provided an economic and effective way to handle these tedious and complicated communications through the Internet.

4.4.2 Stage 2: Cost Optimization

Cost optimization means the best selection among alternative designs, manufacturing processes, manufacturing resources, suppliers and subcontractors, or combinations of these alternatives with the lowest total cost.

For cost optimization, we have developed an optimal algorithm for the best selection of alternative operation routines and suppliers/subcontractors by using a dynamic programming technique (Tu et al. 2007). In general, we can use a commercial computer simulation package, e.g. ProModel (by ProModel Corp., Orem, UT), or Quest and VNC (by Deneb Inc., Dayton, OH), to develop simulation models to solve cost optimization problems. If a manufacturing process/operation or a PD chain consists of random variables or stochastic processes, Markov chains (Zijim 1984) can be used to model the process/operation or the PD chain and, hence, cost optimization heuristics or algorithms can be developed.

It should be mentioned that cost optimization maybe in conflict with other optimizations. An operation routine, for instance, from the lowest cost point of view maybe in conflict with the one from the optimal production schedule point of view (e.g. minimum make-span). This will lead to a decision-making or optimization problem with multiple objectives, i.e. lowest cost vs. minimum make-span. The multi-objective optimization problem can be solved by using rule-based or heuristic/algorithm-based methods, such as analytic hierarchy process (AHP), Pareto optimality, trade-off curves, goal programming (Winston 1994), etc.

4.5 Some Considerations on Cost Estimation

According to the experiences gained from our research and the industrial implementations, by using the two cost estimate methods as discussed in Sect. 4.3, the computer-aided cost analysis module can estimate cost indexes for most planned manufacturing processes/operations. However, there are still some processes/operations that cannot be automatically estimated by the module with either method. In the case where the computer-aided cost analysis module cannot automatically estimate the cost of a planned manufacturing process/operation according to the two cost estimate methods, the module will wait for the user to determine and input the cost indexes for this process or operation under the "Planned" costs (see Fig. 4.3).

For a combined process or operation, if the cost analysis module cannot find the cost indexes for some of its primary processes/operations from the company

Fig. 4.3 The computer interface for determining the cost index of a shop floor operation

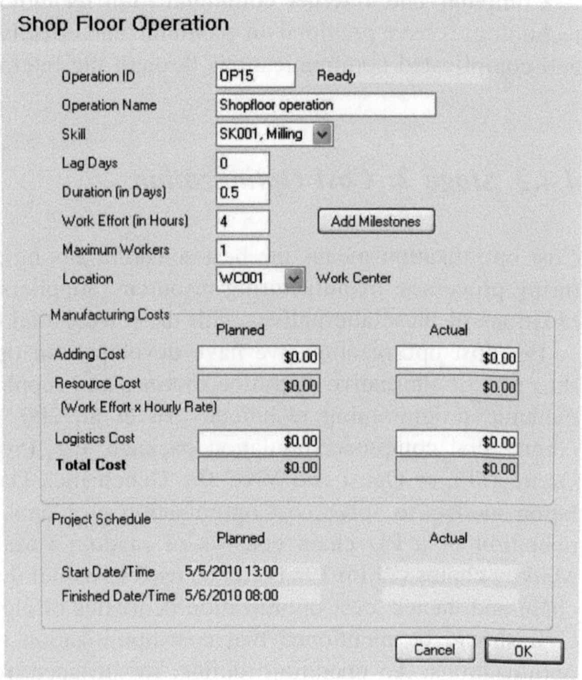

database, the module will wait for manual costing of these primary processes/operations, and then decide the cost index for the combined process/operation.

Another problem for cost estimate is that the cost of a component or subcontracted process/operation is difficult to identify. Under this circumstance, the comparison technique can be applied to identify the price domain for this process/operation.

Let us assume:

OP_a: The problem process/operation;
OP_H: A well-known process/operation which cost, recorded as $C(OP_h)$, is equally or more expensive than the problem process/operation, i.e. $C(OP_a) \leq C(OP_h)$;
OP_L: A well-known process/operation which cost, recorded as $C(OP_L)$, is equally or less expensive than the problem process/operation, i.e. $C(OP_L) \leq C(OP_a)$.

Consequently, the cost domain is determined as:

$$C(OP_L) \leq C(OP_a) \leq C(OP_h)$$

In terms of this price domain, a neural network based algorithm plus some evaluation rules can be used to infer a most likely cost of the component or the subcontracted process/operation (Tu and Jiang 1997).

The third problem in cost estimation and optimization is a conflict between the optimal (or lowest) cost and the reliability of a supplier or subcontractor.

Sometimes, the cheapest price may imply an unreliable supplier or subcontractor. The manager of an OKP company can solve this problem through choosing proper quality control and marketing policies, e.g. only accept quotations from reliable suppliers.

4.6 Cost Control

During the PD process, cost control is very critical for the success of an OKP product development. To control the cost of an OKP product development, the following terms are defined:

Planned Cost (PC): It is a planned cost for a process/operation or part in the PPS of an OKP product.

Planned Budget (PB): It is the total or sum of the planned costs of an OKP product.

$$\mathrm{PB} = \sum_{i=1}^{n} \mathrm{PC}_i + \sum_{j=1}^{m} \mathrm{PC}_j \tag{4.2}$$

where, PC_i is the planned cost of process/operation i, and PC_j is the planned cost of part j.

Actual Cost (AC): It is an actual expenditure of a process/operation or part in the PPS of an OKP product.

Up-To-Date Expenditure (UTDE): It is the sum of the actual costs of the completed processes/operations and the parts in the PPS of an OKP product.

$$\mathrm{UTDE} = \sum_{i=1}^{K} \mathrm{AC}_i + \sum_{j=1}^{L} \mathrm{AC}_j \tag{4.3}$$

where, AC_i is the actual cost of process/operation i which has been completed during the production, and AC_j is the actual cost of part j which has been completed during the production.

Cost Performance Index (CPI): It is defined by the following formula:

$$\mathrm{CPI} = \frac{\sum_{i=1}^{K} \mathrm{PC}_i + \sum_{j=1}^{L} \mathrm{PC}_j}{\sum_{i=1}^{K} \mathrm{AC}_i + \sum_{j=1}^{L} \mathrm{AC}_j} = \frac{\sum_{i=1}^{K} \mathrm{PC}_i + \sum_{j=1}^{L} \mathrm{PC}_j}{\mathrm{UTDE}} \tag{4.4}$$

Cost Variance (CV): It is the variation between the up-to-date expenditure (UTDE) and the sub-total of planned costs of the completed processes/operations and parts in the PPS of an OKP product. It is defined by the following formula:

$$\begin{aligned} \mathrm{CV} &= \mathrm{UTDE} - \left(\sum_{i=1}^{K} \mathrm{PC}_i + \sum_{j=1}^{L} \mathrm{PC}_j \right) = \mathrm{UTDE} - \mathrm{UTDE} * \mathrm{CPI} \\ &= \mathrm{UTDE}(1 - \mathrm{CPI}) \end{aligned} \tag{4.5}$$

If the CPI < 1, the CV is a positive value which is the total overran of the budget. Otherwise, it is a negative value which means a total saving from the planned budget (PB).

Forecasted Budget (FB): It is an expected total cost to complete an OKP product during the production. The value domain of the FB is defined by the following equations:

$$FB_1 = CV + PB \tag{4.6}$$

$$FB_2 = \frac{PB}{CPI} \tag{4.7}$$

The forecasted budget (FB) provides a manager a quantitative indication of cost control against the planned budget during the production.

In some project management textbooks, e.g. Gido and Clements (2009), a concept of cumulative earned value (CEV) has been introduced. The CEV is the value of the total work actually performed. It is calculated by multiplying the percentage of completed processes/operations against the total processes/operations to the planned budget (PB), i.e. CEV = (Percentage of completed processes/operations)·PB. However, in practice, it is difficult to count the percentage of the completed processes/operations against the total processes/operations in a PD project. This is particularly true in OKP since the time estimate of a process/operation is roughly quoted based on the closest past case or experience. On the other hand, it does not always held a one-to-one relation between the percentage of the process/operation and the percentage of PB, i.e. a process/operation may take 10% of the project time but it does not always demand 10% of the PB of the project. Therefore, to be practical, we remove the concept of cumulative earned value (CEV) from this book. Consequently, the CPI and CV as defined above are different from the ones as defined by Gido and Clements (2009). By Gido and Clements, CPI is defined as: CPI = (cumulative earned value)/(cumulative actual cost), and CV = (cumulative earned value) − (cumulative actual cost). The cumulative actual cost is the same as UTDE as defined above. The FB, which is called forecasted cost at completion (FCAC) by Gido and Clements (2009), has the same definition as mentioned above but it uses a different CV to calculate FB_1 and a different CPI to calculate FB_2.

4.7 Case Study

The presented cost estimate and control methods have been coded into the cost analysis module of the OKP management and control software (OKPMCS) system© that is described in Chap. 8. In this section, through a case study of the development of an electrical generator as shown in Fig. 3.4, we demonstrate the principles of the cost estimate and control methods as presented in this chapter.

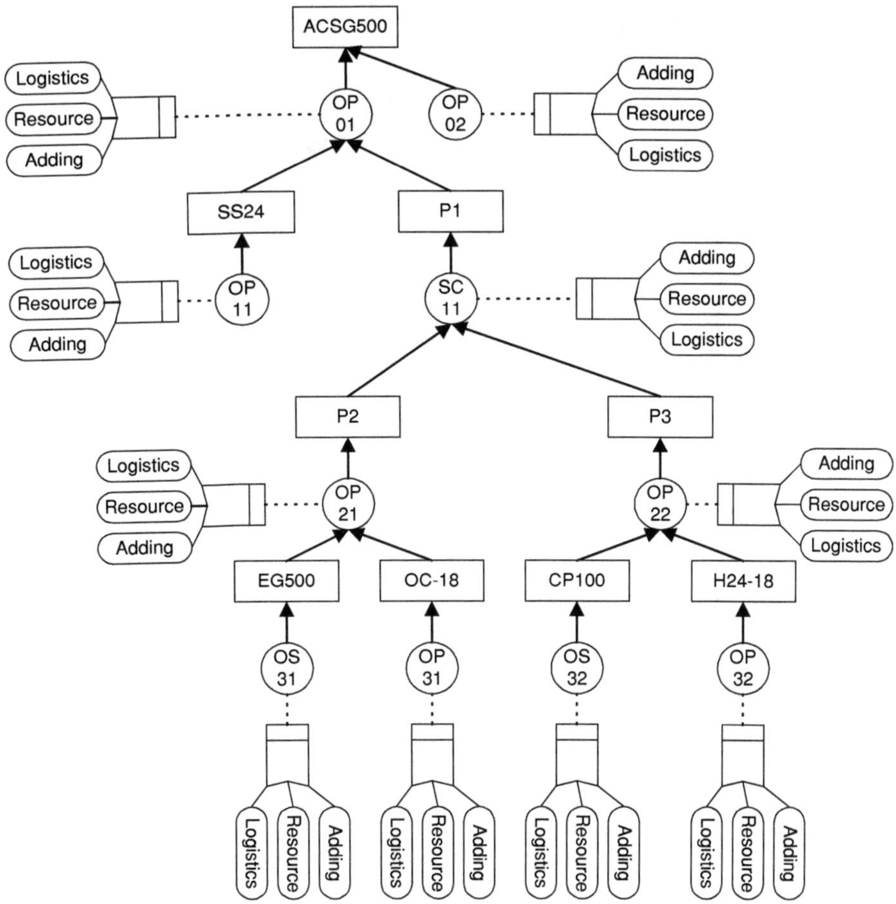

Fig. 4.4 The PPS and cost index of ACSG500

4.7.1 Case Scenario and System Description

Figure 3.4 shows an electrical generator ACSG500. Table 3.1 lists the bill of materials of the electrical generator. The ACSG500 is a specifically ordered or custom product, i.e. OKP product. Its PPS (product production structure) and cost index are jointly shown in Fig. 4.4.

From Fig. 4.4, it can be seen that the ACSG500 are planned to be made by using two outsourcing operations (OS31 and OS32), one subcontracting operation (SC11), and seven shop floor operations (OP01, OP02, OP11, OP21, OP22, OP31 and OP32). These operations are drawn in the PPS from the product level (i.e. Level 0) down to the feature/component level (i.e. Level 3), which were noted in Sect. 3.4.1 already. For the convenience of the reader, we re-copy the

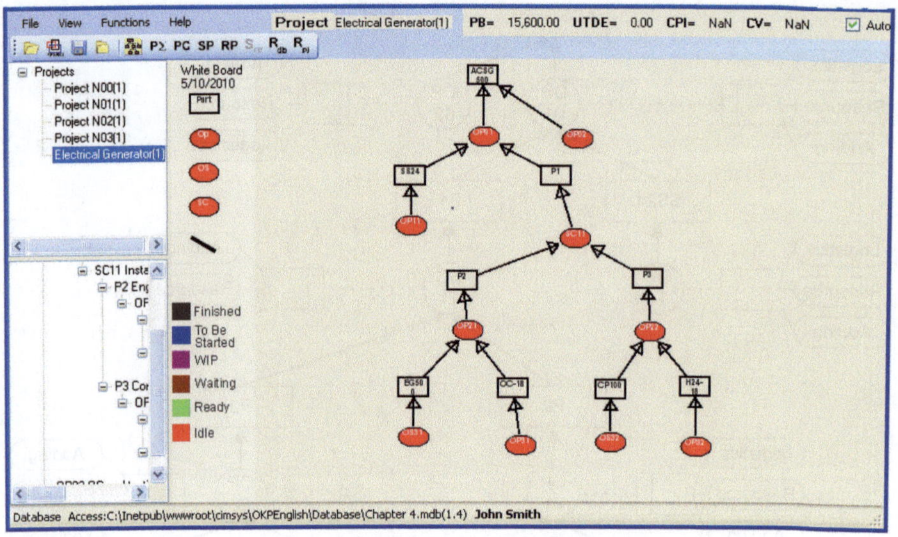

Fig. 4.5 Screen snapshot of the PPS graphic interface for the case product

specifications of the operations as follows: OS31: Purchase EG500 (Electrical Generator Engine, see Table 3.1) from "Custom Part Supplier#2".

OP31: Build OC-18 (Outer Casing 18 cm) casing for the engine.

OS32: Purchase CP100 (Control Panel).

OP32: Build H24-18 (CP Housing 24 cm × 18 cm) for CP (Control Panel).

OP21: Install the engine (EG500) in the casing (OC18). It formulates the intermediate assembly part P2 (EG500 and OC18).

OP22: Install the control panel (CP100) in the housing (H24-18). It formulates the intermediate assembly part P3 (CP100 and H24-18).

OP11: Build SS24 (Steel Stand 24 cm) for ACEG500.

SC11: Install and wire the control panel (CP100) to the engine (EG500). It formulates the intermediate assembly part P1 (P2 and P3).

OP01: Assemble SS24 (Steel Stand 24 cm) to P1. It formulates the final product ACSG500 (electrical generator).

OP02: Quality control and testing.

The interface of the OKPMCS system for this case study is shown in Fig. 4.5. As shown in Fig. 4.5, the planned budget (PB), up-to-date expenditure (UTDE), cost performance index (CPI) and cost variance (CV) are displayed on the top bar of the interface. These cost parameters will be calculated by the OKPMCS system automatically if a user ticks "Auto". These parameters give a production manager a real-time view of the cost planning and control of a project.

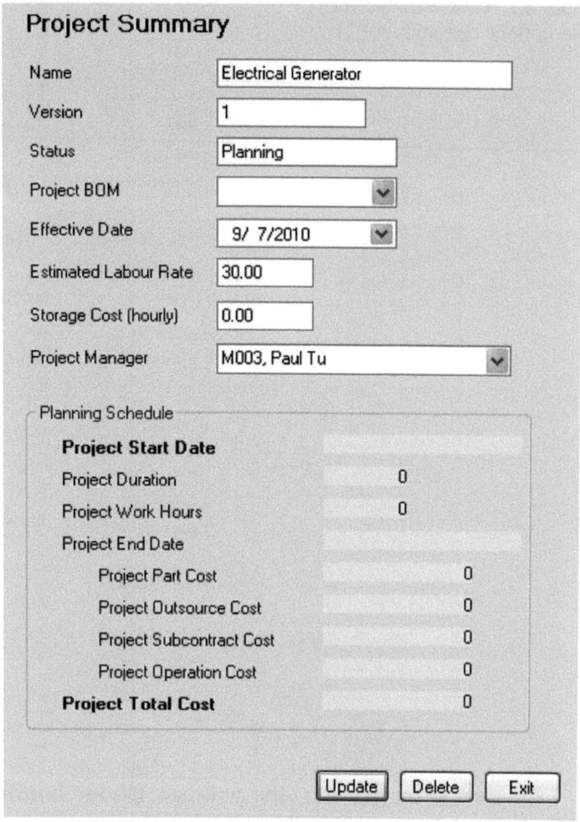

Fig. 4.6 Screen snapshot of project specification for the electrical generator

4.7.2 Cost Estimate for Making the Case Product

Using the graphic interface of the OKPMCS system as shown in Fig. 4.5, a user can easily draw and modify the PPS of the case product. Accordingly, the OKPMCS system will update the data relation in the database. As mentioned in Sect. 3.4.1, the PPS consists of the BOM (bill of material) and the BOO (bill of operation) of a product. Therefore, the changes of product design (i.e. changes of BOM) and production routine (i.e. changes of BOO) in OKP can be easily handled by the OKPMCS system. With a given PPS as shown in Fig. 4.4, the cost estimate is carried out as shown in Fig. 4.6.

First of all, for the seven shop floor operations, the company can estimate the cost of these operations according to the past experiences or through using a computer simulation system as mentioned in Sect. 4.4 to estimate a total operation hours (i.e. Work Effort as shown in Figs. 4.2 and 4.3) for each of these operations. Then the company needs to define an average labour hourly rate for a specific project, e.g. the project for developing the case product (i.e. electrical generator ACSG500) as shown in Fig. 4.6.

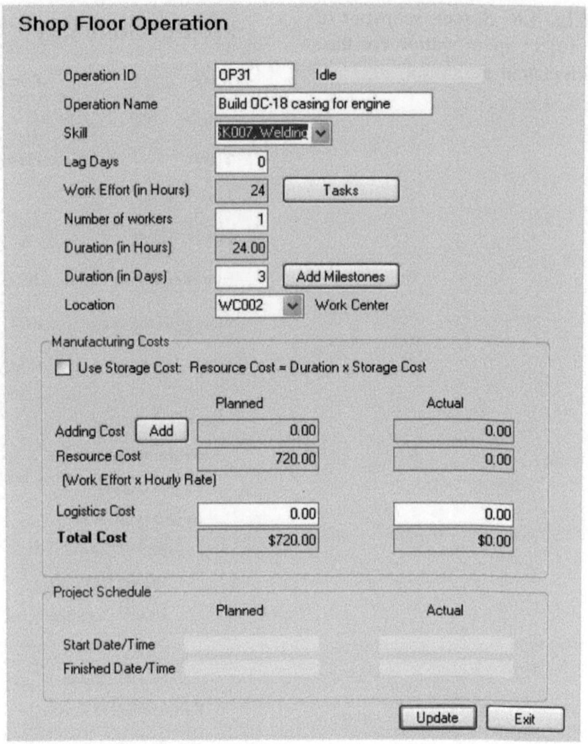

Fig. 4.7 Cost estimate for shop floor operation OP31

As shown in Fig. 4.6, the average labour hourly rate is defined in text box “Estimated Labour Rate”, i.e. $30. Based on the Work Effort and Estimated Labour Rate, the resource cost of an operation can be calculated (or estimated) by the cost analysis module as follows:

$$\text{Resource Cost} = \text{Work Effort} \times \text{Estimated Labour Rate} \tag{4.8}$$

If we double-click OP31 in the PPS as shown in Fig. 4.5, for instance, the resource cost of OR31 is estimated by the computer cost analysis module as shown in Fig. 4.7.

In fact, the cost estimate of OP31 follows the variant cost estimate method as mentioned in Sect. 4.3.2. As shown in Table 3.1, the part OC-18 is in the stock, there is no “Logistics Cost” for OP31. The operation OP31, to build OC-18 casing for engine (see Fig. 4.7), is obviously not a difficult shop floor operation. Therefore, we do not need to input an “Adding Cost” to buffer the uncertainty for carrying out OP31. Likewise, the cost estimate for the other shop floor operations can be done by using the cost analysis module of the OKP management and control software system©. These cost estimates are summarized in Table 4.1.

To estimate outsourcing operations OS31 and OS32 for purchasing parts EG500 and CP100 as well as the costs or prices of EG500 and CP100, we need to communicate with suppliers and look for quotations. To estimate costs of OS31

Table 4.1 Cost estimate of shop operations for making ACSG500

Operation ID	Logistics cost	Resource cost	Adding cost
OP31	0	24 h × $30/h = $720	0
OP32	0	24 h × $30/h = $720	0
OP21	0	40 h × $30/h = $1200	0
OP22	0	40 h × $30/h = $1200	0
OP11	0	8 h × $30/h = $240	0
OP01	0	16 h × $30/h = $480	0
OP02	0	4 h × $30/h = $120	0
Total	Logistics cost + resource cost + adding cost = $4680		

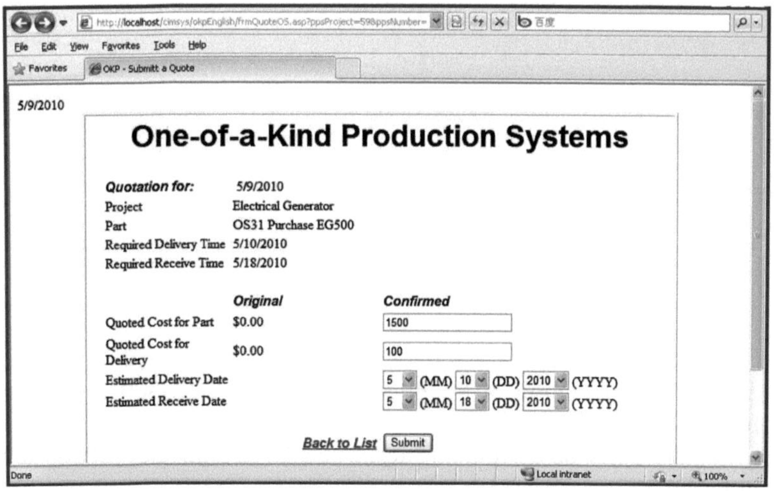

Fig. 4.8 Web-based communication interface with the supplier for quoting EG500

and EG500, by using the PPS Manager module (see Sect. 8.3) of the OKPMCS system, the company manager can send a request e-mail to a supplier, i.e. Custom Part Supplier#2 or simply recorded as Supplier#2, to seek for a quotation of EG500. After receiving the request e-mail which provides Supplier#2 a web link, user ID and a password, Supplier#2 can use a web-based interface as shown in Fig. 4.8 to send a quotation to the company.

Using the interface as shown in Fig. 4.8, Supplier #2 can quote EG500 as $1500 to plus $100 delivery cost. This electronic quotation can be received by the company manager immediately. After the company manager receives this electronic quotation, he/she can set the costs of OS31 as shown in Fig. 4.9.

As shown in Fig. 4.9, in addition to the logistics cost (i.e. delivery cost) as quoted by Supplier#2, the manager also set a "Resource Cost" of $60 to cover the work effort to process this purchasing order and an "Adding Cost" of $40 to cover the communication and overhead cost associated with this purchasing operation.

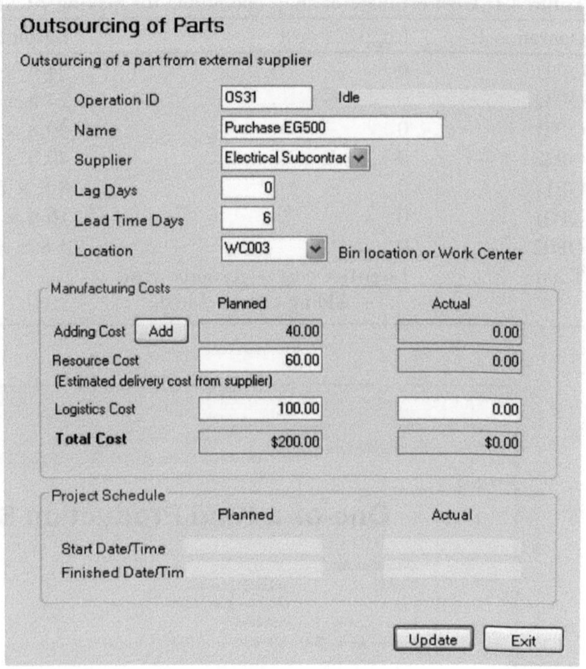

Fig. 4.9 Screen snapshot of cost estimate of outsourcing operation OS31

In terms of the quotation of Supplier#2, the cost of EG500 is estimated in Fig. 4.10.

Likewise, the costs of outsourcing operation OS32 and part CP100 are estimated in Fig. 4.11.

To estimate subcontracting operation SC11, the manager needs to select a subcontractor and seek for a quotation from the selected subcontractor. Let us assume the selected subcontractor is called Electrical Subcontractor. By using a web-based communication interface which is similar to the one as shown in Fig. 4.8 but with a file loading function for the subcontractor to send files or drawings, the manager can seek for a quotation from Electrical Subcontractor. The Electrical Subcontractor provides the quotation to the company as:

$$\text{Total cost} = 80\,\text{h} \times \$80/\text{h} = \$6400$$

Consequently, the cost of SC11 is set in the OKPMCS system as shown in Fig. 4.12.

For the parts in the stock, i.e. OC-18, H24-18 and SS24, the costs can be directly taken from the inventory record, which are OC-18 ($300), H24-18 ($200) and SS24 ($120). Therefore, in terms of the cost estimates of the shop floor operations, outsourcing operations and subcontracting operation as well as the outsourcing parts and in-stock parts, the total cost of the project for making the ACSG500 is estimated or summarized by the OKPMCS system in Fig. 4.13.

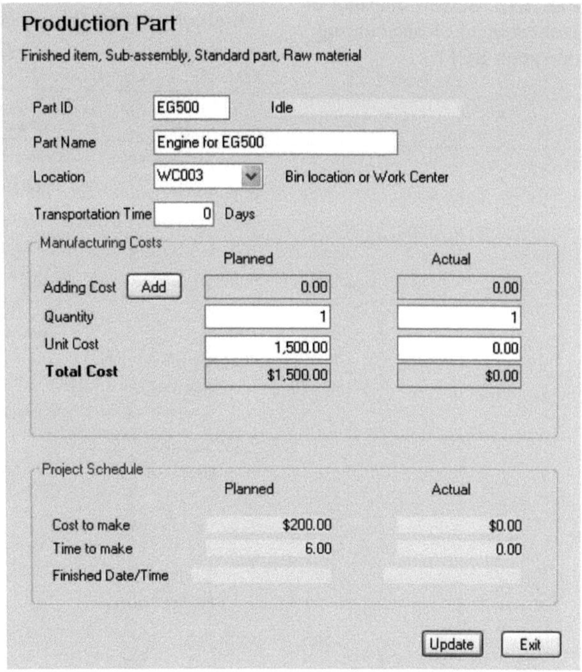

Fig. 4.10 Screen snapshot of cost estimate of part EG500

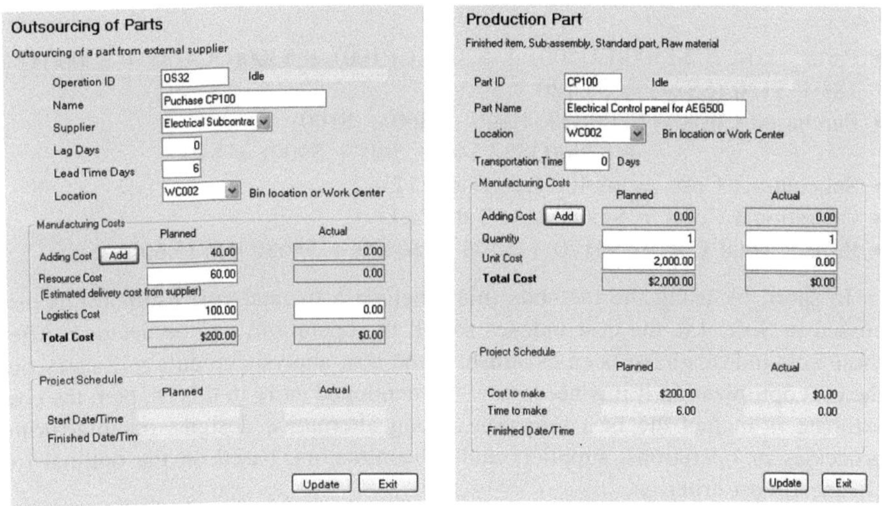

Fig. 4.11 Screen snapshots of cost estimates of OS32 and CP100

The OKPMCS system automatically calculate the total cost of $15,600 according to the generative cost estimate method as described in Sect. 4.3.1. The value of the total cost is approved below:

Fig. 4.12 Screen snapshot of cost estimate of outsourcing operation SC11

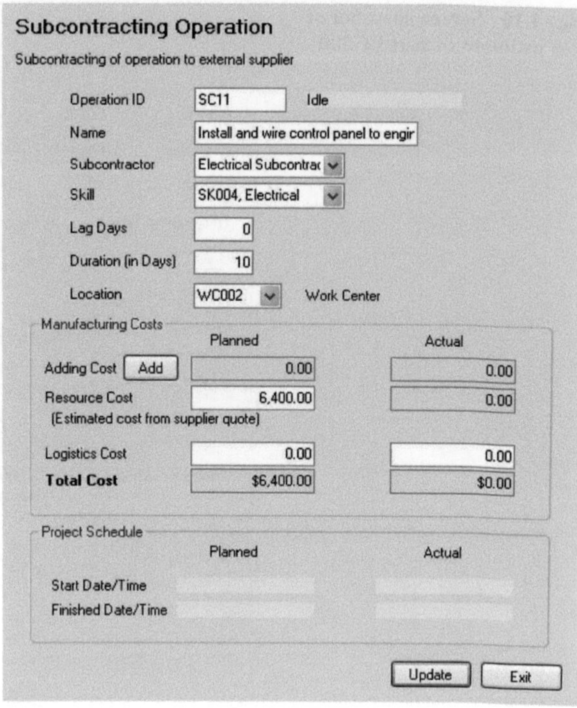

- Parts cost = \$1500(EG500) + \$2000(CP100) + \$300(OC-18) + \$200(H24-18) + \$120 (SS24) = \$4,120
- Purchased Costs = \$200(OS31:\$40 + \$60 + \$100) + \$200(OS32:\$40 + \$60 + \$100) = \$400
- Subcontract Costs = \$6,400 (see Fig. 4.12)
- Operational Costs = \$4,680 (see Table 4.1)
- Project Total Cost = \$4120 + \$400 + \$6400 + \$4680 = \$15,600

In short, by using the methods (may include a manual cost estimate) as discussed in Sect. 4.3, the cost indexes for all the operations can be decided. After these cost indexes have been established, the cost analysis module will carry out the cost optimization if it is necessary. As mentioned early in this chapter, the cost estimate is in fact the best selection among alternative designs, manufacturing processes or operations, suppliers and subcontractors, based on the optimal (or cheapest) cost criterion.

Tu et al. (2007) describe a cost estimate case for making a sheet metal part. This cost estimate case can be carried out under three scenarios, (1) alternative manufacturing processes/operations, (2) optional designs, and (3) different production routines. With the cost estimate methods and the cost analysis module of the OKPMCS system, the cost of this sheet metal part can be estimated (i.e. cost estimate) and minimized (i.e. cost optimization) as follows:

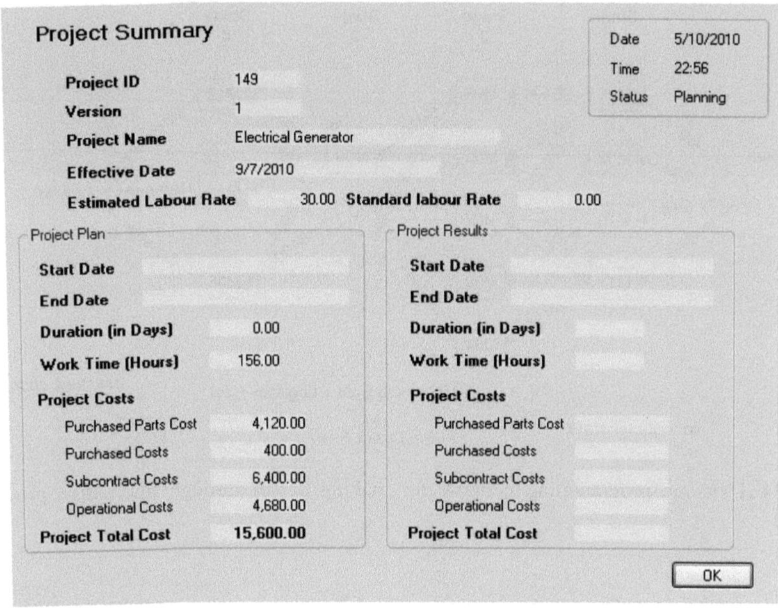

Fig. 4.13 Project summary for making electrical generator ACSG500

1. Alternative manufacturing processes/operations
   The sheet metal part has a number of machining features which can be carried out in two alternative ways: (1) use a CNC press machine to punch all the features, or (2) use a laser cutting machine to cut the two hexagons first and then use a press machine to punch the rest of the holes and rectangles. The first alternative has the advantage of completing all processes in one machine, which saves the material handling cost between the two machines, but it will need a special punching tool to punch the two hexagons. The tool may be searched from suppliers. The second alternative can have the company avoid acquiring a special punching tool, but a material handling cost or logistics cost between the laser-cutting machine and the press machine is implied, particularly if this laser-cutting machine is in a subcontractor's plant. Under this circumstance, if we ignore other technical factors and only use the cost factor as a reference to make a selection between these two alternatives, this can be automatically determined by the cost analysis module of the OKPMCS system in the following way:

   - Use the PPS graphic interface as shown in Fig. 4.5 to modify the PPS in terms of alternative manufacturing processes/operations.
   - According to the generative cost estimate and variant cost estimate methods, the cost analysis module estimates costs and decides the cost indexes of all the optional PPSs.

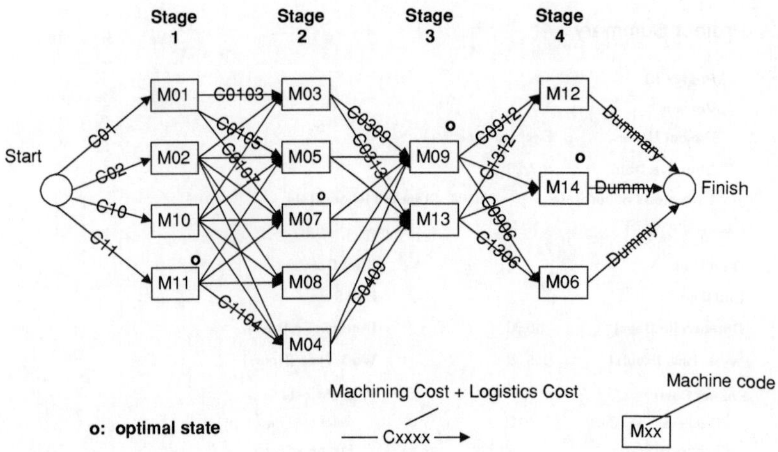

Fig. 4.14 Dynamic programming network for making a part through alternative production routines

- The module calculates the total costs for all the optional PPSs. Hence, a decision can be made according to the cheapest total cost of a PPS which includes the BOM and BOO for making this case sheet metal part at the lowest total cost.

2. Optional designs
   The case part can be designed in optional forms, e.g. including alternative components and consequently different parts or intermediate assembly parts. Accordingly the PPS of making this sheet metal part can be modified and a number of optional PPSs in terms of the different designs or BOMs are modelled with the PPS graphic interface of the OKPMCS system. Following the three steps as mentioned above for dealing with alternative manufacturing processes/operations, the cheapest solution can also be found using the OKPMCS system.
3. Different production routines
   The sheet metal part may be made by using alternative machines and through different production routines, as shown in Fig. 4.14. The nodes in Fig. 4.14 are different machines and the paths between the "Start" node and the "Finish" node mean different production routines to make the case part. The value of a link between a pair of nodes is a cost of which is equal to the machining cost on the next machine plus the logistics cost between these two machines. In this way, we establish a network diagram for applying the dynamic programming technique to find the minimum cost production routine, i.e. the shortest path between the "Start" node and the "Finish" node. This dynamic programming cost optimization method can be coded into a computer software system.

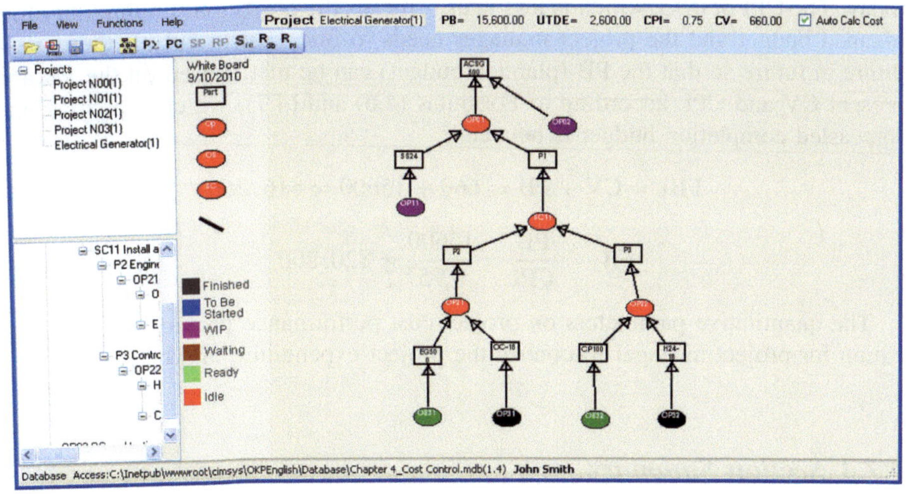

Fig. 4.15 Screen snapshot of cost tracing control parameters of making ACSG500

This dynamical programming optimization method can also be applied to select the cheapest PD (product development) chain among a number of alternatives if we simply think that the machines, as shown in Fig. 4.14, are in the different suppliers' or subcontractors' plants. For a more complicated case on a manufacturing shop floor or in a PD chain, the computer simulation technology, e.g. ProModel or Quest, can be employed for the cost estimation and optimization. Tu et al. (2000b) report an application of computer simulation technology to estimate and optimally control the cost of making injection moulds/tools in a tool-making shop floor.

4.7.3 *Cost Tracing and Control for Making Electric Generator ACSG500*

In the previous section, it describes how to estimate cost for various operations of making the case product ACSG500 and therefore it determines the Planned Budget (PB) of \$15,600 to make ACSG500. When the project starts, the project manager must control the project properly in order to complete this project within the PB.

To help the project manager to trace and control the cost of a project, the cost analysis module of the OKPMCS system automatically calculates the PB, UTDE, CPI, CV and FB if "Auto Calc Cost" as shown in Fig. 4.15 is ticked.

Figure 4.15 shows that the project has run for 3 days, i.e. it starts on September 7, 2010. The UTDE (up-to-date expenditure) of \$2600 means the project has spent \$2600 so far. The CV (cost variance) of \$660 means an overrun of \$660 up to today. The CPI of 0.75, which is less than 1, means the project is running out of its planned budget and the project manager needs to find way to cut off the expenditure in future so that the PB (planned budget) can be met. Based on the PB and present CV and CPI, according to Formulas (4.6) and (4.7), we can calculate the forecasted completion budget is between:

$$FB_1 = CV + PB = 660 + 15600 = \$16,260 \tag{4.6}$$

$$FB_2 = \frac{PB}{CPI} = \frac{15600}{0.75} = \$20,800 \tag{4.7}$$

The quantitative parameters on project cost performance provide a clear indication for project manager to control the project expenditure and budget.

4.7.4 Section Summary

In Sect. 4.7, through a simple case product ACSG500, the feasibility, principle and general procedure of applying the presented cost estimate and control methods, cost index, cost optimization algorithms and the cost analysis module of the OKPMCS system which includes a cost analysis module are demonstrated and validated.

In a large number of published research papers, the mass customization has been understood as a manufacturing paradigm by which a master manufacturing company actively involves suppliers in a whole product development cycle through the use of e-commerce technology to achieve both customization and economy of scale. The presented computer module and system, methods, PPS, cost index and algorithms for cost estimate, control and optimization are e-commerce technology based and can be applied in mass customization. If the case product ACSG500 is replaced by a customized sheet metal part, as far as the corresponding PPSs and cost indexes are derived, which may consist of alternative manufacturing processes/operations, designs and/or production routines, the OKPMCS system can be used to find the optimal solution, i.e. the lowest cost product production solution.

When the OKPMCS system together with its PPS, cost estimate methods and cost index are widely applied to various manufacturing businesses in one-of-a-kind production, some perceived difficulties or limitations can be envisaged. First of all, the cost data in the cost index may be dynamic instead of static, i.e. these data will change over time. The resources cost in the cost index can be affected by the production schedule or the utilization of a critical and expensive machine tool in a company. A production schedule in a manufacturing company, particularly in an OKP company (Tu et al. 2000a, 2002b), is kept open (or changes) until a predetermined last time period, i.e. planning time fence (Krajewski and Ritzman 1999). This will lead the cost data in the cost index to keep changing until the late stage in a PD cycle. This dynamic change of cost data can be coped with in the iterative computation power of the computer system. In fact, to adopt the concepts of prototype-based incremental and concurrent product development method (Sects. 3.2 and 3.3) in the cost estimate and optimization procedure is to allow a certain uncertainty or data change in earlier product design and planning stages of a PD cycle in one-of-a-kind production. However, if the data fluctuate too much and keep changing until the last minute when an order is to be issued to a supplier or subcontractor, this will result in the decision-making process becoming too complex to be handled by the system. This problem can be solved by controlling the degree of suppliers' or subcontractors' involvement in a PD cycle or adopting the adaptive production scheduling and control system that will be discussed in Chap. 7.

The second problem of wide application of the presented cost analysis module is that some cost data may be non-linear and it cannot be proportionally distributed to different parts or components. The ordering cost, for instance, may be fixed regardless of the number of parts to be ordered from a supplier. This problem can be solved by adding the ordering cost to the logistics cost of a part in a batch and set the null value of the ordering cost for the rest of the parts in the batch. In this way, the cost analysis module can ignore the non-linear cost data and correctly estimate cost. However, the batch of the parts must be recorded in the PPS instead of in the cost index. The other way to solve this problem is to add a sub-data-class under the logistics cost, which may be called transaction, to store these non-linear cost data in the cost index. However, this may cause the cost index structure to become tedious or complicated.

The third problem is the interdependence between the cost estimates of some processes/operations. A selection of alternative machining processes on the component level in a PPS may affect the cost estimate of an assembly operation on the assembly level. This problem can be solved by adding the alternative paths in the cost index hierarchical structure to include all possible options of process/operation selection.

4.8 Chapter Conclusions

In this chapter, we present the cost analysis module of the OKP management and control software system© to automate the cost estimation and control process for quickly quoting and controlling the cost of product development in OKP through the use of cost estimate methods and computer technologies. By following the concept of the prototype-based incremental and concurrent product development method, it is able to provide in-time cost estimate and optimization results to the product developers as important decision-making support references in a concurrent approach to product development in OKP. This concurrent approach enables product developers to design the product, plan the processes/operations, schedule production routines, manage logistics, and estimate and optimally control the cost in a feature-by-feature incremental manner. This incremental way of developing a product can help the product developers to clarify the product development life cycle data at earlier product development stages and hence avoid reworks, which often occur in those manufacturing companies that employ the traditionally sequential approach of a product development. The avoidance of reworks in a product development cycle will reduce the product development cycle time and cost significantly and hence improve efficiency of one-of-a-kind production.

The proposed PPS and cost index are generic, and they should be applicable to a wide range of one-of-a-kind production (OKP) companies. The PPS and cost index are modelled in a relational database as a part of the product life cycle data. Hence, they can be accessed by all kinds of computer-aided engineering and management software systems. It provides links or indexes that associate various costs to product features through the manufacturing processes/operations that are used to create these features. In this way of recording the various costs of a product development, cost estimation and optimization in a product development cycle can be performed, semi or fully automatically, by a computer software system like the OKPMCS system. The PPS and cost index, as a part of the product life cycle data stored in the company database, can be further used as important knowledge and references for supporting the new product development. It is very important for manufacturing companies to record its knowledge and past experiences to compete effectively in a global market. This practice is referred as knowledge management. It is interesting to note that the hierarchical structure of the PPS fits very well with a product-oriented supply chain or product development chain in mass customization, which is proposed by Huang and Mak (2000).

The cost estimation and optimization methods presented in this chapter are effective. They can be easily understood by and implemented in a manufacturing company. The OKPMCS system runs on intranets and the Internet. This makes the system be able to run in a distributed computer network, which is required by e-commerce technologies. The use of relational databases and computer Internet/intranet communication technology to develop these systems makes these systems very compatible with existing commercial computer-aided engineering and management software systems, such as ProEngineer, SmartGroup's PDM (product data management) system (Smart Team 2001), etc.

It is worth to note that the presented cost estimate and optimization methods together with the PPS and cost index have been developed to meet the needs of rapid and economic product development of OKP companies. OKP companies normally compete in market niches and adopt make-to-order or even engineer-to-order business strategies. These OKP companies survive, particularly in developed countries, mainly due to their high flexibility, customization and rapid responsiveness to the market needs. However, they are facing increasing competitive pressure on productivity and production costs from the overseas manufacturing companies. The OKP strategy, technology and system as presented in this book provide promising approaches to reduce production costs and improve productivity of OKP.

However, the cost estimation and optimization methods and the computer cost analysis module may not be applied in traditional mass or large batch size production companies, since the production in these companies is highly repetitive. It is unnecessary or even meaningless to estimate and control production cost according to the PPS and cost index presented in this chapter. This is because the product development and/or production costs are normally quoted in these companies by using a cost knowledge database that stores the market prices and cost data for standard parts and common manufacturing processes/operations from either their own practices in the past or a limited number of long-time cooperative suppliers. In some mass production companies, e.g. electronic appliance manufacturing companies, the unit cost of a product depends very much on the utilization of critical and expensive machine tools. In this case, the cost estimate and control problem becomes an optimal production scheduling problem. This production schedule and the consequently dynamic cost estimate and control problem also exist in one-of-a-kind production. We have developed an adaptive production scheduling and control system in one-of-a-kind production (OKP), which is described in Chap. 7. This adaptive production scheduling and control system is able to change the production schedule to adaptively cope with unexpected uncertainties in OKP, e.g. customer order changes, supplier delivery date changes, production capacity changes due to machine breakdowns, etc. It sends the updated cost rates of manufacturing resources in the PPSs of being produced products in the shop floors according to production schedule changes and the utilization of manufacturing resources. Accordingly, the cost analysis module changes cost estimates and the optimization decision regularly. In this chapter, the manufacturing resource cost rates are considered as constants and stored in the company database. These rates can be changed automatically by the OKPMCS system through updating the PPSs which are stored in the company database. Finally, the cost estimation, optimization and control methods, the PPS, cost index and the cost analysis module should not be perceived as universal solutions for the PD process of all kinds of products although it can be widely applied in OKP companies. In some manufacturing companies where, for instance, the products include emerging technologies, the product cost is mainly determined by the adoption of a particular kind of emerging technology. Due to the rapid development of or uncertainties with these emerging technologies, the product cost may change drastically. This makes it difficult to apply the cost estimate and control methods, as presented in this paper, in these manufacturing companies.

References

Becker J, Prischmann M (1993) Supporting the design process with neural network complex application of cooperating neural networks and its implementation. J Inf Sci Technol 3:79–95

Bode J (1998) Neutral networks for cost estimation. J Cost Eng 40:25–30

Bode J (2000) Neutral networks for cost estimation: simulations and pilot applications. Int J Prod Res 38(6):1231–1254

Casey J (1987) Digitizing speeds cost estimating. Am Mach Autom Manuf 131(1):71–73

Cawthorne-Nugent M, Da Luz Vieira J, Watson PA (1989) An intelligent knowledge-based system for cost estimating in the make-to-order environment. Comput Aided Eng 6(4):121–127

Gido J, Clements JP (2009) Successful project management, 4th edn. Thomson South-Western, Florence

Goldberg J (1987) Rating cost-estimating software. Manuf Eng 98(2):37–41

Huang GQ, Mak KL (2000) Modelling the customer–supplier interface over the world-wide web to facilitate early supplier involvement in the new product development. Proc Inst Mech Eng B Eng Manuf 214:759–769

Jiao JX, Ma QH, Tseng MM (2003) Towards high value-added products and services: mass customization and beyond. Technovation 23(10):809–821

Kals H, Brinke ET, Lutters E, Streppel T(1999) Integrated cost estimation based on information management. University of Ternte, Laboratory of Design, Production and Management. http://opm.wb.utwente.nl/staff/erik/phd/wgp99.pdf. Accessed Dec 2009

Kamarthi SV, Cohen PH, Demetter EC (1993) CMES: cost minimization and estimation system for sheet metal parts. Int J Flex Autom Integr Manuf 1(1):69–79

Kawada M, Johnson DF (1993) Strategic management accounting—why and how. Manage Acc 75(2):32–38

Koltai T, Lozano S, Guerrero F, Onieva L (2000) A flexible costing system for flexible manufacturing systems using activity based costing. Int J Prod Res 38(7):1615–1630

Krajewski LJ, Ritzman LP (1999) Master production scheduling: operations management, 5th edn. Addison-Wesley, Reading, p 725

Marx WJ, Mavris DN, Schrage DP (1998) Knowledge based system integrated with numerical analysis tool for aircraft life cycle design. Artif Intell Eng Des Anal Manuf 12(3):211–229

Monden Y, Lee J (1993) How Japanese auto makers reduce costs. Manage Acc 75(2):22–26

Niemand HP, Wolbold M (1994) Target costing: state of the art review. Manuf Compet Front 18(3):30–40

O'Grady P, Chuang WC (2001) Research issues in e-commerce and product development. Int J Cybern Syst 32:775–796

Ostwald P (1984) Cost estimating, 2nd edn. Prentice Hall, Englewood Cliffs

Ou-Yang C, Lin TS (1997) Developing an integrated framework for feature-based early manufacturing cost estimation. Int J Adv Manuf Technol 13(9):618–629

Pine BJ (1993) Mass customization. Harvard Business School Press, Boston

Poirier CC, Reiter SE (1996) Supply chain optimization: building the strongest total business network. Berrett-Koehler Publishers, Inc, San Francisco

Smart Team (2001) SmarTeam users guide, version 4.0. Smart Solutions Ltd, Kfar Saba. www.smarteam.com. Accessed June 2009

Smunt TL (1999) Log-linear and non-log-linear learning curve models for production research and cost estimation. Int J Prod Res 37(17):3901–3911

Tseng MM, Jiao JX (1996) Design for mass customization. Ann CIRP 45(1):153–156

Tu YL, Jiang ZB (1997) A neural network based approach for the quotation of a product in one-of-a-kind production. In: Proceedings of 1st international conference on engineering design and automation, March 18–21, Bangkok, Thailand, pp 307–310

Tu YL, Kam JJ (2004) Manufacturing network for rapid tool/die making. In: Proceedings of the 2nd international workshop on supply chain management and information systems (SCMIS2004), July 7–9, 2004, Hong Kong, PR China

Tu YL, Chu XL, Yang WY (2000a) Computer aided process planning in virtual one-of-a-kind production. Comput Ind 41:99–110

Tu YL, Xie SQ, Fung RYK (2000b) Rapid product development in a virtual production environment. In: Proceedings of IFAC symposium on manufacturing, modelling, management and control (MIM 2000), 12–14 July, Patras, Greece, pp 63–68

Tu YL, Kam JJ, Fung RYK, Tang JF (2002a) Resource deployment in earlier product development stages. In: Proceedings of 4th international symposium on tools and methods of competitive engineering, TMCE 2002, April 22–26, Wuhan, P.R. China, pp 499–509

Tu YL, Kam JJ, Xie SQ (2002b) Rapid one-of-a-kind product development in a global and virtual manufacturing environment. In: Proceedings of international conference on advanced production management systems (APMS 2002), 8–13 September, Eindhoven, The Netherlands

Tu YL, Xie SQ, Fung RYK (2007) Product development cost estimation in mass customization. IEEE Trans Eng Manage 54(1):29–40

Wang CH, Bourne DA (1995) Using features and their constraints to aid process planning of sheet metal parts. In: Proceedings of IEEE international conference on robotics and automation, pp 1020–1026

Wasti SN, Liker JK (1999) Collaborating with suppliers in product development: a US and Japan comparative study. IEEE Trans Eng Manage 46(4):444–461

Winbourne JP, Toolsie GM (1991) Computer-aided tool cost estimating (CATE). In: Proceedings of the 1991 ASME international computers in engineering conference and exposition, computers in engineering (New York), vol 1, pp 617–621

Winston WL (1994) Operations research: applications and algorithms, 3rd edn. Duxbury Press, Belmont, pp 772–818

Zhang YF, Fuh JYH (1998) Neural network approach for early cost estimation of packaging products. Comput Ind Eng 34(2):433–450

Zijim H (1984) The optimality equations in multichain denumerable state Markov decision processes with the average cost criterion: the unbounded cost case. CQM Note 22. Centre for Quantitative Methods, Philips BV Eindhoven

Chapter 5
Knowledge Mapping

As mentioned in Chaps. 1–3, there are a number of differences between the one-of-a-kind production (OKP) and traditional manufacturing modes. From a computer aided production management and control point of view, the data on product design (i.e. bill of material or BOM), process planning and production routines (i.e. bill-of-operation or BOO), production resources and information (bill of information or BOI) are much more extensive in OKP than the ones demanded in traditional manufacturing modes, e.g. batch production and mass production. In traditional manufacturing systems, a set of such data can be used for a large number of repeat or identical products. In OKP, every product demands a set of data. Therefore, to mass-produce OKP products, an efficient method for a computer system to generate the necessary data is required. To develop a such system, a knowledge mapping method to capture product designs, manufacturing processes and information requirements for one-of-a-kind production is needed. We simply call this method as knowledge mapping method.

Knowledge mapping requires a graphical interface to define various components of the production plant, processes, products, and information as well as their relationships. Products made by an OKP company are grouped by product families. However, the concept of product family in OKP is different from the one in traditional manufacturing systems. It means a kind of products which are designed and manufactured in terms of or by modification of a base product (or template). The variations within an OKP product family could be infinitive and the difference among the products within an OKP product family could be very large. An OKP product family contains configurations that define a set of variations. Each set of variations is defined by engineering constraints, input variables, attributes, operations and components. These configurations are used by a computer module, which is called information generator to be discussed in Chap. 6, to produce the BOM (bill-of-material), BOO (bill-of-operation) and BOI (bill-of-information) for every customer order or customized product. The knowledge mapping process captures the source of information and stores it in the company database for reuse by the modules, e.g. the information generator. The knowledge mapping process is

Y. Tu and P. Dean, *One-of-a-Kind Production*,
DOI: 10.1007/978-1-84996-531-6_5, © Springer-Verlag London Limited 2011

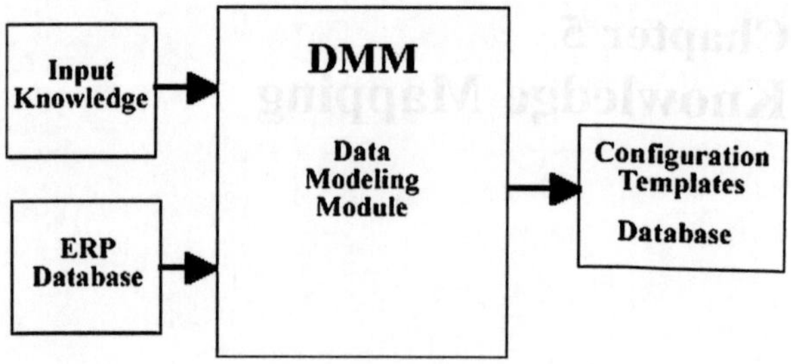

Fig. 5.1 Conceptual illustration of data modelling

in fact a data modelling and knowledge express process. The concept of the data modelling is illustrated in Fig. 5.1.

5.1 Scope of Knowledge Mapping Process

The scope of the knowledge mapping process includes the following functions:

1. Input, expression and maintenance of basic elements such as products, components, raw materials, work centres, operations, tasks, reports, etc.;
2. Expression, creation and maintenance of relationships among these elements that constitute configurations;
3. Output view of configurations for verification purposes;
4. Permanent storage of information in a relational database.

For the case studies in this book, Microsoft Access database is used to hold the various elements in tables, along with a set of tables that maintain the relationships. However in industrial implementations of the methods and computer modules as presented in this book, the Microsoft SQL Server is used. It is not difficult or a big job to transfer the Access database to the SQL Server. The database and the data tables can be reused by other non-object oriented programs and it is easy to integrate with ERP (enterprise resource planning) systems and computer aided design systems, e.g. Solid Works.

Input of different elements in the knowledge mapping process can be entered altogether initially, or as they are needed when establishing relationships. This standard process creates a database. It is also possible to import some or all of these elements from an ERP system.

The initial set of elements is defined as follows:

- *Product*: configurations, attributes, components, and raw materials;
- *Process*: operations and tasks;
- *Information*: constraints, variables, control instructions, and reports;
- *Structure*: plant, production lines, work centres, equipment, and resources.

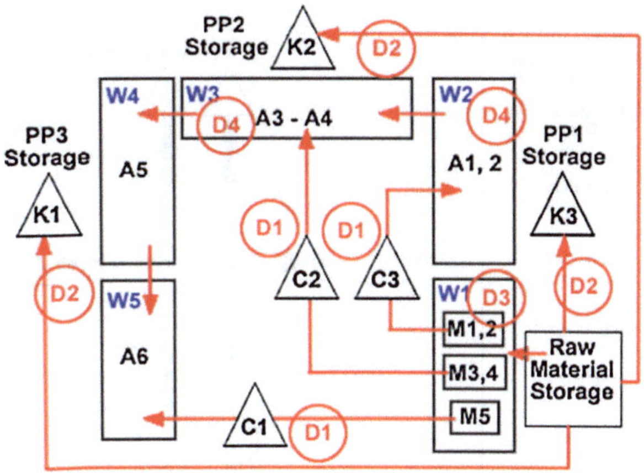

Fig. 5.2 A sample production process and its relevant information flow

Figure 5.2 shows an example of a production process and its relevant information flow. In Fig. 5.2, W indicates work centres which include W1, W2, W3, W4 and W5; A means assembly operations, i.e. A1, A2, A3, A4, A5 and A6; M notes machine operations which are M1, M2, M3, M4, and M5; K are Kanban storages for parts, i.e. K1, K2 and K3; C represents moveable storages (i.e. cart, pallet or conveyer) which include C1, C2 and C3; D1 is the Cart (or moveable pallet) delivery report, D2 is the Kanban parts storage report, D3 is the cutting machine download file, and D4 is the optional part installation and purchase order report; Finally PP is for purchased parts, i.e. PP1, PP2 and PP3. In Sect. 2.5, the basic elements of data structure for the configurations were outlined.

To represent the process (or operation) and its information flow as an example shown in Fig. 5.2, a symbolic form of a workflow template has been created in Fig. 5.3.

By using the workflow template as defined in Fig. 5.3, a production unit, which could be a shop floor or a flow production line, can be modelled. Figure 5.4 shows an example of a production unit that includes five work centres (WC). In fact, this production unit is taken from a custom window and door manufacturing company in Calgary, Alberta, Canada. It is formulated for producing various custom windows. Each of the work centres is defined with resources, operations, information, means of transportation of parts, and workflows. This production unit is configured for producing a specific kind (or family) of custom windows, i.e. OKP products. In practice, OKP requires frequent changes of configuration of a production unit in order to achieve high production efficiency according to the customer orders.

The necessary parts for a customized product are defined by the BOM of the product and the operation routings (or bill of operation (BOO)) are determined by the relationships between work centres and operation instructions. Information defines the parameters and tooling requirements for each operation. For example, if a part requires a cutting operation, then the information provides the dimensions

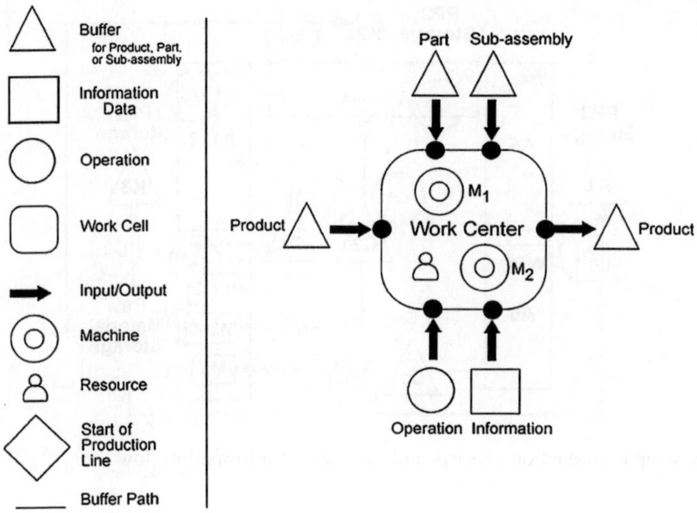

Fig. 5.3 Workflow template

of the cut and the necessary cutting tool. If the part requires a drilling operation, consequently the information defines the size, depth and location of the drill in tool-base of a CNC machining centre.

Information can take a number of forms. It depends on the operation, how the operation is performed (e.g. by machine or by human), and the purpose of the information. For example, the information can be a means of identification in the form of a printed tag with or without a bar code, or it can be an instruction in the form of a printed document, a download file to a CNC machining centre, or a display on a screen for an operator to read. Figure 5.4 illustrates an example of the configuration of a production unit. As shown in Fig. 5.4, WC means work centres which include WC1, WC2, WC3, WC4 and WC5; A notes assembly operations, including A1, A2, A3, A4, A5 and A6; M represents machining operations, i.e. M1, M2, M3, M4 and M5; K indicates Kanban storages for parts, including K1, K2 and K3; C notes moveable storages (i.e. cart, pallet or conveyer), which include C0, C1, C2, and C3; D1 is the cart (or moveable pallet) delivery report, D2 is the Kanban parts storage report, D3 is the cutting machine download file, and D4 is the optional part installation and purchase order report; Finally S indicates Main storage or warehouse, i.e. S1. The information for the configuration of a production unit as shown in Fig. 5.4 is defined in Table 5.1.

It should be noted here, the configuration of the production unit as shown in Fig. 5.4 produces OKP products with similar or close structures of PPS (product production structure). However, the operations (circles/ovals) and parts (boxes) in the PPS could completely different. This means an operation to be carried out in work centre WC1 for producing an OKP product could be completely different from the one for producing another OKP product, and the purchasing parts (PPs) or raw materials (RMs) are also quite different from

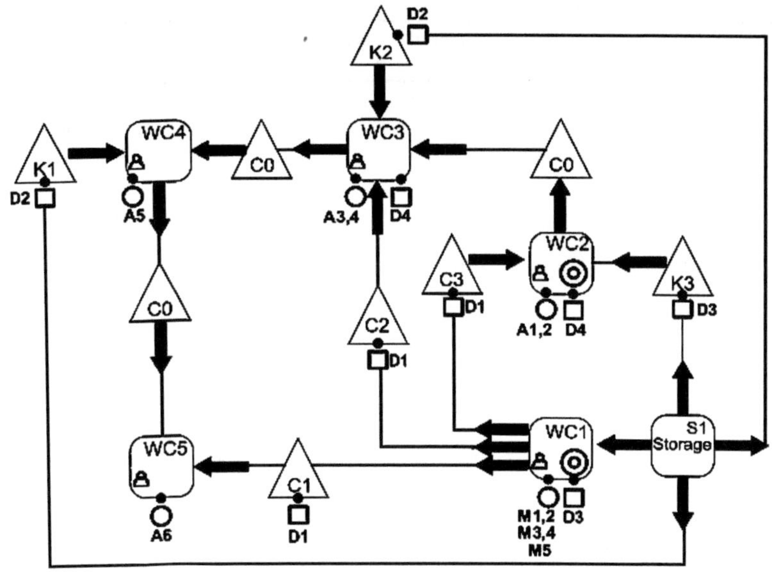

Fig. 5.4 Example of the configuration of a production unit

one product to the other. Therefore, PP1 may not mean the same or a type of parts, and likewise RM1 may not mean the same thing or same type. However, OKP products that can be produced on the configuration as shown in Fig. 5.4, may have the same number of circles/ovals and boxes which are linked by the same set of arrow lines since these OKP products are designed through modifications of the same base product (or template), i.e. they are in the same product kind or family.

Once a base product and the configuration of a production unit are defined in terms of the template as shown in Fig. 5.4 and the information as shown in Table 5.1, a specific set of data templates associated with the configuration will be formulated according to the generic data structures as described in Sect. 2.5. These templates include the BOM, BOO and BOI templates. Consequently, we can create a database that contains the tree structure of these templates and the conditions for addition or removal of optional paths or branches of the templates (see Sect. 2.5). The customer specifications for a product will determine instances, values and instructions of these template classes and variables. By applying customer specifications and optimization algorithms (e.g. for minimum production cost or maximum customer satisfaction) to the templates, the conditional tree structures are "pruned" and a PPS structure is produced for each of the customized products to be produced on the configuration of production unit or production line. As described in Sect. 3.4.1, the PPS of a product contains parts, operations and workflows for producing a customized product. It is in fact an integrated flowchart of BOM and BOO for the product production (see examples in Chap. 4).

Table 5.1 Information definition

Work centre	Data type	Material	Cart #	To WC	Machine #	Display
WC1	D1—cart delivery report	IP2, IP3	C1	W2	–	Run no, cart #, slot #
	D1—cart delivery report	IP5, IP6	C2	W3	–	Run no, cart #, slot #
	D1—cart delivery report	SC3	C3	W5	–	Run no, cart #, slot #
S1	D2—part storage report	PP1, PP2, PP4	–	K1, K2, K3	–	Part # for item # and location
W1	D3—download file	RM1, RM2	–	–	M1, M2	Machine parameters
W1	D3—download file	RM3, RM4	–	–	M3, M4	Machine parameters
W1	D3—download file	RM5		–	M5	Machine parameters
W2	D4—optional part purchase and installation	PP1	–	–	–	Part # for item # and sequence #
W3	D4—optional part purchase and installation	PP2	–	–	–	Part # for item # and sequence #

PP purchased part (PP1, PP2, PP3); *RM* raw material (RM1, RM2, RM3, RM4, RM5); *SC* sub-component (SC1, SC2, SC3); *IP* intermediate part (IP1 to IP6)

5.2 Graphical Interface

To establish the data set of BOM, BOO and BOI for the configuration of a production unit, the graphical interface needs to be developed. The interface brings together the BOM, BOO and BOI, and establishes all interdependent relationships between the classes and variables of the BOM, BOO and BOI data structures. The functionalities of the graphical interface are described by using it to process BOM, BOO and BOI templates respectively in the following sub-sections.

5.2.1 Processing BOM

The products, which are designed according to the same base product and are processed in a specific configuration of a production unit, should have the similar structure of a BOM template. The BOMs of various customized products based on this base product can be viewed as sub-trees of the BOM template. The BOM template can be imported or created using the graphical interface according to the following steps.

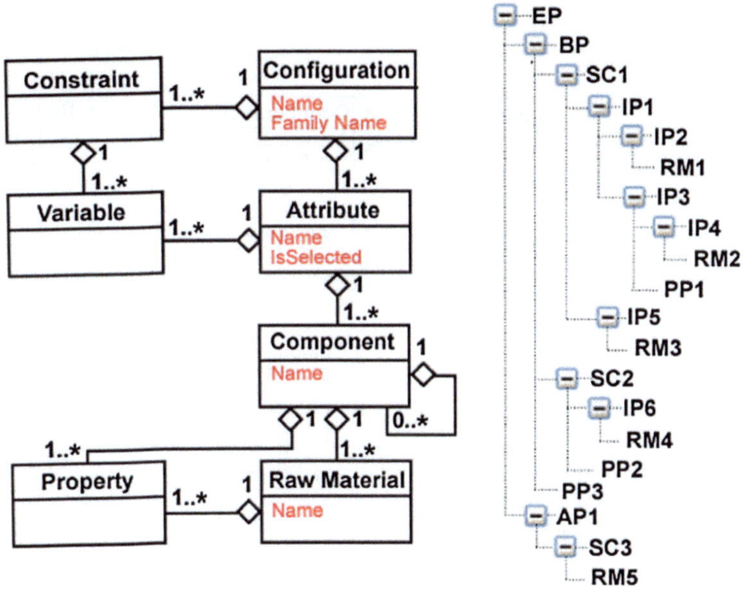

Fig. 5.5 The tree view and the relevant classes and attributes of BOM template

5.2.1.1 Step 1: Import or Create a BOM

Creating a BOM requires entering parts and subcomponents of an end product. An end product could be a final product or a transition product state, e.g. a part or intermediate assembly (see Sect. 3.4.1). The graphical interface will then create a tree view of the product. Figure 5.5 shows an example of a BOM template in tree view form alongside the class diagram for the configuration and associated classes. The BOM template is used by the computer system to automatically generate the PPS of a customized product.

In the example as shown in Fig. 5.5, EP has two attributes BP and AP1. The BP attribute has two components SC1 and SC2 and a purchased part PP3. The SC1 component is made up of two components IP1 and IP5. The IP1 component has 2 components IP2 and IP3, while IP1 is made from raw material RM1.

As each element of the BOM template is added, the product property information is required. For the configuration, a product family name or the name of the base product is provided. For an attribute, its name is required along with a default value for the "IsSelected" (Boolean) field. In the case of a base product attribute, the default value is TRUE, whereas for all the other attributes it is FALSE. The properties of a product are the values or instances of the attribute variables or classes. If a computer aided customer interface is used to capture customer requirements or specifications of the ordered product, these attribute variables are

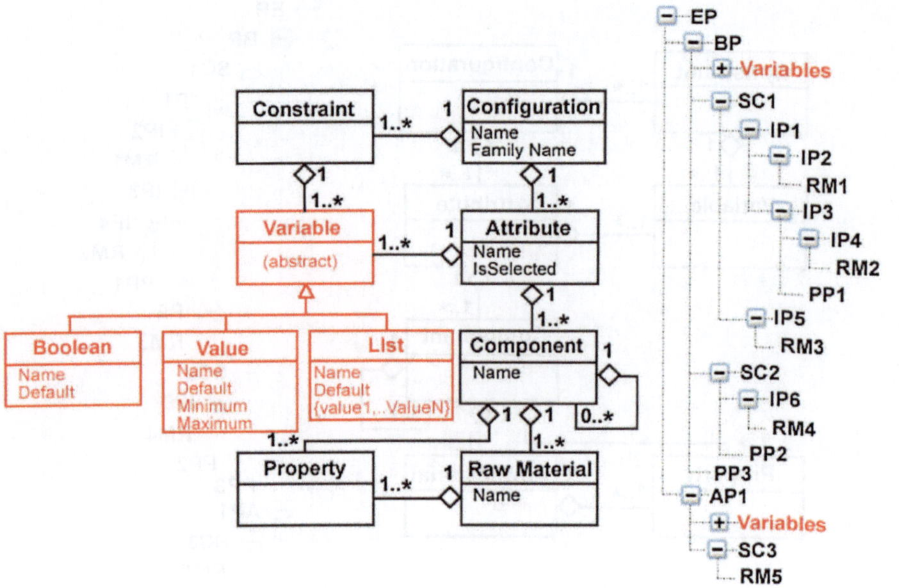

Fig. 5.6 Adding the variables to the BOM tree view

the data entrances or the required inputs by which the customer can specify their requirements of the product.

5.2.1.2 Step 2: Identify and Add Variables to the Attributes in the BOM

These variables will be used to capture specifications from the customer during the order entry. They will also be used to resolve engineering constraint rules which are applied next.

Variables shown in Fig. 5.6 can be one of three types—Boolean, Value or List. For example, an optional part can be selected with a Yes/No variable, a type of material can be selected from a list containing {WOOD, METAL, GLASS}, and a value can be a dimension such as LENGTH or WIDTH. An attribute can have any number of variables, but they must be inclusive so that values for the variables specify a unique variation of a product.

5.2.1.3 Step 3: Identify and Add the Constraint Rules

To manage and control the production of an OKP product, some constraints, as examples shown in Fig. 5.7, need to be identified and added to the end product, intermediate assemblies, parts, components and raw materials in the tree view of the BOM template as shown in Fig. 5.6.

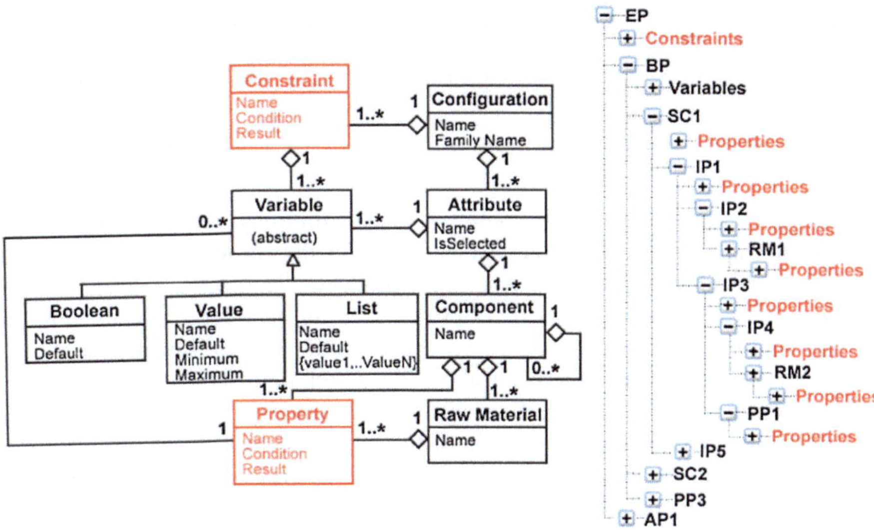

Fig. 5.7 Adding constraints and properties to the BOM tree view

Table 5.2 Conditions of the constraint rules

Name	Condition	Result
CR1	IF MATERIAL = "WOOD"	THICKNESS = 0.5
CR2	IF MATERIAL = "METAL"	THICKNESS = 0.375
CR3	IF MATERIAL = "GLASS"	THICKNESS = 0.25

As shown in Fig. 5.7, the constraint rules are added at the End Product (EP) level so that they can be used anywhere in the BOM tree by various properties. A Constraint Rule for producing a kind of customized windows may be:

IF MATERIAL = "WOOD": THICKNESS = 0.5 (mm)
THEN a Property of a Sub-component, raw material or purchased part must be:
LENGTH = WIDTH − (2 * THICKNESS)

where WIDTH means the width of a customized window, it is an Input Variable, and its value is defined by the customer during order entry. Likewise, the MATERIAL and THICKNESS are also customer controlled variables. If the MATERIAL has three options: WOOD, METAL, and GLASS, the conditions of three constraint rules are identified in Table 5.2.

The three steps as mentioned above can be repeated as many times as required until all the end products in the product family have been specified or designed.

The end product has at least one attribute which is called the Base Product (BP). It can have multiple optional attributes determined by the design. The base product and any additional attributes can have variables that control allowable product variations. These can be variables such as dimensions, colors, materials and

Table 5.3 Part number conditions

Length of piece	Raw Material Part #
IF LENGTH < LEVEL1	WOODOAK-24
IF LEVEL1 <= LENGTH < LEVEL2	WOODOAK-48
IF LENGTH >= LEVEL2	WOODOAK-72

Initial values are: LEVEL1 = 24 and LEVEL2 = 48

optional parts. These variables become the specification variables that a customer must define for the ordered product. Constraint rules define other variables used in the manufacturing process to make custom products. They also determine what operations take place and the instructions that go with the operations. In addition, the part numbers and the properties of purchased parts or raw materials are defined. In some cases, the raw material may be a bulk item such as a sheet (metal, wood, or glass). The BOM template will identify the part number and the constraint rules will determine dimensions of the piece required. In other cases, raw material may come in multiple forms (e.g. different types and sizes). Then, the constraint rules will determine the part number required based on the customer requirements for material and size. For example, the customer may have specified the product to be made from oak wood and to be 23″ long. A constraint rule would specify the part number based on IF MATERIAL = "OAK WOOD" and LENGTH < 24: PART NUMBER = "WOODOAK-24".

If any pre-specified value in the application varies from time to time, it can be set up as a global variable. When it does change, there is only one place to make the change and the new value is propagated throughout the system via the use of that variable. For example, if a part number for a raw material is dependent on a range of lengths as it does for wood, the method as described in Table 5.3 can be used.

The method shown in Table 5.3 can be used for constant rules as well by giving it a variable name and a fixed value.

5.2.2 *Processing BOO and Establishing the PPS for a Customized Product*

In this section, we show how the PPS of a customized product is established by using a computer system through the identification and specification of relationships between the customized product and its BOM template and between the customized product and the manufacturing process and production routines, i.e. BOO.

5.2.2.1 Step 1: Import or Create the BOO

Creating the BOO requires entering the operations and tasks that make the end product. The interface will then create a tree view of production. One or more operations are grouped in work centres. These work centres contain the resources

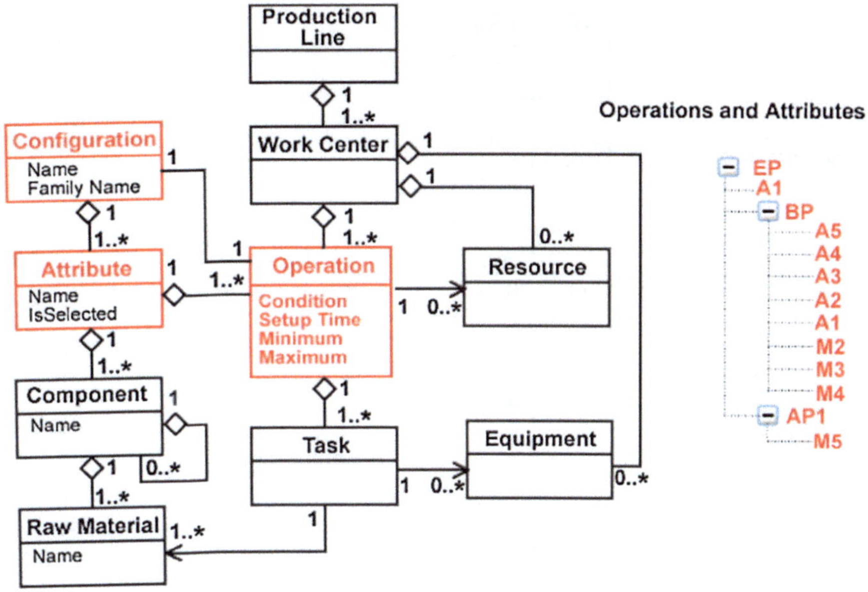

Fig. 5.8 Operations added to the tree view of the BOO

necessary to complete operations assigned to each work centre. These resources include people, equipment and machines. Figure 5.8 shows an example of BOO in tree view form beside the class diagram for the Operation and associated classes, which is a part of the PPS of a customized product.

The example as shown in Fig. 5.8 shows operations being added first to the tree view while making relationships between Operations and Attributes. The end product has at least one operation that combines all the attributes to make the finished product. To add the operation, certain properties have to be defined, which include conditions (when this operation is required), setup time and minimum and maximum number of workers required to perform this operation.

Operation A1 assembles the base product together with attribute AP1 (if selected). The Base Product has operations A1–A5 and M1–M4, while the attribute AP1 has operation M5.

5.2.2.2 Step 2: Add the Tasks

After adding operations according to the work flow diagram as shown in Fig. 5.8, tasks are added next to the tree view structure of the BOO. This is shown in Fig. 5.9.

Tasks and task properties are added. A task may be defined as "Lineal Cutting" and it can be used in many different operations. When it is added to an operation in this tree view, the specific information is required and it will be different each time when it is used. Information includes: a number that indicates which task is performed, a condition that determines when the task is required, and an expression

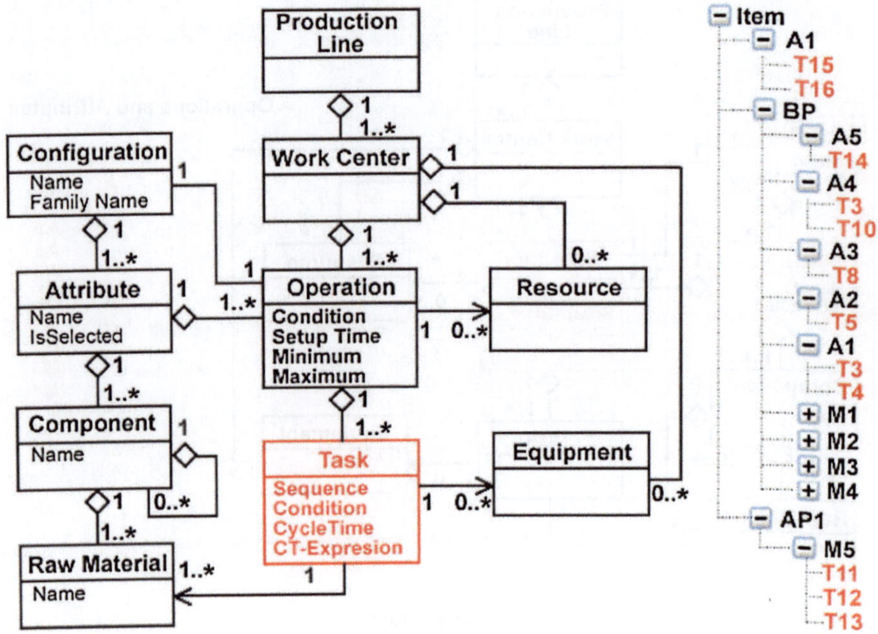

Fig. 5.9 Tasks are added to the tree view of the BOO

that will calculate the cycle time based on one or more of the variables. Since the variables have already been established in the BOM, any of them can be included in the expression. For example, if the cycle time is dependent on the length of the material being cut, then the expression is:

$$\text{Cycle Time (CT)} = (10 + (\text{LENGTH} * 2))$$

5.2.2.3 Step 3: Add Work Centres, Resources and Equipment in the PPS

Work centres are grouped under a production line that has an identification, name and performance factor. As each work centre is created with an identification and name, resources and equipment can be added to a work centre. The resource definition includes an hourly rate for costing purposes. The machine definition includes an identification number, specification to indicate whether it is automatic or not, and a change-over or setup time. For a highly automatic CNC machining centre, the change-over or set-up time can be viewed as zero since it is normally very short. Figure 5.10 illustrates the addition of work centre, resource and equipment into the PPS.

Now the material storage information can be added into the PPS in forms of storage centres, online storage (Kanbans) and moveable storage (conveyer, cart or pallet). Information is maintained in the storage class which is associated with raw

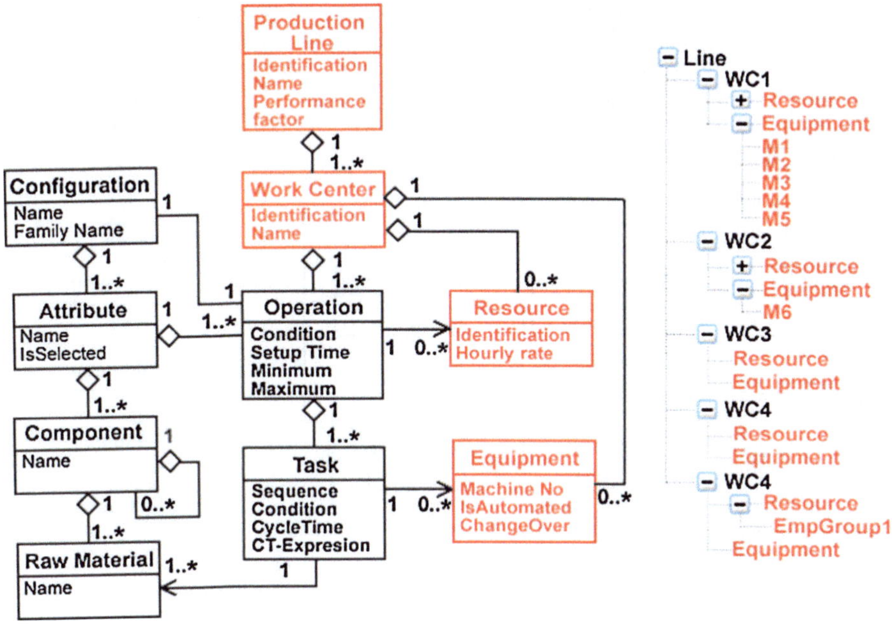

Fig. 5.10 Work centres, resources and equipment are added into the PPS

materials and components (see Fig. 5.11). With input/output pointers and path lines, the movement of material can be defined.

5.2.2.4 Step 4: Define the Work Flows in the PPS

The work flows among the work centres need to be defined in the PPS. In order to do this, a graphic work flow definition interface as shown in Fig. 5.12 is used. The interface has four sections. Section 1 is shown on the left part of Fig. 5.12 under Tools that contains work flow definition symbols. Section 2 is shown on the middle part of Fig. 5.12 and it is a blank screen or whiteboard on which a flow production line is created with various elements such as work centres, operations, machines and resources. Section 3 shows the tree views of the PPS. On this section, the BOM, BOO or Line can be viewed and utilized. Section 4 is Property Window as shown on right-bottom part of Fig. 5.12 and it shows properties of the selected item in the interface.

To define the work flows, we need to place work centres on the whiteboard according to the layout of a production unit or production line. If the work centres have been defined in Step 3, these work centres can be dragged from the tree view on to the whiteboard. The other way of defining a work centre is to drag a work centre symbol from the Tools (or Sect. 1 of the interface) and drop it to the whiteboard. By double-clicking this symbol, a property dialog box will open and the information about the work centre can be input.

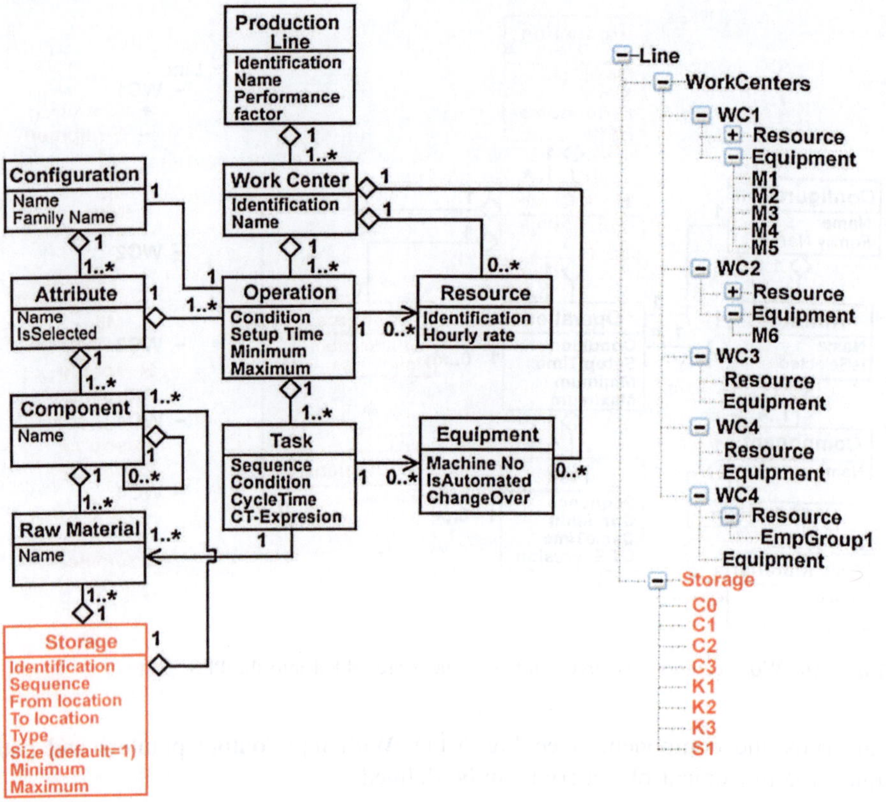

Fig. 5.11 Definition of material storages in the PPS

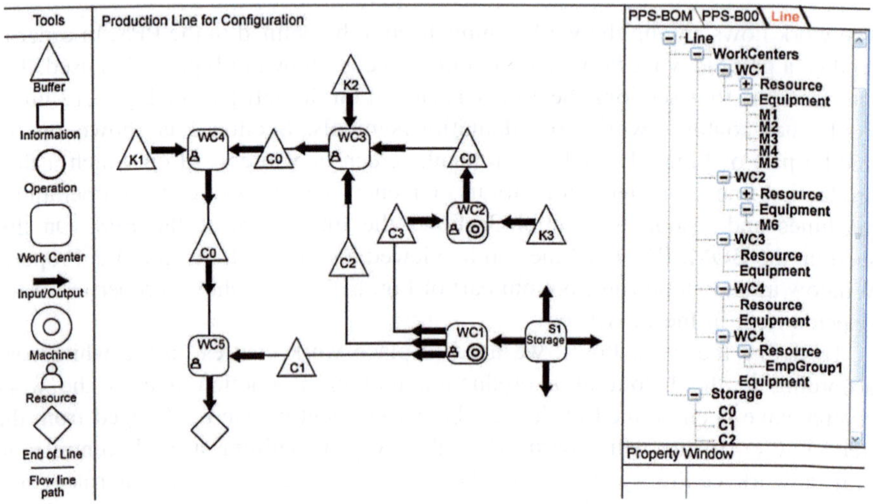

Fig. 5.12 Graphic work flow definition interface

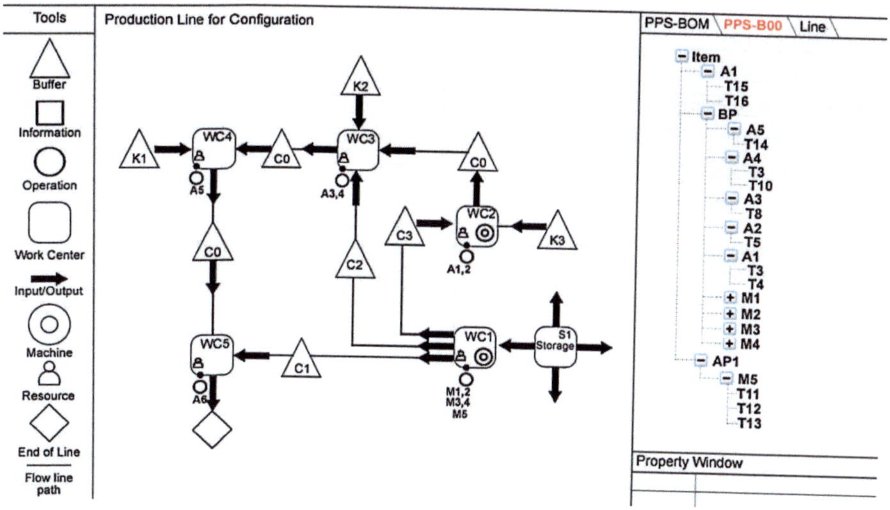

Fig. 5.13 Assign operations to work centres

This double-clicking function and the dialog box can also be used at anytime to update the definition of a work centre. Likewise, the resources and machines can be added to the whiteboard by dragging the symbol from the Tools or double-clicking the icon under the existing work centre from the Tree View of the PPS (Sect. 3 of the interface).

5.2.2.5 Step 5: Assign Operations to Work Centres and Specify Tasks in Operations

In this step, operations are grouped and assigned in various work centres and the tasks for each operation are specified. Operation assignment is simply done by dragging an operation as defined in Step 2 and adding it to the work centre. Multiple operations can be added to a work centre. As operations are dragged to the work centre in this manner, tasks are also added automatically. The operational content of any work centre can be viewed by double-clicking the operational icon attached to the work centre, and a dialog box will open to show the content.

Figure 5.13 continues the example from Fig. 5.12 and shows where the operations have been added. The new or additional operations can be added by dragging the operation symbol from the Tools or operation icon from the PPS-BOO tree view in Sect. 3, and then add it to a work centre. By double-clicking the operation, the operation dialog box will open for inputting or updating the operation information. Figure 5.14 shows the operation dialog box.

The dialog box as shown in Fig. 5.14 allows for the definition of an operation and the selection of tasks that comprise the operation. Tasks are selected from a list. Tasks can include both value-added and none value-added work. Tasks can

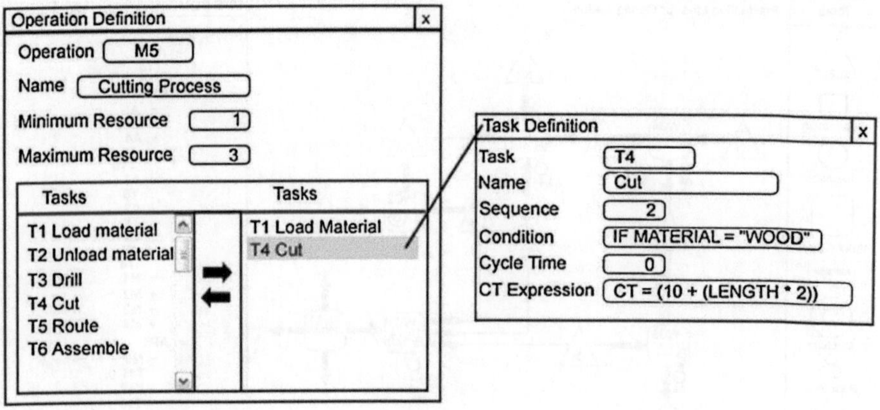

Fig. 5.14 The operation and task dialog boxes for operation and task definitions

include set-up time, clean-up time and time for quality check and statistical process control (SPC) functions. When a task is added to the operation by moving it from the left side to the right side, the task dialog box opens automatically so that the specific use of the task in this operation can be fully defined. The same task can be specified in any number of times. However, the cycle time of a task can be different when the task is used in different operations for processing different materials.

The resource allocation and production lead-time control are carried out by the production scheduling and control module, which will be discussed in Chap. 7.

5.2.2.6 Step 6: Assign Materials to the PPS

Once all the operations and tasks have been defined in the PPS of a customized product, the materials, which are required by this customer order, need to be added into the PPS according to the BOM template of this customized product. To do this, the PPS-BOM tree view is selected and the various components, sub-components, raw materials and purchase parts of the BOM template are shown in Fig. 5.15.

For a type of OKP products which can be produced in a large number daily, e.g. a custom window or door, the required materials are determined, at the stage when customer order is entered through a computer customer interface, according to customer specifications, technical requirements and safety or trade codes (e.g. construction code for making windows and doors). In other words, the number of different types of materials is limited but the number of dimensions or sizes could be infinitive. For this type of OKP products, the BOM template consists of all the possible types of materials and purchased parts for this type of OKP products. Furthermore, this type of OKP products is often produced according to a standard production routine, i.e. on a flow production line. Although every product passes the same work centres on this flow line, the different product requires operations

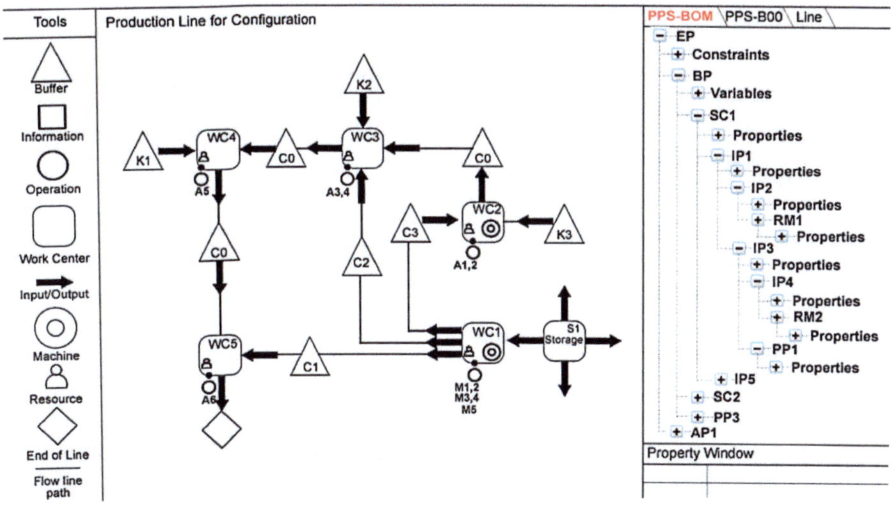

Fig. 5.15 The PPS-BOM tree view

that consist of different tasks. For this type of OKP products, the materials for a specific customized product can be automatically assigned by the computer system. We describe this automatic assignment process as follows.

Let us consider a work centre from a flow line as modelled in Fig. 5.15. This work centre is used to produce an intermediate part of a specific customized product. In other words, it carries out a sub-structure of the PPS of the product, i.e. a sub-PPS. Based the BOM template and the standard production routine defined by the BOO, the computer system can identify a generic structure of this sub-PPS as shown on the left side of Fig. 5.16. In Fig. 5.16, M1 through M7 are parts in a BOM; OP1, OP2 and OP3 are the three operations performed in the work centre; C1, C2 and C3 are moveable storages or buffers; K1 and K2 are purchased parts Kanban storages; T1 through T7 are operational tasks; and 01 and 02 are machines.

The work centre as shown in Fig. 5.16 assembles intermediate part M7. It consists of three operations OP1, OP2 and OP3, and each of these operations requires at least one resource. Materials come to the work centre by the ways like on a cart, in a buffer or from a part storage bin which will be selected when we define the work flows in Step 4. According to customer specifications, technical requirements and safety or trade codes, the computer system can automatically specify the generic sub-PPS into the detailed sub-PPS or an instance of the generic sub-PPS that is shown on the right side of Fig. 5.16. This automatic knowledge mapping process for generating an instance of the generic sub-PPS is illustrated in Fig. 5.17.

It should note that for a complicated OKP product that is produced in weeks or even longer period, this automatic material assignment process may not be applicable since the BOM template cannot include all the possible materials and parts. The BOM template of a complicated OKP product is a major part of the PPS

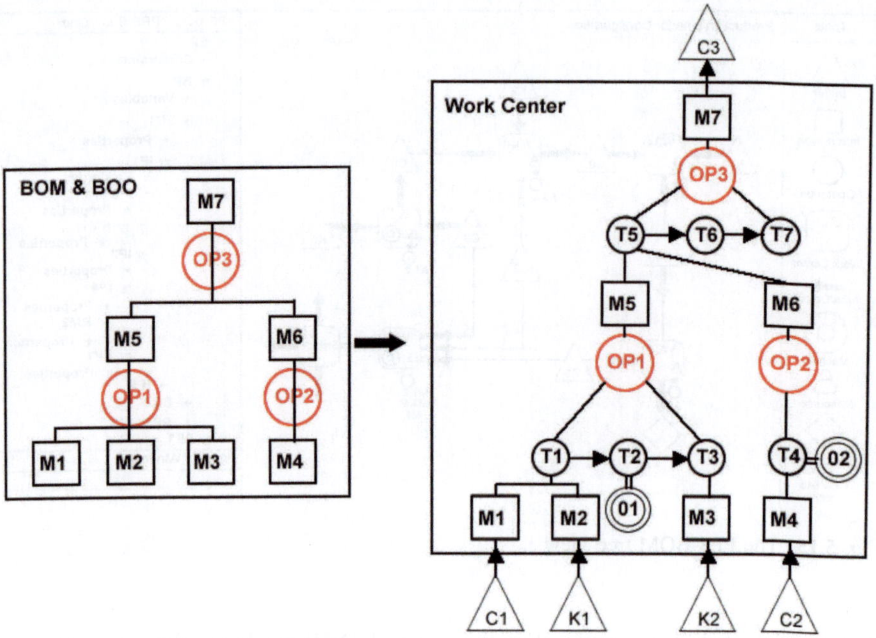

Fig. 5.16 The generic sub-PPS in a work centre and its instance for a specific customized product

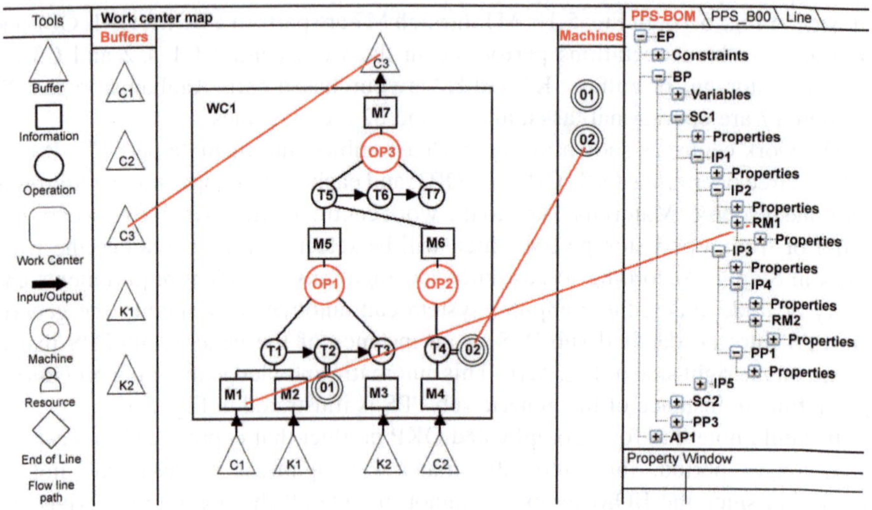

Fig. 5.17 The knowledge mapping process for generating an instance of the generic sub-PPS

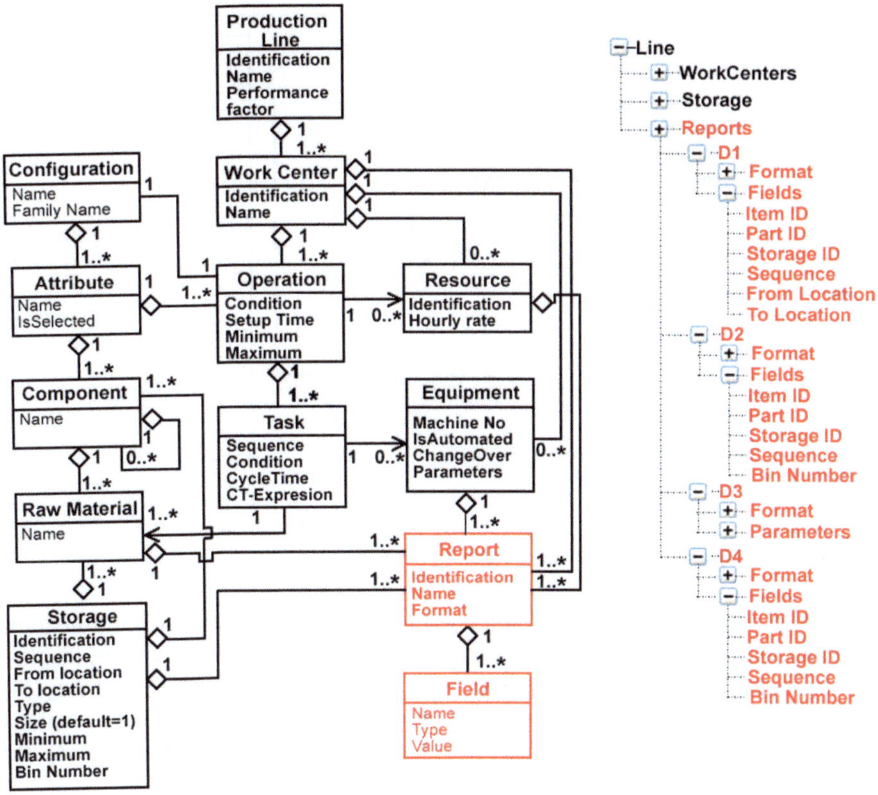

Fig. 5.18 Attaching the report to an object in the PPS

of a prototype (see Sect. 3.2). The PPS of a complicated OKP product is normally specified manually using the prototype-based incremental and concurrent product development methods as described in Sects. 3.2 and 3.3 under help of this knowledge mapping computer module.

5.2.3 Processing BOI

The bill-of-information (BOI) represents the major reporting requirement in a computer aided production management and control software system. It defines the required fields for the reports rather than a report generator as commonly known in information technology.

The basic reports from a computer system are defined with the information content, which is taken from various objects and variables in the system. Once the required fields are defined, the report can be attached to any object in the PPS of a product as illustrated in Fig. 5.18.

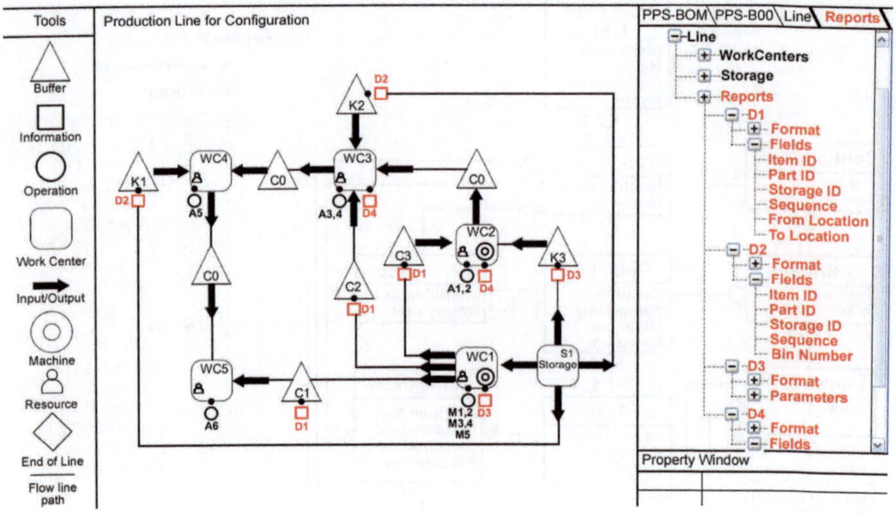

Fig. 5.19 Attaching the reports to the objects in the PPS

Various reports from a computer system are created with an identification, name and format. The format determines that the report is a printed report, display, download or other types. The required fields of a report are specified and they can be drawn from any class property or variable in the system. A report is generally created at the time when an application of the system is being setup. However, the reports can be created anytime when they are needed.

Once report fields are defined, the graphic work flow definition interface can be used to attach the report to any of the objects in the configuration. Figure 5.19 demonstrates the attachment of reports D1–D4 to the case used in previous section.

As shown in Fig. 5.19, the moveable carts are attached with the D1 cart report which provides information to the cart loader, the deliverer, and the recipient. The D2 storage reports are attached to bins K1, K2, and K3 which help with either getting the parts from the main warehouse or ordering them from the supplier. Work centre WC1 has a number of machines that require download files to automatically perform the change-over and processing tasks. The D3 download-file report is attached to these machines. Finally, work centres WC2 and WC3 have optional custom parts added to some products in the production run, and have report D4, named Optional Purchase Parts report, attached to these work centres. This ensures the timely acquisition of parts and the assignment of the right parts to the right custom product.

This completes the first phase of the definition of the data structure, including classes and their properties. Additional classes will be discussed in Chaps. 5 and 6 along with static and dynamic behaviours as the information generator and monitoring modules are specified. In the next section, this first phase will now be specified using real-time process algebra (RTPA).

5.3 Data Structure and Classes Specification

The data classes and the relevant properties for the knowledge mapping process for generating PPS as discussed in previous section need to be specified formerly so that the readers with computer programming knowledge can further understand the data structure of the PPS in a computer system. Therefore, we define the data structure with the unified modelling class diagram as shown in Fig. 5.20.

In terms of real-time process algebra (RTPA), we specify the data modelling module of the PPS as follows. For the RTPA notations please refer to Appendix C.

§DMM ≙ DMM.Architecture // Class definitions
|| DMM.StaticBehaviors // Class behaviours
|| DMM.DynamicBehaviors // Event handling

1. DMM.Architecture ≙ CLM : **ST** // Data Modelling Module
|| Events : **S**
|| Status : **BL**

DMM.Architecture.CLMST ≙ Configuration: **ST** |[1..*]> // Classes
||< Attribute: **ST** | [1..*]>
||< Component: **ST** | [1..*]>
||< RawMaterial: **ST** | [1..*]>
||< Constraint: **ST** | [1..*]>
||< Variable: **ST** | [1..*]>
||< Property: **ST** | [1..*]>
||< ProductionLine: **ST** | [1..*]>
||< WorkCenter: **ST** | [1..*]>
||< Operation: **ST** | [1..*]>
||< Task: **ST** | [1..*]>
||< Resource: **ST** | [1..*]>
||< Equipment: **ST** | [1..*]>
||< Storage: **ST** | [1..*]>
||< Report: **ST** | [1..*]>
||< DataField: **ST** | [1..*]>

DMM.Architecture.CLM.Configuration**ST** ≙ ConfigurationID**S**:
// Configuration class
(<Name:**S**>
<FamilyName:**S**>
< Attributes:= $\mathop{R}\limits_{i=1}^{n}$ <Attribute(i**N**)>)

DMM.Architecture.CLM.Attributes**ST** ≙ AttributeID**S**:
// Attribute class
(<Name:**S**>
<IsSelected:**BL** | **T**=Selected ∧ **F**=Not Selected |default=FALSE>

< Variables:= $\overset{n}{\underset{i=1}{R}}$ <Variable(iN)>

< Operations:= $\overset{n}{\underset{i=1}{R}}$ <Operation(iN)>

< Components:= $\overset{n}{\underset{i=1}{R}}$ <Component(iN)>)

DMM.Architecture.CLM.Component**ST**≙Component**IDS**:
// Component class

(<Name:**S**>

< Components:= $\overset{n}{\underset{i=1}{R}}$ <Component(iN)>

< Properties:= $\overset{n}{\underset{i=1}{R}}$ <Property(iN)>

< RawMaterials:= $\overset{n}{\underset{i=1}{R}}$ <RawMaterial(iN)>

<StorageID:S>)

DMM.Architecture.CLM.RawMaterial**ST** ≙ RawMaterial**IDS**:
// Raw material class

(<Name:**S**>
<PartNumber:**S**>

< Properties:= $\overset{n}{\underset{i=1}{R}}$ <Property(i**N**)>

< Reports:= $\overset{n}{\underset{i=1}{R}}$ <Report(iN)>)

DMM.Architecture.CLM.Constraint**ST** ≙ Constraint**IDS**:
// Constraint class

(<Name:**S**>
<ConfigurationID:**S**>
<Condition:**S**>
<Result:**S**>

< Variables:= $\overset{n}{\underset{i=1}{R}}$ <Variable(i**N**)>)

DMM.Architecture.CLM.Variable**AB** ≙ Variable**IDS**:
// Variable class

(<Name:**S**>
<Type:**S** | "value" ∧ "list" ∧ "Boolean">
<Value:**O** > // Object depending its type

DMM.Architecture.CLM.ValueVariable**ST** ≙ Variable**IDS**:
//Inherits Variable - Value

(<Value:**R**>
<DefaultValue:**R**>
<Minimum:**R**>
<Maximum:**R**>)

DMM.Architecture.CLM.ListVariable**ST**≜VariableID**S**:
//Inherits Variable - List
(<Value:**S**>
<DefaultValue:**S**>
< ListValues:= $\mathop{R}\limits_{i=1}^{n}$ <ListValue(i**N**)>)

DMM.Architecture.CLM.BooleanVariable**ST** ≜ VariableID**S**:
//Inherits Variable - Boolean
(<Value:**BL**>
<DefaultValue:**BL**>)

DMM.Architecture.CLM.Property**ST** ≜ PropertyID**S**:
// Property class
(<Name:**S**>
<ConfigurationID:**S**>
<Condition:**S**>
<Result:**S**>
< Variables:= $\mathop{R}\limits_{i=1}^{n}$ <Variable(i**N**)>)

DMM.Architecture.CLM.ProductionLine**ST** ≜ ProductionLineID**S**:
// Production Line
(<Name:**S**>
<PerformanceFactor:**R**>
< WorkCenters:= $\mathop{R}\limits_{i=1}^{n}$ <WorkCenter(i**N**)>)

DMM.Architecture.CLM.WorkCenter**ST** ≜ WorkCenterID**S**:
// Work Centre
(<Name:**S**>
< Operations:= $\mathop{R}\limits_{i=1}^{n}$ <Operation(i**N**)>
< Resources:= $\mathop{R}\limits_{i=1}^{n}$ <Resource(i**N**)>
< Equipments:= $\mathop{R}\limits_{i=1}^{n}$ <equipment(i**N**)>
< Reports:= $\mathop{R}\limits_{i=1}^{n}$ <Report(i**N**)>)

DMM.Architecture.CLM.Operation**ST** ≜ OperationID**S**:
// Operation class
(<Name:**S**>
<Condition:**S**>
<SetupTime:**R**>
<MinimumOps:**N**>
<MaximumOps:**N**>
<AttributeID:**S**>
<ResourceID:**S**>

< Tasks:= $\overset{n}{\underset{i=1}{R}}$ <Task(i**N**)>)

DMM.Architecture.CLM.Task**ST** ≙ TaskID**S**:
// Task class

(<Name:**S**>
<Sequence:**N**>
<Condition:S>
<CycleTime:**R**>
<CT-Expression:S>
< Equipments:= $\overset{n}{\underset{i=1}{R}}$ <Equipment(i**N**)>
< RawMaterials:= $\overset{n}{\underset{i=1}{R}}$ <RawMaterials(i**N**)>)

DMM.Architecture.CLM.Resource**ST** ≙ ResourceID**S**:
// Resource class

(<Name:**S**>
<HourlyRate:**R**>
< Reports:= $\overset{n}{\underset{i=1}{R}}$ <Report(i**N**)>)

DMM.Architecture.CLM.Equipment**ST** ≙ EquipmentID**S**:
// Equipment class

(<MachineNo:**S**>
<IsAutomated:**BL** | **T**=Automated ∧ **F** = Manual>
<SetupTime:**R**>
<MachineParameters:**S**>
< Reports:= $\overset{n}{\underset{i=1}{R}}$ <Report(i**N**)>)

DMM.Architecture.CLM.Storage**ST** ≙ StorageID**S**:
// Storage class

(<Sequence:**N**>
<FromLoaction:**S**>
<ToLocation:**S**>
<Type:**S** | Cart ∧ Buffer ∧ Kanban ∧ Storage ∧ Conveyer>
<Size:**N** | default = 1>
<Minimum:**N**>
<Maximum:**N**>
<BinNumber:**N**>
< Components:= $\overset{n}{\underset{i=1}{R}}$ <Component(i**N**)>
< RawMaterials:= $\overset{n}{\underset{i=1}{R}}$ <RawMaterial(i**N**)>

$$\text{Reports:=}\ \mathop{R}_{i=1}^{n} \text{<Report(iN)>)}$$

DMM.Architecture.CLM.Report**ST** ≙ Report**IDS**:
// Report class

(<Name:**S**>
<Format:**S**>

$$\text{< DataFields:=}\ \mathop{R}_{i=1}^{n} \text{<DataField(iN)>)}$$

DMM.Architecture.CLM.DataField**ST**≙DataField**IDS**:
// Field class

(<Name:**S**>
<Type:**S**>
// All values converted to string format
<Value:**S**>

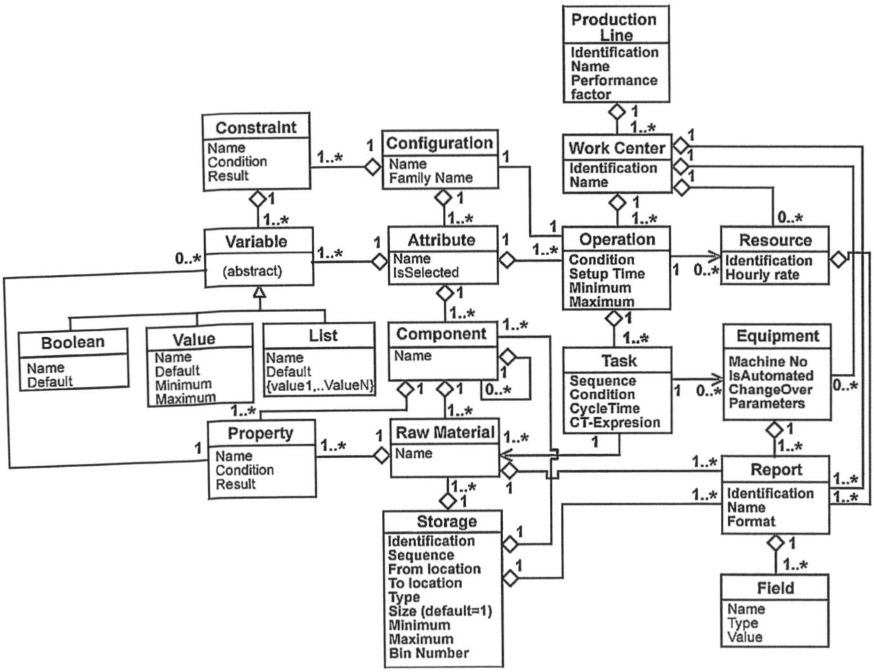

Fig. 5.20 Class definitions of the PPS data structure

Static and dynamic behaviours relate to the capture of knowledge and mapping functions. Output from the mapping process can either be an object file from which the class structure and the current state can be restored, or the data can be output to a relational database. When the data are re-read by a computer system, the class structure has to be re-established by creating all objects with the data from the database.

5.4 Chapter Summary

In this chapter, we present a knowledge mapping technique for computer aided generating the necessary product design and production data for mass-producing OKP products. The emphasis is placed on a computerized process of generating these data and integration of the data in the data structure called product production structure (PPS). The combination of BOM and BOO has been researched for quite a few years and several methods have been proposed in other references. However, from our best knowledge, none of these methods has been implemented in a computer aided production management and control systems, such as MRP II or ERP systems. In this chapter, we not only provide the concept and methods for integrating product design and production data, but also expose the importance of including information requirements in the form of BOI in the PPS. Only with these applicable methods and data structure in place, can information generation occur and a one-piece flow production line be implemented, that is required if we want to mass-produce OKP products.

Chapter 6
Information Generation

After the knowledge mapping method as described in Chap. 5, this chapter presents an information generation module (IGM) for generating the necessary product design and production data as well as the manufacturing instructions for producing an OKP product. This IGM is efficient and scalable and it is based on client/server computer architecture.

In order to manufacture a custom product, the system uses a library of templates from the company database. A customer uses an interface to select the product family and answer questions that will eventually define all the information required to create a custom product. The interface validates these answers through accessing the company data by means of a request and then passes the answers to the IGM via the company database. Figure 6.1 shows a conceptual diagram for this process based on client/server computer architecture.

The top-level specification of the IGM can be expressed in RTPA as follows:

$$\begin{aligned} \S\text{IGM} \triangleq\ & \text{IGM.Architecture} \\ & \|\ \text{IGM.StaticBehaviors} \\ & \|\ \text{IGM.DynamicBehaviors} \end{aligned} \tag{6.1}$$

The initial state of the server application of the IGM is established through loading the program. When the Configuration database is read, it initializes configuration classes. Then the server application reverts to a waiting state to wait for a request that comes either from the shop floor control module (i.e. PPSShopFloor module, see Sect. 8.4) for some information, or from the data modelling module (DMM) to test a change or new configuration. To avoid the tedious and complicated description of the program coding details, we focus our discussion in this chapter on the process aspect of manufacturing and generation of the PPS data.

Y. Tu and P. Dean, *One-of-a-Kind Production*,
DOI: 10.1007/978-1-84996-531-6_6, © Springer-Verlag London Limited 2011

6.1 IGM Architecture

The basic function of the IGM is to take a customer's specification, select an appropriate configuration (for producing a product family) and generate the data structures for product variants within the product family. Data structure will be made up according to the BOM, BOO and the BOI templates as discussed in Chap. 5. This data structure can be output in a number of forms and contains all the information necessary to plan, produce and monitor manufacturing processes for making a custom product. This is different from the process in mass production, where information for large batches of products is rolled up into summary form and provides no details. In OKP, the detailed information is necessary for each of the products because the product is unique and processed one-at-a-time or in a very small batch size.

The architecture of the IGM, at the top level, specifies the list of names of the component logic modules (CLMs), events and status in RTPA as follows:

$$
\begin{array}{lll}
\textbf{IGM.Architecture} \triangleq & \text{CLM} : \mathbf{ST} & \text{// Relational Database} \\
& \| \ \text{Configuration DB} : \mathbf{ST} \mid [1] & \\
& \| \ \text{InputDemand} : \mathbf{ST} \mid [1..*] & \\
& \| \ \text{InputItem} : \mathbf{ST} \mid [1] & \\
& \| \ \text{DemandDB} : \mathbf{ST} \mid [1] & \\
& \| \ \text{Request} : \mathbf{S} \mid [1] & \\
& \| \ \text{Response} : \mathbf{S} \mid [1] & \\
& \| \ \text{OutputInfor} : \mathbf{ST} \mid [1..*] & \\
& \| \ \text{ConfigModel} : \mathbf{ST} \mid [1..*] & \\
& \| \ \text{Events} : \mathbf{ST} & \\
& \| \ \text{Status} : \mathbf{ST} & \\
\text{IGM.Architecture.Events}\mathbf{ST} \triangleq & @\text{ServerInitial}\mathbf{S} & \\
& \| \ @\text{ShutDownServer}\mathbf{S} & \\
\text{IGM.Architecture.Status}\mathbf{ST} \triangleq & \textcircled{s}\text{ConfigModelsCreated}\mathbf{BL} & \\
& \| \ \textcircled{s}\text{ValidRequest}\mathbf{BL} & \\
& \| \ \textcircled{s}\text{BOOGenerated}\mathbf{BL} & \\
& \| \ \textcircled{s}\text{BOOOutputCompleted}\mathbf{BL} & \\
& \| \ \textcircled{s}\text{ImportDemandCompleted}\mathbf{BL} & \\
& \| \ \textcircled{s}\text{UpdateDemandCompleted}\mathbf{BL} & \\
& \| \ \textcircled{s}\text{ItemInterface}\mathbf{BL} & \\
& \| \ \textcircled{s}\text{ServerShutDown}\mathbf{BL} &
\end{array}
\tag{6.2}
$$

The RTPA expression (6.2) specifies inputs and outputs, main events that occur while the server is active, and the statuses. Some of the statuses can change when the server starts running.

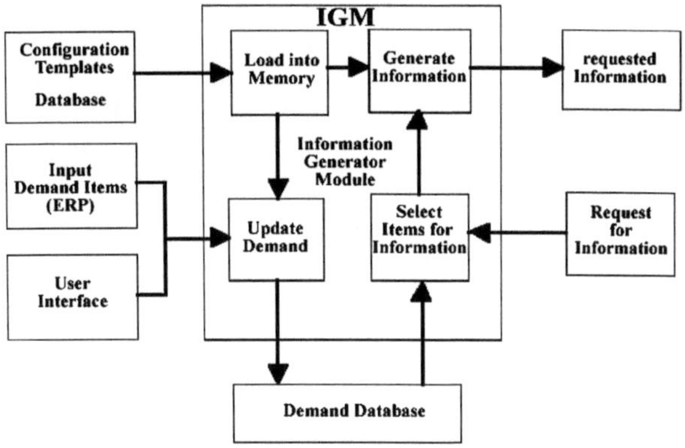

Fig. 6.1 Diagram of information generating process

6.1.1 Input Definitions

There are five types of input to the IGM

1. Configuration Database
2. Input Demand Items
3. Input Items
4. Demand Database
5. Request for Information

Configuration Database holds all the dada templates for a configuration for producing a product family. It is produced and maintained by the data modeling module (DMM) in knowledge mapping process. Since an OKP company normally has a number of production units/lines and produces a number of different kinds (or families) of products, the Configuration database in fact holds all the data templates for all the configurations. When the IGM starts, the data templates of all the configurations are loaded into memory. These data templates can be stored in memory because of their compact structure. To simplify the description in the rest of this chapter, we will call the set of data templates associated with a configuration simply as the configuration.

The benefit of keeping the configuration in memory is that the IGM processes much faster than when the configurations are retrieved from the Configuration database. When the configurations are in memory, a single or small batch of products can be processed. The specification for the Configuration database is given in RTPA expression (6.3) as follows:

$$
\begin{aligned}
&\text{IGM.Architecture.ConfigurationDB}\mathbf{ST} \triangleq \text{ConfigurationDB}\mathbf{S} : \\
&\qquad (\langle \text{TaskList} : \mathbf{ST} \mid [1..*] \rangle, \\
&\qquad \langle \text{AttributesVariables} : \mathbf{ST} \mid [1..*] \rangle, \\
&\qquad \langle \text{VariableList} : \mathbf{ST} \mid [1..*] \rangle, \\
&\qquad \langle \text{GlobalVariables} : \mathbf{ST} \mid [1..*] \rangle, \\
&\qquad \langle \text{WorkCenterOperations} : \mathbf{ST} \mid [1..*] \rangle \\
&\qquad) \\
&\text{IGM.Architecture.ConfigurationDB.TaskList}\mathbf{ST} \triangleq \text{TaskList}\mathbf{S} : \\
&\qquad (\langle \text{ProductID} : \mathbf{S} \rangle, \\
&\qquad \langle \text{ConfigID} : \mathbf{S} \rangle, \\
&\qquad \langle \text{AttributeID} : \mathbf{S} \rangle, \\
&\qquad \langle \text{isBase} : \mathbf{BL} \rangle, \\
&\qquad \langle \text{isSelected} : \mathbf{BL} \rangle, \\
&\qquad \langle \text{WorkCenterID} : \mathbf{S} \rangle, \\
&\qquad \langle \text{OperationID} : \mathbf{S} \rangle, \\
&\qquad \langle \text{OperationCondition} : \mathbf{S} \rangle, \\
&\qquad \langle \text{TaskSequence} : \mathbf{N} \mid 1 \leq \text{TaskSequence}\mathbf{N} \leq 9999 \rangle, \\
&\qquad \langle \text{TaskID} : \mathbf{S} \rangle, \\
&\qquad \langle \text{TaskCondition} : \mathbf{S} \rangle, \\
&\qquad \langle \text{TaskExpression} : \mathbf{S} \rangle) \\
&\text{IGM.Architecture.ConfigurationDB.AttributeVariables}\mathbf{ST} \triangleq \text{AttributeVariables}\mathbf{S} : \\
&\qquad (\langle \text{ProductID} : \mathbf{S} \rangle, \\
&\qquad \langle \text{ConfigID} : \mathbf{S} \rangle, \\
&\qquad \langle \text{AttributeID} : \mathbf{S} \rangle, \\
&\qquad \langle \text{VariableID} : \mathbf{S} \rangle, \\
&\qquad \langle \text{Type} : \mathbf{S} \rangle, \\
&\qquad \langle \text{Value} : \mathbf{RT} \rangle, \\
&\qquad \langle \text{Default} : \mathbf{RT} \rangle, \\
&\qquad \langle \text{Min} : \mathbf{N} \mid 1 \leq \text{Min}\mathbf{N} \leq 9999 \rangle, \\
&\qquad \langle \text{Max} : \mathbf{N} \mid 1 \leq \text{Max}\mathbf{N} \leq 9999 \rangle) \\
&\text{IGM.Architecture.ConfigurationDB.VariableList}\mathbf{ST} \triangleq \text{VariableList}\mathbf{S} : \\
&\qquad (\langle \text{ProductID} : \mathbf{S} \rangle, \\
&\qquad \langle \text{ConfigID} : \mathbf{S} \rangle, \\
&\qquad \langle \text{AttributeID} : \mathbf{S} \rangle, \\
&\qquad \langle \text{VariableID} : \mathbf{S} \rangle, \\
&\qquad \langle \text{ListValue} : \mathbf{S} \rangle) \\
&\text{IGM.Architecture.ConfigurationDB.GlobalVariables}\mathbf{ST} \triangleq \text{GlobalVariables}\mathbf{S} : \\
&\qquad (\langle \text{GlobalVariableID} : \mathbf{S} \rangle, \\
&\qquad \langle \text{Type} : \mathbf{S} \rangle, \\
&\qquad \langle \text{Value} : \mathbf{S} \rangle) \\
&\text{IGM.Architecture.ConfigurationDB.WorkcenterOperations}\mathbf{ST} \\
&\triangleq \text{WorkCenterOperations}\mathbf{S} : \\
&\qquad (\langle \text{WorkCenterID} : \mathbf{S} \rangle, \\
&\qquad \langle \text{OperationID} : \mathbf{S} \rangle, \\
&\qquad \langle \text{Min} : \mathbf{N} \mid 1 \leq \text{Min}\mathbf{N} \leq 9999 \rangle, \\
&\qquad \langle \text{Max} : \mathbf{N} \mid 1 \leq \text{Max}\mathbf{N} \leq 9999 \rangle)
\end{aligned}
\tag{6.3}
$$

Input Demand Items maybe imported from an ERP (enterprise resource planning) system or typed by an operator into the application either in real time or off-line. There is a specific format required for items to be accepted. An interface may be required to ensure that input demand records conform to the specification as listed below by RTPA expression (6.4) and will be discussed in detail later in this chapter.

$$
\begin{aligned}
&\text{IGM.Architecture.InputDemand}\mathbf{ST} \triangleq \text{InputDemand}\mathbf{S}\,(<\text{OrderNo}:\mathbf{N} \mid 1 \le \text{OrderNo}\mathbf{N} \le 999999>,\\
&\qquad <\text{LineNo}:\mathbf{N} \mid 0 \le \text{LineNo}\mathbf{N} \le 999>):\\
&\qquad (<\text{ConfigID}:\mathbf{S}>,\\
&\qquad <\text{ProductionLine}:\mathbf{S}>\\
&\qquad <\text{ScheduleDate}:\mathbf{D}>,\\
&\qquad <\text{SequenceNo}:\mathbf{N} \mid 0 \le \text{SequenceNo}\mathbf{N} \le 9999>,\\
&\qquad <\text{CompletedDate}:\mathbf{DT}>,\\
&\qquad <\text{ConfigString}:\mathbf{S} \mid\\
&\qquad \mathop{R}_{i=1}^{NofAttributes\mathbf{N}} [\text{ATT}(\text{name}(i\mathbf{N}),\\
&\qquad \mathop{R}_{j=1}^{NoVariables\mathbf{N}} [\text{Var}(\text{name}(j\mathbf{N}) = \text{Value}(j\mathbf{N})]|] >\\
&\qquad)
\end{aligned}
\tag{6.4}
$$

Input Items are normally input via a computer aided customer interface (Tu et al. 2003, Hong et al. 2008) and they enable individual demand items to be specified and input into the application. The computer aided customer interface could be simple, e.g. a parametric based customer interface for specifying a customized jewellery box (see Fig. 6.2), or a complicated one by which the computer can interpret or transform the customer requirements into the product specifications and design parameters. Tu et al. (2003) employ QFD (quality function deployment) method to interpret and transform customer requirements into the technical specifications and attributes of a product, and linear programming technique to find the parameters of a customized electrical appliance in order to maximize the customer's satisfaction. Hong et al. (2008) used GA (generic algorithm) to find the design configuration and parameters of a customized window for the maximization of a comprehensive design index value under a given budget. No matter which type of customer interface is used, it must make sure that the customer interface is compatible with the data structure of the configuration. The values of the attribute variables of the configuration have to be either input by a user if a parametric based customer interface is used, or converted and derived from the customer requirements if a complex customer interface is used. By the computer aided customer interface, the user can specify product variants within a product family. The Input Items of the IGM is specified in RTPA expression (6.5).

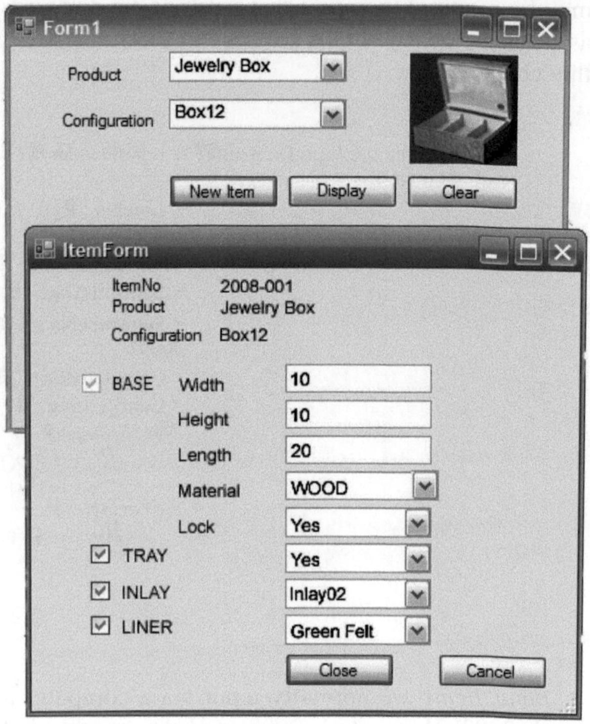

Fig. 6.2 A simple parametric customer interface

$$
\begin{aligned}
\text{IGM.Architecture.InputItem}\mathbf{ST} \triangleq \text{InputItem}\mathbf{S}\,(<\text{ItemID}:\mathbf{S}>):\ & \\
& (<\text{ItemConfigID}:\mathbf{S}>, \\
& <\text{ItemConfigString}:\mathbf{S}\mid \\
& \mathop{R}_{i=1}^{NofAttributes\mathbf{N}} [\text{ATT}(\text{name}(i\mathbf{N}), \\
& \mathop{R}_{j=1}^{NoVariables\mathbf{N}} [\text{Var}(\text{name}(j\mathbf{N}) = \text{Value}(j\mathbf{N})]|] > \\
&)
\end{aligned}
\quad (6.5)
$$

Demand Database stores all the demand items that an OKP company has produced in the past, is producing now and will produce in future. From this database, the IGM is able to analyze history of demand against the company's production capacity, monitor the WIP (work-in-progress) orders, and predict and plan future resource requirements. All of these functions require details on each customized product in order to effectively manage its production in an OKP company. The Demand database is defined by RTPA expression (6.6).

$$
\begin{aligned}
&\text{IGM.Architecture.DemandDB}\mathbf{ST} \triangleq \text{DemandDB}\mathbf{S}\,(<\text{OrderNo}:\mathbf{N}\mid 1 \leq \text{OrderNo}\mathbf{N} \leq 999999>,\\
&\qquad <\text{LineNo}:\mathbf{N}\mid 0 \leq \text{LineNo}\mathbf{N} \leq 999>):\\
&\qquad (<\text{ConfigID}:\mathbf{S}>,\\
&\qquad <\text{ProductionLine}:\mathbf{S}>,\\
&\qquad <\text{ScheduleDate}:\mathbf{D}>,\\
&\qquad <\text{SequenceNo}:\mathbf{N}\mid 0 \leq \text{SequenceNo}\mathbf{N} \leq 9999>,\\
&\qquad <\text{CompletedDate}:\mathbf{DT}>,\\
&\qquad <\text{StdMins}:\mathbf{N}\mid 0 \leq \text{StdMins}\mathbf{N} \leq 9999>,\\
&\qquad <\text{ConfigString}:\mathbf{S}\mid \mathop{R}\limits_{i=1}^{NofAttributes\mathbf{N}} [\text{ATT}(\text{name}(i\mathbf{N})), \mathop{R}\limits_{j=1}^{NoVariables} [\text{Var}(\text{name}(j\mathbf{N})) = \text{Value}(j\mathbf{N})]||]>\\
&\qquad)
\end{aligned}
\tag{6.6}
$$

Requests for Information can be made by the IGM for a single item or any group of items from the Demand database. The historical items can be requested to perform analysis. The WIP items are requested to monitor the progress of production, and the future orders are requested for planning and forecasting. The requests for information are specified in RTPA expression (6.7).

$$
\begin{aligned}
&\text{IGM.Architecture.Request}\mathbf{ST} \triangleq \text{Request}\mathbf{S}\,(<\text{RequestType}:\mathbf{S}\mid \text{Request}\mathbf{S} = \text{“Import”} \wedge \text{Request}\mathbf{S} = \text{“Generate”} \wedge \text{Request}\mathbf{S} = \text{“Input”} \wedge \text{Request}\mathbf{S} = \text{“Stop”}>):\\
&(<\text{Parameters}:\mathbf{S}\mid \mathop{R}\limits_{i=1}^{NofParameters\mathbf{N}} (\text{ParameterName}\mathbf{S} = \text{Value}\mathbf{S}>)
\end{aligned}
\tag{6.7}
$$

6.1.2 Output Definitions

The IGM has three outputs which are defined as follows:

1. Generated Information
2. Demand Database
3. Response

Generated Information is based on the bill-of-information (BOI). A part or all of the BOI can be requested. This output is in the form of data files or in a table or tables added to the Demand database with the BOO (bill of operation). The information can be further passed to or accessed by the shop floor control module (see Chap. 8). RTPA expression (6.8) specifies the output record set of the BOO for a demand item.

$$
\begin{array}{l}
\text{IGM.Architecture.OutputInfo}\mathbf{ST} \triangleq \text{OutputBOO}\mathbf{S}\,(<\text{OrderNo}:\mathbf{N} \mid 1 \le \text{OrderNo}\mathbf{N} \le 999999>,\\
\qquad <\text{LineNo}:\mathbf{N} \mid 0 \le \text{LineNo}\mathbf{N} \le 999>):\\
\qquad (<\text{ProductionLine}:\mathbf{S}>,\\
\qquad <\text{ScheduledDate}:\mathbf{DT}>,\\
\qquad <\text{SequenceNo}:\mathbf{N} \mid 0 \le \text{SequenceNo}\mathbf{N} \le 9999>,\\
\qquad <\text{StdMins}:\mathbf{N} \mid 0 \le \text{StdMins}\mathbf{N} \le 9999>,\\
\qquad <\text{WorkCenterID}:\mathbf{S}>,\\
\qquad <\text{OperationID}:\mathbf{S}>,\\
\qquad <\text{TaskID}:\mathbf{S}>,\\
\qquad <\text{TaskStd}:\mathbf{R} \mid 0 \;\; \text{TaskStd}\mathbf{R} \;\; 9999.99>\\
\qquad)
\end{array} \tag{6.8}
$$

Demand Database stores the Input Demand Items which can be imported from an ERP system. The IGM outputs the generated data to the Demand database. The standard processing time of a customized product or part is either estimated by the IGM according to the historical records or directly input by the company manager. The IGM will also output other information to the Demand database, such as the date and time when each item is completed at the shop floor. The Demand database as the output of the IGM is specified by RTPA expression (6.9).

$$
\begin{array}{l}
\text{IGM.Architecture.DemandDB}\mathbf{ST} \triangleq \text{DemandDB}\mathbf{S}\,(<\text{OrderNo}:\mathbf{N} \mid 1 \le \text{OrderNo}\mathbf{N} \le 999999>,\\
\qquad <\text{LineNo}:\mathbf{N} \mid 0 \le \text{LineNo}\mathbf{N} \le 999>):\\
\qquad (<\text{ConfigID}:\mathbf{S}>,\\
\qquad <\text{ProductionLine}:\mathbf{S}>,\\
\qquad <\text{ScheduleDate}:\mathbf{D}>,\\
\qquad <\text{SequenceNo}:\mathbf{N} \mid 0 \le \text{SequenceNo}\mathbf{N} \le 9999>,\\
\qquad <\text{CompletedDate}:\mathbf{DT}>,\\
\qquad <\text{StdMins}:\mathbf{N} \mid 0 \le \text{StdMins}\mathbf{N} \le 9999>,\\
\qquad <\text{ConfigString}:\mathbf{S} \mid\\
\qquad \mathop{R}\limits_{i=1}^{NofAttributes\mathbf{N}} [\text{ATT}(\text{name}(i\mathbf{N}), \mathop{R}\limits_{j=1}^{NoVariables} [\text{Var}(\text{name}(j\mathbf{N}) =\\
\qquad \text{Value}(j\mathbf{N})]|] >\\
\qquad)
\end{array} \tag{6.9}
$$

Response by the server to the client request takes the form of a message that indicates the result of the request, either a success or failure. It is specified below in RTPA expression (6.10).

$$
\text{IGM.Architecture.Response}\mathbf{ST} \triangleq \text{Response}\mathbf{S}:(<\text{Response}:\mathbf{S}>) \tag{6.10}
$$

6.1.3 Process Definitions

There are many processes which take place in the server application as it imports demand data from an ERP system, including updating standard times and generating various output information sets. The server has to be initialized, and the initiation includes the storage of configurations in memory. Then the server waits for a client request. In this chapter, we focus our discussion on importing and updating demand data, which is the main source of information, and the generation of a BOO for a selected group of items. RTPA expression (6.11) defines the static behaviours of the server application.

$$
\begin{aligned}
\textbf{IGM.StaticBehaviors} \triangleq\ & \text{ServerInitial}\ (<\mathbf{I}::(\,)>, <\mathbf{O}::(\,)>) \\
& \|\ \text{ShutDownServer}\ (<\mathbf{I}::\ \text{Response}\mathbf{S}>, <\mathbf{O}::(\,)>) \\
& \|\ \text{ProcessRequest}\ (<\mathbf{I}::\ \text{Request}\mathbf{S}>, <\text{O}::\ \text{Response}\mathbf{S}>) \\
& \|\ \text{ValidateRequest}\ (<\mathbf{I}::\ \text{Request}\mathbf{S}>, <\text{O}:: \\
& \textcircled{s}\text{ValidParameters}\mathbf{BL}>) \\
& \|\ \text{CreateConfigModels}\ (<\mathbf{I}::(\,)>, <\mathbf{O}:: \\
& \textcircled{s}\text{CreatedConfigModels}\mathbf{BL}>) \\
& \|\ \text{ImportDemand}\ (<\mathbf{I}::\ \text{Request}\mathbf{S}>, <\mathbf{O}::\ \text{Response}\mathbf{S}, \\
& \textcircled{s}\text{ImportDemandCompleted}\mathbf{BL}>) \\
& \|\ \text{GenerateBOO}\ (<\mathbf{I}::\ \text{Request}\mathbf{S}>, <\mathbf{O}::\ \text{Response}\mathbf{S}, \\
& \textcircled{s}\text{BOOGenerated}\mathbf{BL}>) \\
& \|\ \text{ItemInterface}\ (<\mathbf{I}::\ \text{Request}\mathbf{S}>, <\mathbf{O}::\ \text{Response}\mathbf{S}>)
\end{aligned}
\qquad (6.11)
$$

6.2 Input Demand Specification

Input demand specifications are based on the selected product family and the related configuration. When a user has made a selection, the user interface can generate all the questions that will help to make a complete specification. The computer system defines a format for specifications that is flexible, so that any product family and group of configurations can use this format to specify the variants. This format is also used when importing demand items from an ERP system. The format includes a configuration string (CS). It is defined by the following RTPA expression:

$$
\mathop{R}\limits_{i=1}^{NofAttributes} (ATT(name_i) = TRUE/FALSE) \mathop{R}\limits_{j=1}^{NofVariables} (, VAR(name_j) = value_j)| \qquad (6.12)
$$

Where

ATT = an attribute
VAR = a variable

An example of a configuration string for a jewelry box would be:

(ATT(**BASE**)=TRUE,VAR(WIDTH)=10,VAR(HEIGHT)=10,VAR(LENGTH)=20,VAR(MATERIAL)=WOOD,LOCK=YES|ATT(**TRAY**)=TRUE,VAR(TRUE)=TRUE|ATT(**LINER**)=TRUE,VAR(LINER)=GREENFELT|ATT(**INLAY**)=TRUE,VAR(INLAYNO)=INLAY02|)

Where
BASE attribute has 5 variables:

WIDTH = value
HEIGHT = value
LENGTH = value
MATERIAL = list (WOOD, METAL, GLASS)
LOCK = Boolean (Yes/No)

TRAY attribute IsSelected (TRUE/FALSE) and 1 variable

TRUE = Boolean (YES/NO) – activates the option

LINER attribute IsSelected (TRUE/FALSE) and 1 variable

LINER = list (GREENFELT, BLACKFELT, READFELT)

INLAY attribute IsSelected (TRUE/FALSE) has 1 variable

INLAYNO = list (INLAY01, INLAY02, INLAY03)

When input demand is imported from an ERP system, the format of demand items has to be the same. If not, demand from the ERP is initially processed through an interface that will format the data correctly.

6.3 User Interface

A parametric based customer interface for a jewellery box is shown in Fig. 6.2. A user selects a product family and then the associated configuration for producing the products within this family. A description or image of the product and the configuration is shown to help in the selection process. When the selection has been made, the interface sends a request to the server application which uses the variables in the configuration to construct and display a specification form as an example shown in Fig. 6.2. As mentioned before, the values of the variables in the configuration are either directly input by the user or converted and derived by a more complex customer interface. The numerical variables can have a range of valid values from a minimum to maximum with a starting default value. The range of valid values creates validation rules for checking the inputs, so that the final specification is a feasible product variant from a manufacturability aspect. When confirmation of the specifications is finished, the IGM will create a configuration string as defined above and add it to the Demand database.

By using the customer interface, the configurations as modelled in the knowledge mapping process by using the data modelling module (DMM) can be tested.

Order identification includes a customer number, order number and line number for each product. Orders will be categorized in some form, e.g. customer type, location, geographic area or market. This type of information is helpful in forecasting process for planning production demand.

6.4 Creation of Configuration Model

One of the central functions of the IGM is to create configurations (objects) in the memory of the server so that the IGM can process information and generate output. Specifications of the configuration class are shown in Appendix D1. To create configurations in memory, it is necessary to create classes and establish their relationships. When this is done, they can be reused repeatedly as each demand item in the group is processed. An input item determines which configuration to use. The server then makes a copy of it in memory and applies all the values of the variables taken from the input item to construct the precise structure of the custom item. When this is completed, the structure (a group of related objects representing the BOM, BOO and BOI of an item) can be used to generate multiple outputs as requested. In this chapter, the full BOO with standard times will be output either for a single item or for a group of items. This demonstrates the efficiency and flexibility of the IGM. Through the following stages, the configurations can be generated.

Stage 1

In this stage, the IGM creates a basic tree structure of Configurations, Attributes, Operations and Tasks.

- 1.1. Process the TaskList record set in sequence of Product Family, Configuration, Attribute, Operation, and Task sequence number. The TaskList record has been defined in RTPA expression (6.3) in Sect. 6.1.1 and it includes TaskList(ProductID, ConfigID, AttID, OpID, TaskSeq, TaskID).
- 1.2. For the first record of the configuration, create an object of the configuration class using the combination of ProductID and ConfigID as a key. The configuration class is expressed in RTPA as follows:

$$
\begin{aligned}
&\text{IGM.Architecture.Configuration}\mathbf{ST} \triangleq \text{ConfigID}\mathbf{S}\text{:}\\
&\quad (\langle \text{ProductID:}\mathbf{S}\rangle,\\
&\quad \langle \text{ProductName:}\mathbf{S}\rangle\\
&\quad \langle \text{ConfigID:}\mathbf{S}\rangle,\\
&\quad \langle \text{ConfigName:}\mathbf{S}\rangle\\
&\quad \langle \text{Attributes:}\mathbf{S} \mid \mathop{R}_{i=1}^{\textit{NofAttributes}\mathbf{N}} \langle \text{Attribute(i}\mathbf{N}\text{)}\mathbf{ST}\rangle)
\end{aligned}
\qquad (6.13)
$$

- 1.3. Create an object of the attribute class and add it to the configuration object using AttID as a key, if this attribute object is not in this configuration's list.

Please refer to RTPA expression (6.14) for details.

$$\begin{array}{l}
\text{IGM.Architecture.Attributes}\mathbf{ST} \triangleq \text{AttributeIDS:} \\
\quad (<\text{Name:}\mathbf{S}> \\
\quad <\text{IsSelected:}\mathbf{BL} \mid \mathbf{T}\text{=Selected} \wedge \mathbf{F}\text{=Not Selected} \mid \text{default=FALSE}> \\
\quad <\textbf{\textit{Variables:}}\,\boldsymbol{ST} \mid \mathop{R}\limits_{i=1}^{NofVariables\boldsymbol{N}} <\textbf{\textit{Variable(i}}\boldsymbol{N}\textbf{\textit{)}}> \\
\quad <\textbf{\textit{Operations:}}\,\boldsymbol{ST} \mid \mathop{R}\limits_{i=1}^{NofOperations\boldsymbol{N}} <\textbf{\textit{Operation(i}}\boldsymbol{N}\textbf{\textit{)}}> \\
\quad <\textbf{\textit{Components:}}\,\boldsymbol{ST} \mid \mathop{R}\limits_{i=1}^{NofComponents\boldsymbol{N}} <\textbf{\textit{Component(i}}\boldsymbol{N}\textbf{\textit{)}}>)
\end{array} \tag{6.14}$$

- 1.4. Create an object of the Operation class and add it to the attribute object using OpID as a key, if this operation is not in this attribute's list. The Operation class is defined in RTPA expression (6.15).

$$\begin{array}{l}
\text{IGM.Architecture.Operation}\mathbf{ST} \triangleq \text{OperationID}\mathbf{S}\text{:} \\
\quad (<\text{Name:}\mathbf{S}> \\
\quad <\text{Condition:}\mathbf{S}> \\
\quad <\text{SetupTime:}\mathbf{R}> \\
\quad <\text{MinimumOps:}\mathbf{N}> \\
\quad <\text{MaximumOps:}\mathbf{N}> \\
\quad <\text{AttributeID:}\mathbf{S}> \\
\quad <\text{ResourceID:}\mathbf{S}> \\
\quad <\text{Tasks:=} \mathop{R}\limits_{i=1}^{NofTasks\boldsymbol{N}} <\text{Task(i}\mathbf{N}\text{)}>)
\end{array} \tag{6.15}$$

- 1.5. Create an object of the Task class and add it to the operation object using TaskSeq and TaskId as keys. There should not be any duplicates of the TaskSeq and TaskID combination. See RTPA expression (6.16).

$$\begin{array}{l}
\text{IGM.Architecture.Task}\mathbf{ST} \triangleq \text{TaskID}\mathbf{S}\text{:} \\
\quad (<\text{Name:}\mathbf{S}> \\
\quad <\text{Sequence:}\mathbf{N}> \\
\quad <\text{Condition:}\mathbf{S}> \\
\quad <\text{StdMins:}\mathbf{R}> \qquad \text{// Cycle Time} \\
\quad <\text{Expression:}\mathbf{S}> \\
\quad <\text{Equipments:=} \mathop{R}\limits_{i=1}^{NofEquipment\boldsymbol{N}} <\text{Equipment(i}\mathbf{N}\text{)}> \\
\quad <\text{RawMaterials:=} \mathop{R}\limits_{i=1}^{NofRawMaterials\boldsymbol{N}} <\text{RawMaterials(i}\mathbf{N}\text{)}>)
\end{array} \tag{6.16}$$

- 1.6. Repeat Steps 1.3, 1.4 and 1.5 until the next Product Family and Configuration ID, then go back to Step 1.1.

Stage 2
In this stage, the IGM adds attribute variables to the structure.

- 2.1. Process the AttributeVariables record set (for any sequence but no duplication is allowed). The AttributeVariable record is defined in RTPA expression (6.3) in Sect. 6.1.1 and it contains the ProductFamily, ConfigID, AttID and VariableID.
- 2.2. Add the variable to the attribute object using the VarID as a key. Variables are set with default values between the minimum and the maximum values where applicable. This type of information is used to create the run-time interface for entering item specifications as mentioned in Sect. 6.3.
- 2.3. Repeat Step 2.2 for each record in the set.

Stage 3
In this stage, the IGM adds the content of variables that are lists. The following steps are used to create the user interface.

- 3.1. Process the VariableList record set (for any sequence but no duplication is allowed). The VariableList record is defined in RTPA expression (6.3) in Sect. 6.1.1 and it contains the ProductFamily, ConfigID, AttID and VariableID.
- 3.2. Add the list value to the variable list in the relevant attribute object using the key data.
- 3.3. Repeat Step 3.2 for each record in the set.

Through the three stages as mentioned above, the IGM creates the tree structure for the BOO as shown in Fig. 1.4 in Chap. 1. For the complete set of data templates, i.e. BOM, BOO and BOI, the other parts would be added into the database by the IGM in a similar manner.

Multiple configurations can be constructed in memory and are only limited by the amount of memory available to the server application. In a client/server environment, this produces a very responsive application, as opposed to a client based program that reads the database each time it needs to structure a demand item to generate information. RTPA expression (6.17) is the complete specification for storing configurations in the memory of the server application.

```
IGM.CreateConfigModel ( < I:: ( ) >, < O:: ⓢCreatedConfigModelsBL >) ≙
{
→ConfigDB = RelDataBase( < I::( ) >, < O::( ) >)
→ConfigDB.Connect( < I::("Configuration.mdb" ) >, < O::
ⓢRelDBConnectedBL >)

→ (? ⓢRelDBConnectedBL = F
        → ⓢCreatedConfigModelsBL:= F
        →ResponseS := "Failed to connect to Configuration Database"
        →⊗              // Exit Configuration Model Function
   )
```

```
→SQLCmdS := "Select * From TaskList"
→TaskListRS := ConfigDB.Execute( < I::SQLCmdS >, < O:: RecordSetRS,
                                  ⓈRelDBExecutedBL >)
→ (? ⓈRelDBExecutedBL = F
        → ⓈCreatedConfigModelsBL:= F
        →ResponseS := "Failed to access Configuration Database - TaskList"
        →⊗                  // Exit Configuration Models Function
)
// TaskList is sorted by ProductID, ConfigID, WorkCenterID, OperationID and
TaskSequenceNo
// The tree structure of each configuration is now built, add attributes, operations
and tasks in this
//  part of the Framework application
→ ConfigsST := CollectionST( ) (or ArrayListST)
    NofTaskListRecordsN
→              R              (→ ConfigKeyS := TaskList(iN)RS.ProductIDS &
              i=1
TaskList(iN)RS.ConfigIDS
                               (? ConfigKeyS ≠ LastConfigKeyS
                                      → c:ST := ConfigST( )
                                      → cST.ProductIDS := Task-
                                      List(iN)RS.ProductIDS
                                      → cST.ConfigIDS := Task-
                                      List(iN)RS.ConfigIDS
                                      → ConfigsST.add(cST, ConfigKeyS)
                                      → LastConfigKeyS := ConfigKeyS
                               |?~
                                      → cST := ConfigsST(ConfigKeyS)
                               )
                              → AttKeyS := TaskList(iN)RS.AttributeIDS
                               (? AttKeyS ≠ LastAttKeyS
                                      → a:ST := AttributeST( )
                                      → aST.AttributeIDS := Task-
                                      List(iN)RS.AttributeIDS
                                      → aST.isSelectedBL := Task-
                                      List(iN)RS.isSelectedBL
                                      → cST.addAttribute(aST, AttKeyS)
                                      → LastAttKeyS := AttKeyS
                               |?~
                                      → aST := cST.Attributes(AttKeyS)
                               )
                              → OpKeyS := TaskList(iN)RS.WorkCenterIDS &
                              TaskList(iN)RS.OperationIDS
                               (? OpKeyS ≠ LastOpKeyS
                                      → Op:ST := OperationST( )
                                      → OpST.OperationIDS := Task-
                                      List(iN)RS.OperationIDS
```

```
                                        → OpST.ConditionS := Task-
                                        List(iN)RS.OperationConditionS
                                        → aST.addOperation(OpST, OpKeyS)
                                        → LastOpKeyS := OpKeyS
                                  |?~
                                        → OpST := aST.Operation(OpKeyS)
                                  )
                                → TaskKeyS := TaskList(iN)RS.SequenceN & Task-
                                List(iN)RS.TaskIDS
                                 (? TaskKeyS ≠ LastTaskKeyS
                                        → Tk:ST := TaskST( )
                                        → TkST.TaskIDS := TaskList(iN)RS.TaskIDS
                                        → TkST.SequenceN := Task-
                                        List(iN)RS.SequenceN
                                        → TkST.ConditionS := Task-
                                        List(iN)RS.TaskConditionS
                                        → TkST.ExpressionS := Task-
                                        List(iN)RS.ExpressionS
                                        → OpST.addTask(TkST, TaskKeyS)
                                        → LastTaskKeyS := TaskKeyS
                                  )
                                )
// The variables are added to the Attributes (along with the Operations)
→SQLCmdS := "Select * From Variables"
→VariableRS := ConfigDB.Execute(SQLCmdS)RS
→ (? ⓈRelDBExecutedBL = F
        → ⓈCreatedConfigModelsBL:= F
        →ResponseS := "Failed to access Configuration Database - Variables"
        →⊗                  // Exit Configuration Models Function
   )
```

$$\rightarrow \quad \overset{NofVariable\mathrm{Records}\boldsymbol{N}}{\underset{i=1}{R}} \quad (\rightarrow \text{c:}\mathbf{ST} := \text{Configs(Variable(i}\mathbf{N}\text{)}\mathbf{RS}\text{.ProductID}\mathbf{S}\ \&$$

```
Variable(iN)RS.ConfigIDS)ST
                                → a:ST :=
                                cST.Attributes(Variable(iN)RS.AttributeIDS)ST
                                 → v:ST := VariablesST
                                 → vST.NameS := Variable(iN)RS.VariableIDS
                                 → vST.TypeS := Variable(iN)RS.TypeS
                                 → vST.ValueRT := Variable(iN)RS.ValueRT
                                 → vST.DefaultRT := Variable(iN)RS.DefaulRT
                                 → vST.MinN := Variable(iN)RS.MaxN
                                 → vST.MaxN := Variable(iN)RS.MinN
                                 → aST.addVariable(vST, vST.NameS)
                                )
// The variable Lists are added to the Variables of type LIST
→SQLCmdS := "Select * From VariableList"
```

→VarList**RS** := ConfigDB.Execute(SQLCmd**S**)**RS**
→ (? ⓈRelDBExecuted**BL** = **F**
→ ⓈCreatedConfigModels**BL**:= **F**
→ResponseS := "Failed to access Configuration Database – Variable List"
→⊗ // Exit Configuration Models Function
)

$$\rightarrow \mathop{R}_{i=1}^{NofVariableRecords\mathbf{N}} (\rightarrow c{:}\mathbf{ST} := \text{Configs(VarList(i}\mathbf{N}\text{)}\mathbf{RS}\text{.ProductID}\mathbf{S}\ \&$$

VarList(i**N**)**RS**.ConfigID**S**)**ST**
→ a:**ST** := c**ST**.Attributes(VarList(i**N**)**RS**.AttributeID**S**)**ST**
→ lv:**ST** := a**ST**.Variable(VarList(i**N**)RS.VarID**S**)**ST**
→ lv**ST**.add(VarList(i**N**)**RS**.VarListValue**S**)
)
→ ⓈCreatedConfigModels**BL**:= **T**
} (6.17)

In the following section, the process of information generation from a configuration will be described. This process can be performed on a single item of a selected configuration, or a group of items with a mixture of configurations.

6.5 Basic Processes of Information Generation

In this section, a brief description of the server application of the IGM is presented, which includes the structure and five of the main functions: Initialize Server, Import Data, Generate Information, Item Interface and Stop Server. There are many ways to implement a server function and it is dependent on the operating environment. The basic concept of the server application will be described, but the main focus will be placed on the functionality in response to specified requests from client programs. The two client programs are the data modelling module and the shop floor control module.

6.5.1 Server Initialization

The server application is initialized as it is loaded into the memory of the computer and then it enters on a waiting state until it receives a request from a client program to perform one of its services. At any time, a request can be sent to shut down the server application or it can be interrupted through the server console to shut it down immediately. The main process of the server is specified by RTPA expression (6.18)

```
IGM.ProcessDeployment ≙
    { // Server in waiting state
    @ServerInitialS
    ↳ ( ServerInitial
                @ServerShutDown=T
            ↳       R                ProcessRequest ( < I:: RequestS >, <
                   ≥1
            O:: ResponseS >)
            ↳ ShutDownServer (< I:: ResponseS >, < O:: ( ) >)
        )
    || // Interrupt Shut Down
    ↳ @Interrupt
            →ResponseS := "Server interrupted – shutting down!"
            →OutputDevice.Write ( ResponseS )
            →⊠
    }                                                                   (6.18)
```

Initialization of the server includes loading of configurations from the Configuration database into memory and creating an object of the Expression class. This expression object will be used throughout the running of the server to evaluate various expressions and constraints. The initialization behaviour is defined by RTPA expression (6.19).

```
IGM.ServerInitial ( < I::( ) >, < O::( ) >) ≙
    {
    ↳ CreateConfigModels ( < I::( ) >, < O:: ⓢCreatedConfigModelsBL >)
    → (?ⓢCreatedConfigModelsBL = F
            →ResponseS := "Invalid Models"
            →ⓢShutDownServerBL = T
            ↳ StopServer (< I::ResponseS >, < O::( ) >)
      |?~
            →ⓢShutDownServerBL = F
            ExpST := New Expression( )ST      // Instantiates and Expression
            object to evaluate expressions
      )
    }                                                                   (6.19)
```

The CreateConfigModels has been specified by RTPA expression (6.17).

The full specification of the Expression class can be found in Appendix D4 and has the following static behaviours:

```
Expression.StaticBehaviors ≙ ExpressionS ( < I:: ( ) >, < O:: ( ) >)
    || Evaluate ( < I:: < ExpressionS >, < O:: ValueRT >)
    || AddVariable ( < I:: < ExpVariableST >, < O::( ) >)
    || SetGlobals (<I :: ( ) >, <O :: ( )>)
    || AddRange ( < I:: VariableArrayST >, < O:: ( )>)
    || Clear ( < I:: ( ) >, < O:: ( ) >)                                (6.20)
```

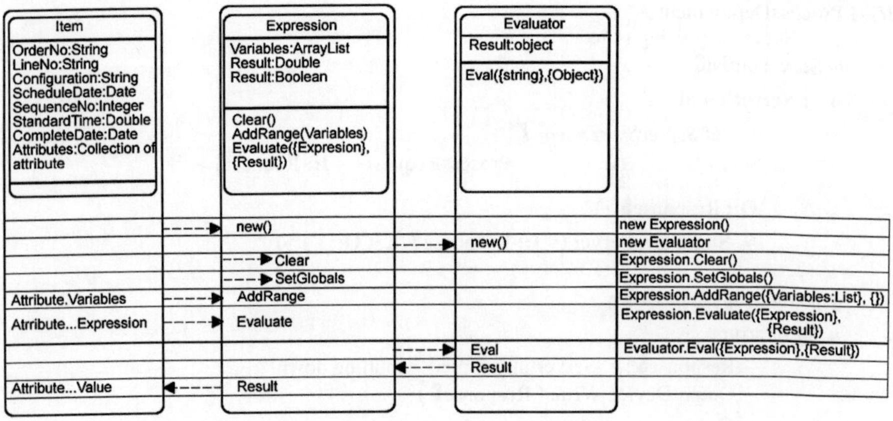

Fig. 6.3 The process and data relations among item, expression and evaluator

The Expression class uses an Evaluator class (available in several forms from a variety of sources on the Internet and in most programming languages) which evaluates expressions that produce either a Boolean or a numerical result. The Evaluator will evaluate expressions with run-time variables, and the Expression object creates a variable array before calling the Evaluate function. Other behaviours of the Expression class are clearing, setting global variables and adding run-time variables to its array.

The Expression class holds all variables associated with the expression and the Evaluator performs the evaluation. The Evaluator returns an object with the result. The Expression class determines if the result is either a Boolean or a numerical value, then passes the result back to the calling class. This process is illustrated in Fig. 6.3.

An expression can either be a Boolean expression or a mathematical expression. Both have a single result returned.

Example 1 Item (e.g. a jewellery box) in Fig. 6.2 has a variable called MATERIAL which can have one of three values {WOOD, GLASS, METAL}. The customer specification sets the value of MATERIAL. The IGM wants to know which material has been specified. Item object adds the MATERIAL variable with its value to the Expression class with the behaviour AddRange(Variables). Then Item object asks Expression to evaluate the expression MATERIAL = WOOD. Depending on the initial value of the variable MATERIAL passed by Item, Item will receive the result of either TRUE or FALSE.

Example 2 Item object has two variables called LENGTH and WIDTH which can have values in the range of 10–50 in. The customer specification sets the values of LENGTH and WIDTH. The IGM wants to know the area of the item. Item object adds the two variables LENGTH and WIDTH with their values to the Expression class using AddRanges(Variables). Item object asks Expression to evaluate the expression LENGTH * WIDTH and then Item object will receive the result as a numerical value.

6.5.2 Processing Requests

To process a request, the server application of the IGM first validates the request parameter. There are four types of requests: import, generate, input and stop. In the first three types of request, the server application completes each request, sends a response, and stays on the waiting state to wait for the next request. If the request is the fourth type (stop), a response is sent back to the client to acknowledge the request to terminate the server and the server shuts down. RTPA expression (6.21) defines this process.

```
IGM.ProcessRequest ( < I:: RequestS >, < O:: ResponseS >) ≙
    {
    → (? RequestS = "Import"
            ↳ ImportDemand (< I:: RequestS >, < O:: ResponseS >)
       |? RequestS = "Generate"
            ↳ GenerateBOO (< I:: RequestS >, < O:: ResponseS >)
       |? RequestS = "Input"
            ↳ ItemInterface (< I:: RequestS >, < O:: ResponseS >)
       |? RequestS = "Stop"
            →ResponseS := "User request to shut down server"
            ↳ ShutDownServer (< I:: ResponseS >, < O:: ( ) >)
       |?~
            →ResponseS := "Invalid request from user"
       )
    }                                                          (6.21)
```

6.5.3 Import Demand Items

Request parameters are validated first. These requests require a schedule date to select items. The schedule date can be imported from the production scheduling and control module of the OKP management and control software (OKPMCS) system© or an ERP system. The transfer of data can be performed by a conventional SQL command or a stored procedure in the database. Each requested service retrieves records from the Demand database.

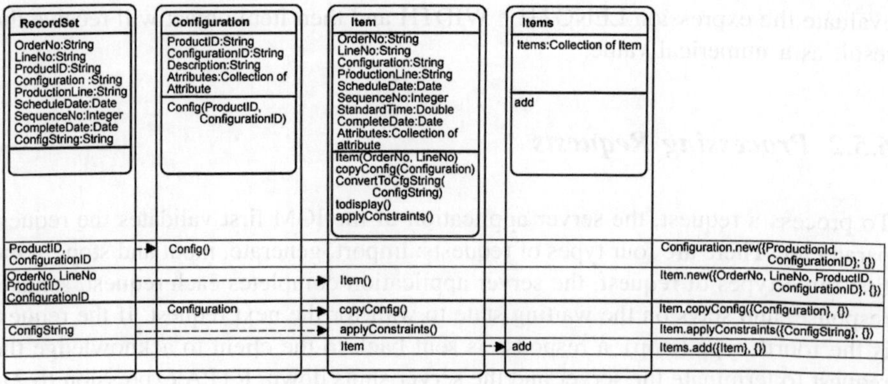

Fig. 6.4 Process of importing demand data items

The import data service uses Import Data, Configuration and Item classes. The Import Data class is used to create a relational database object connecting to the Demand database. The SQL command is then set using request parameters. The SQL command is executed by the Import Data object and produces a record set. This process is illustrated in Fig. 6.4.

For each record in the data set, a Configuration object is created based on configuration ID. A new Item object is created and the Configuration object is copied to the Item object, setting all default values for variables. The configuration string in the input record is then used to evaluate all of the conditions and variables. The expression object described in Sect. 6.5.1 is used. Constraint conditions for the item (i.e. a manufacturing product or part) are evaluated, all the internal variables are calculated, and the data fields are accordingly assigned values. This includes calculating the standard time to manufacture items. The item is then added to the Items class which is a collection (or array list).

When the Items collection has been successfully built, the update function of the import data object is called to update the imported data with the standard time calculated by the configuration model and run-time item variables. This function is specified by RTPA expression (6.22) below:

IGM.ImportDemand (< **I::** Request**S** >, < **O::** Response**S**,
ⓢImportDemandCompleted**BL** >) ≙
{
↳ ValidateRequest (< **I::** Request**S** >, < **O::** ⓢValidRequest**BL** >)
→ (?ⓢValidRequest**BL** = **F**
→ⓢImportDemandCompleted**BL** = **F**
→ResponseS := "Invalid Request Parameters"
→⊗ // Exit Import Function
)
→ Id:**ST** := ImportData(< **I::** ( ) >, < ⓢRelDBConnected**BL** >)
→SQLCmd**S** := "Select * From Items where Items.ScheduleDate =" & Im-

```
portDateD
→ (? ⓈRelDBExecutedBL = T
    →DemandItemRS := IdST.Execute( < I::SQLCmdS >, < O:: Record-
    SetRS,
                        ⓈRelDBexecutedBL >)
    → (? ⓈRelDBExecutedBL = F
            →ⓈImportDemandCompletedBL = F
            →ResponseS := "Failed to access Demand Database"
            →⊗                  // Exit Import Function
    )
    NofItemsN
    →      R      (→ c:ST = config(DemandItemRS(iN).ProductIDS &
          i=1
                              DemandItemRS(iN).ConfigIDS)
              → i:ST := ItemST(DemandItemRS(iN).OrderNoN & De-
            mandItemRS(iN).LineNoN)
              → iST.copyConfig(cST)
              →
            iST.ApplyConstraints(DemandItemRS(iN).ConfigStringS)
              → iST. ConfigStd( )
              → ItemsST.add(iST)
             → ⓈImportDemandCompletedBL := T
            )
→ (? ⓈImportDemandCompletedBL = T
    ↳ Id.UpdateDemand ( < I:: ItemsST >, < O::
    ⓈDemandUpdateCompletedBL >)
|?~
    →ⓈUpdateDemandUpdatedBL = F
)
}                                                              (6.22)
```

Calculation of the standard time (cycle time) is performed by the item class and uses the Attribute, Operation, Task and Expression classes. This process is described in Fig. 6.5.

To calculate the standard time (cycle time) for an item, it is necessary to view the model like branches on a tree. An item is made up of a number of attributes and an attribute is made by a number of operations. For each attribute selected in the custom specification, the time for operations needs to be accumulated. An operation has a condition to be satisfied before it can be included in the accumulation. An operation is made up of one or more tasks. The task conditions must be satisfied and then the time for the task is calculated and accumulated to calculate the total time for the operation.

$$ItemStd = \sum_{i=1}^{NofAttributes=isSelected} \sum_{j=1}^{NofOperations=True} \sum_{k=1}^{NofTask=True} t_{ijk} \tag{6.23}$$

Where

NofAttributes = number of selected attributes (isSelected)

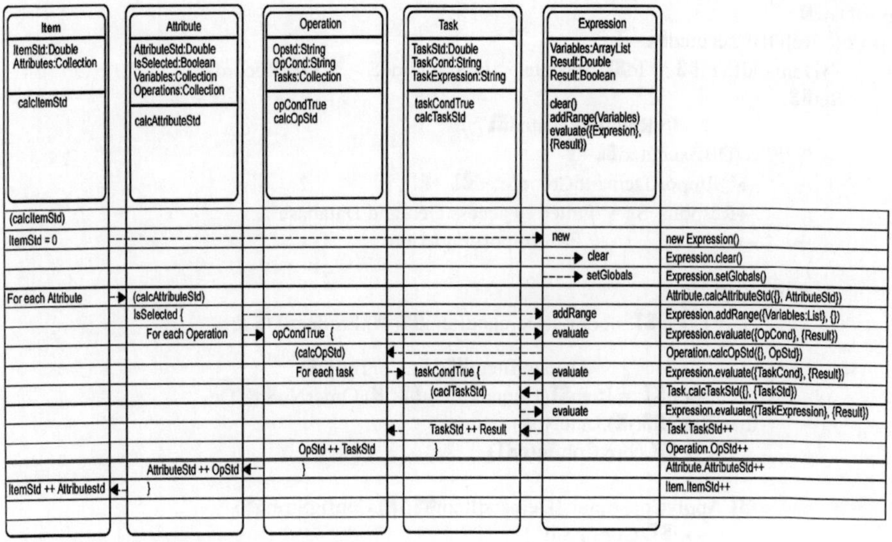

Fig. 6.5 Calculate the cycle time for an item

NofOperations = number of operations with an operation condition that is TRUE
NofTasks = number of tasks with an task condition that is TRUE

The specifications for these standard time calculations in RPTA are shown in expressions (6.24) and (6.25).

ITEM.CalcStdTime (< **I**:: ( ) >, < **O**:: StdTime**R** >) ≙
{
→ StdTime**R** := 0
→ Exp.Clear ( )
→ Exp.setGlobals ( )
→ $\overset{NoAttributes\mathbf{N}}{\underset{i=1}{R}}$ (→ Exp.AddRange(Attributes(i**N**).Variables**ST**)
→ (? Attributes(i**N**).isSelected = **T**
→ $\overset{NoOperations\mathbf{N}}{\underset{j=1}{R}}$ StdTime**R** := stdTime**R** + Attributes(i**N**).Operation(j**N**).calcOpStd( )**R**
)
)
(? Exception = **T**
→ ⓈStdTimeCalculated**BL** := **F**
|?~
→ ⓈStdTimeCalculated**BL** := **T**
)
} (6.24)

OPER.calcOpStd (< **I**:: Variable**ST** >, < **O**::OpStdMins**R** >) ≙
{
→ OpStdMins**R** := 0
→ (? OpCond(Condition**S**) = **T**

→ $\mathop{R}\limits_{i=1}^{NofTasks\mathbf{N}}$ (? TaskCond(Task(i**N**).Condition**S**) = **T**

→ OpStdMins**R** := OpStdMins**R** +
Exp.Evaluate(Task(i**N**).TaskExpresion**S**)**R**
)
} (6.25)

6.5.4 Generation of Bill of Operation (BOO)

In this section, the basic function of generating the BOO for an item is presented. The Import Data, Configuration and Items classes will be used. These classes are defined in RTPA expression (6.22).

In the request, the request parameters, schedule date and a range of production sequence numbers determine the selection of items from the Demand database. When the item collection is completed successfully, the output information function is called to output the BOO. RTPA expression (6.26) defines the output of the BOO via this function.

ID.OutputBOO (< **I**:: Items**ST** >, < **O**:: <ⓢBOOGenerated**BL** >) ≙
{
→ $\mathop{R}\limits_{i=1}^{NofItems\mathbf{N}}$ (→OrderNo**N** := Item(i**N**).OrderNo**N**

→OrderLine**N** := Item(i**N**).OrderLine**N**
→ProductionLine**S** := Item(i**N**).ProductionLine**S**
→ScheduleDate**D** := Item(i**N**).ScheduleDate**D**
→SequenceNo**N** := Item(i**N**).SequenceNo**N**
→StdMins**R** := Item(i**N**).StdMins**R**

→ $\mathop{R}\limits_{i=1}^{NofAttributes\mathbf{N}}$ (? Items.Attribute(i**N**)**ST**.isSelected**BL** = **T**

→ $\mathop{R}\limits_{j=1}^{NofOperations\mathbf{N}=\mathbf{T}}$ (→ OperationID**S** := Attribute(i**N**).Operation(j**N**).OpID**S**

→ WorkCenterID**S** := Attribute(i**N**).Operation(j**N**).OpID**S**

→ $\mathop{R}\limits_{n=1}^{NofTasks\mathbf{N}=\mathbf{T}}$ (→ TaskID**S** := Attribute(i**N**).Operation(j**N**).Task(n**N**).Name**S**

→ TaskStdMinsR := Attribute(i**N**).Operation(j**N**).Task(n**N**).StdMins**R**
)
)
)
)
→ SQLCmd**S** = "Insert into ItemBOO (OrderNo, LineNo, ProductionLine, ScheduleDate, SequenceNo, StdMins, WorkCenterOperation,Task, TaskStdMins) Values (" & OrderNo**N** & ", " & LineNo**N** & ", "' & ProductionLine**S** & "', '" & ScheduleDate**D** & "', '" & StdMins**R** & "' , '" & WorkCenterID**S** & "', '" & OperationID**S** & "', '" & TaskID**S** & "', " & TaskStdMins**R** & ")"
→DemandDB**ST**.Execute(< **I**::SQLCmdS >, < **O**::ⓈRelDBexecuted**BL** >)
→ (? ⓈRelDBExecuted**BL** = **F**
→ⓈBOOGenerated**BL** = **F**
→Response**S** := "Failed to access Demand Database"
→⊗ // Exit Import Function
)
→ ⓈBOOGenerated**BL** := **T**
} (6.26)

An alternative way of outputting the BOO directly to the Demand database is to create a file output similar to the format of the Comma Separated Variables (.csv) or XML format (.xml). There are several standard routines available to create these types of output files. Consequently, NewReport class is described in Fig. 6.6.

To output a report from a request, two classes are used. The first is the NewReport class and the second is the OutputReport class. To generate the OutputReport object, the NewReport first sets the heading of the report which is a list of the NAMES of fields in the selected report. Then, for each Item object in the Items collection, the report object corresponding to the ReportID specified in the request

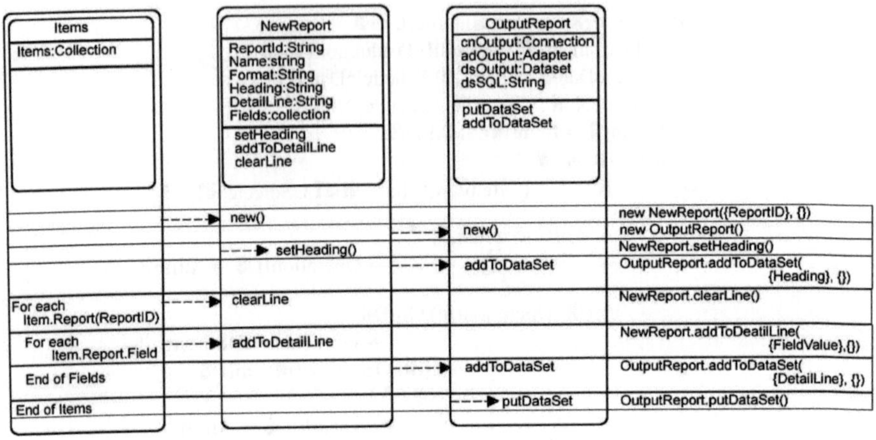

Fig. 6.6 NewReport class

is accessed. Each field is added to a detail line (fields separated by a deliminator, e.g. a comma), until the last field is added. The detail line is added to the OutputReport. When all the items in the collection have been processed, the OutputReport is written to the database. In XML format, heading names are used as tag field definitions as in <Name>Value<\Name>. Consequently this format uses many of the features and functions of the XML environment.

6.6 Chapter Summary

In this chapter, an information generation module (IGM) is presented. In terms of the data templates associated with a configuration, i.e. BOM, BOO and BOI which are created by the data modelling module in the knowledge mapping process, and the customer specifications of a product within the product family, the IGM is able to automatically generate the necessary production data for producing the product. In this chapter, only the generation of BOO, which is similar to the generations of BOM and BOI, is described as a case to demonstrate the principles and the procedure of the IGM.

The IGM is based on a server application that responds to requests from client programs for information. Central to information generation is the configuration. Based on the description made in Chap. 5, the configuration means a layout of a production unit and set of data templates (i.e. BOM, BOO and BOI) for producing OKP products under a product family. The product family in OKP means a base product by which a very large number of variants can be developed to meet customer requirements. Customers can specify their orders through a computer aided customer interface. According to the configuration and the customer specifications, the IGM can produce relevant material, resource and manufacturing information for many customized products under a product family.

The IGM is a novel and key technology for OKP companies to achieve both high customization and high production efficiency, i.e. mass customization. Without this technology, it would be very difficult to implement the OKP strategy and technology (e.g. as one-piece flow production line) for OKP companies that produce a large number of customer orders daily, e.g. a windows and doors manufacturing company in Calgary produces about 2000 custom windows and doors. As McCarthy (2004) stated, very few work has developed and proposed models for understanding how to mass customize. The IGM, from a system and information aspect, provides a possible solution to the problem.

References

Hong G, Hu L, Xue D, Tu YL, Xiong YL (2008) Identification of the optimal product configuration and parameters based on individual customer requirements on performance and costs in one-of-a-kind production. Int J Prod Res 46(12):3297–3326

McCarthy IP (2004) Special issue editorial: the what, why and how of mass customization. Prod Plan Control 15(4):347–351
Tu YL, Fung RYK, Tang JF, Kam JJ (2003) Computer aided customer interface for rapid product development. Int J Adv Manuf Technol 21:743–753

Chapter 7
Adaptive Scheduling and Control of One-of-a-Kind Production

As a typical manufacturing paradigm, one-of-a-kind production (OKP) challenges production scheduling and control differently than mass production. High throughput in OKP is a typical example of mass customization, which is one of the important strategies in the current economy (Blecker and Friedrich 2006) where the objective is to maximize the customer satisfaction by producing highly customized products with near mass production efficiency. As mentioned frequently in this book, OKP is intensely customer focused such that every product is based on specific customer requirements, and products differ on matters of colors, shapes, dimensions, functionalities, materials, processing times, etc. Consequently, a product that is produced on an OKP flow line is rarely repeated (Wortmann et al. 1997).

To achieve scope of economy, some of the OKP companies are trying to design its production lines as reconfigurable to achieve high flexibility. At meanwhile, they also design the products through the modifications of a base product which can be produced on a particular configuration of the reconfigurable production line. As mentioned in Chap. 5, the case product determines the product family in OKP. Therefore, the product family in OKP is much wider than a product family in traditional batch or mass production company and it can have nearly infinitive number of variations. A typical case product and the reconfigurable production line in a custom window and door manufacturing company in Calgary, Alberta, Canada is illustrated in Fig. 7.1.

A customized window can be designed through modification of the PPS of the base product as shown in Fig. 7.1. The windows can be different dimensions, colours, materials, configurations (e.g. with more or less sections, openings, or just picture windows, etc.) and even different geometric shapes. The customers can select different layers of glaze, hinges, handles or bars, and locks which can be ordered in batch (if these appliances are common or standard) or individually from suppliers according to the customer requirements. The production line is configured or reconfigured to be flexible to produce all possible variations from the base product. The products that are designed according to the base product will have the same or similar geometric forms of the PPS. However, the contexts in a circle/oval

Y. Tu and P. Dean, *One-of-a-Kind Production*,
DOI: 10.1007/978-1-84996-531-6_7, © Springer-Verlag London Limited 2011

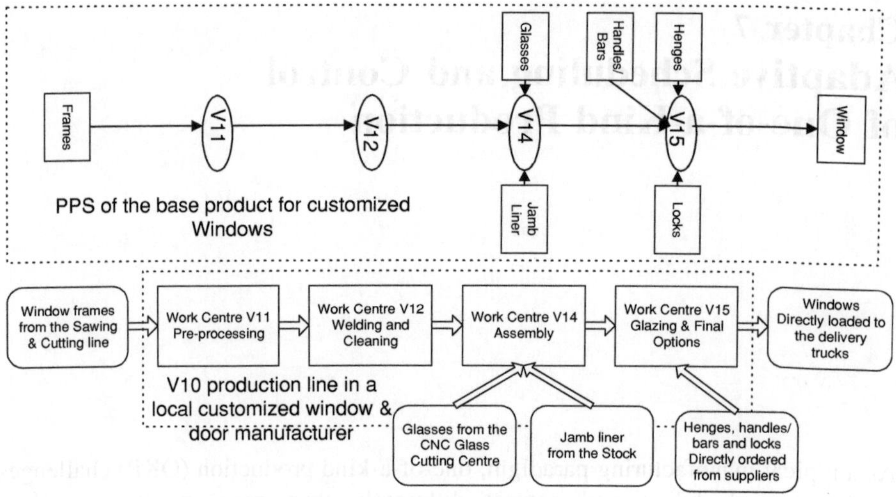

Fig. 7.1 V10 production line and the base product PPS for mass producing customized windows

or a box are quite different. For example, the process and the time of welding and cleaning (in Work Centre V12) a large size window is quite different from the ones of welding and cleaning a small size, different shape or different configuration of window. The detailed data or information for producing different customer orders are automatically generated by the IGM (see Chap. 6). These data are used as instructions for guiding a manual operating work centre or generating NC (numerical control) codes and control instructions for an automatic work entre, e.g. CNC (computer numerical control) glass cutting centre. According to the OKP paradigm as illustrated in Fig. 7.1, OKP products (i.e. customized windows) can be mass produced on a flow line. It is obvious that the products to be produced on this type of OKP flow line are all different and are controlled and produced one by one, which is commonly referred as one-piece flow technology in lean production. However, lean production technique is normally applied in traditional batch or mass production companies where the set of production data is repeated used to manufacture identical products. Therefore, in traditional manufacturing companies, they do not have the pressure or demand for automatically generating production data as we discussed in Chaps. 5 and 6. The ERP systems, which are developed in terms of material requirement planning technique (or MRP I), can be applied in these companies, but are difficultly applied in OKP companies. In the customized window and door manufacturer in Calgary, the company has to run its computer systems to produce about 2000 sets of production data for the 2000 customer orders in the morning of every day. Since the products based on a case product are far different from each other, the OKP is normally viewed as a mix-product mode production.

If a customer order in OKP is viewed as a project, to schedule the production of a customer order in OKP is very similar to scheduling a project. Therefore, OKP is referred as project-based production in this book. However, in project scheduling, the emphasis is placed on how to allocate scarce resources to dependent activities or operations of a project to control the budget or minimize the duration of the project (Brucker et al. 1999; Herroelen et al. 1998). These dependent activities or operations are normally arranged in a hierarchy diagram, commonly known as a project network diagram. In the book, we present the PPS (product production structure) as the production data modelling structure mainly for overcoming the drawbacks of separating BOM (bill of material) and BOO (bill of operation) in the current project management practices and software systems. In Microsoft Project 2007, the project network diagram which can be viewed as a BOO is separated from the BOM. As mentioned in previous chapters, this is also a common logic error in present ERP systems. To gain high production efficiency, OKP companies are suggested to arrange its production in a flow line no matter that the product is complex or simple. In Calgary, Alberta, Canada, for instance, a manufacturing company for producing customized compressors and refrigerators for oil/gas industries arranges its production in a shop floor through six sections (or work centres) that are sequentially linked by means of transportation tools and operation routine sheets. A customized compressor or refrigerator, which is usually complex and a large size, is assembled sequentially through these six sections.

Based on the discussion as mentioned above, the OKP scheduling problem is a combined problem of flow shop scheduling and project scheduling. However, unexpected disturbances frequently happen to affect the daily production on OKP shop floors, such as job insertion (for urgent orders) or cancellation (often due to unreliable supplies), change operation routines (due to uncertain manufacturing process or unexpected work centre breakdowns), operator absence, the breakdown of one of the machines in a work centre, and variations in processing times (due to high customization). To cope with these uncertainties, the OKP must be scheduled and controlled adaptively. The off-line scheduling techniques are not applicable in OKP.

Currently, OKP management primarily uses priority dispatching rules (PDRs) to deal with daily production schedule and control. It is fast and simple to use PDRs to control production online, but PDRs depend heavily on the configuration of shop floors, characteristics of jobs, and scheduling objectives (Goyal et al. 1995), and there is no specific PDR that clearly dominates the others (Park et al. 1997). Moreover, the performance of PDRs is poor on some scheduling objectives (Ruiz and Maroto 2005), and it is especially inconsistent when a processing constraint changes (King and Spachis 1980). Consequently, there is a considerable difference between the scheduled and actual production progress (Ovacik and Uzsoy 1997). When unexpected changes occur and PDRs are used to adaptively control production, production often runs in a chaotic or an "ad hoc fire fighting" manner (Tu 1996a, b).

To adaptively schedule and control OKP shop floors, a fast and efficient scheduling heuristic and close loop production control concept are needed. In this chapter, a state space (SS) heuristic (Li et al. 2011b) is presented. This SS heuristic has been developed by us to support a computer-aided production scheduling and control system for adaptively scheduling an OKP flow line. We compare the optimality of the SS heuristic in terms of minimizing the maximum completion time, to the CDS heuristic (Campbell et al. 1970) and to the NIS heuristic (Thornton and Hunsucker 2004) using case studies based on well-accepted benchmarks, for both traditional flow shop (TFS) and hybrid flow shop (HFS) problems under no pre-emption or no wait processing constraints. Both the CDS and NIS heuristics use Johnson's (1954) algorithm as their core. Across the different case studies, the SS heuristic outperforms the CDS and NIS heuristics. In addition to its self-contained performance, we believe that the SS heuristic—because of its simplicity and computational efficiency—is the most suitable heuristic for adaptively scheduling and control of OKP shop floors.

In addition to the scheduling heuristic, this chapter will also present the adaptive production control structure and strategies, resource planning heuristics, minimum cost project crashing technique, and electronic Gantt chart (eGantt chart) planning technique. These fundamental system structure, strategy, heuristics and technique constitute the shop floor scheduling and control module in the OKP management and control software system© as presented in Chap. 8.

7.1 Review of Flow Shop Scheduling

Research in production scheduling has been carried out for many decades, and there are numerous scheduling methods developed in the literature. In this section, we briefly review flow shop production scheduling methods, and discuss the requirements of heuristics for adaptive production control.

Scheduling is a decision making process of allocating resources to jobs over time to optimize one or more objectives. According to Pinedo (2002), one type of flow shop consists of *m* machines in series, and each job has to be processed on each one of *m* machines in a single direction, which means first on machine 1, then machine 2, and so on. This is typically called a traditional flow shop (TFS). Another type of flow shop where there are *S* work centres in series with a number of machines/operators in parallel in each work centre is a flexible flow shop or hybrid flow shop (HFS). In addition to the difference in flow shop configurations, processing constraints are also different for TFS and HFS. For TFS, if the first in first out (*FIFO*) discipline is applied to jobs in work-in-process (WIP) inventories, then it becomes a no pre-emption or permutation (*prmu*) flow shop problem. For HFS, if the first come first serve (*FCFS*) discipline is applied, then it is still a no pre-emption flow shop problem but the output sequence from each work centre may change. Another processing constraint could be no wait (*nwt*), that is, jobs are not allowed to wait between work centres, which also means there is no intermediate storage. The most common objective of flow shop scheduling is to minimize the maximum completion time or make-span, that is, min ($C_{\max}$). Following the popular three parameter notation, $\alpha/\beta/\gamma$, the above problems can be expressed as:

1. $Fm/prmu/C_{max}$ for m (Fm) machine TFS problems with no pre-emption constraint to minimize make-span,
2. $Fm/nwt/C_{max}$ for m machine TFS problems with no wait constraint to minimize make-span,
3. $FFs/FCFS/C_{max}$ for S work centre HFS problems with *FCFS* constraint to minimize make-span, and
4. $FFs/nwt/C_{max}$ for S work centre HFS problems with no wait constraint to minimize make-span.

Gupta and Stafford (2006) chronologically review flow shop scheduling research in the past five decades since the classic Johnson's algorithm was proposed in 1954 (Johnson 1954). They found that the emergence of NP-completeness theory in the third decade (1975–1984) profoundly impacted the direction of research in flow shop scheduling. That is why heuristics are required to solve large problems. HFS problems emerged in the fourth decade (1985–1994), and various artificial intelligence based heuristics were proposed then. The fifth decade (1995–2004) witnessed the proliferation of various flow shop problems, objective functions, and solution approaches. Although flow shop scheduling has been researched for more than 50 years, there remains a gap between theoretical research and industrial applications (Gupta and Stafford 2006).

Framinan et al. (2004) propose a general framework for the development of heuristics which consists of three phases: index development, solution construction, and solution improvement. Phase I, index development, means that jobs are arranged according to a certain property based on processing times. For example, Campbell et al. (1970) extend Johnson's algorithm and propose the CDS heuristic for an n-job m-machine TFS problem to minimize C_{max}. The CDS heuristic using Johnson's algorithm to arrange jobs is as follows:

> If there is a counter (Ctr) pointing to machine j, for each job i ($i = 1 \ldots n$), the sum of processing times on the machines before the Ctr point (or machine j) is regarded as the processing time on virtual machine 1, and the sum of processing times of job i ($i = 1 \ldots n$) on the rest $m - Ctr$ machines as the processing time on virtual machine 2. Then apply Johnson's algorithm to this virtual 2-machine flow shop problem to get a sequence. As Ctr changes from machine 1 to machine $(m - 1)$, the $(m - 1)$ sequences are generated, and the one with the minimum make-span is the final solution.

In phase II, solution construction, a solution is constructed by a recursive procedure, trying to insert one or more unscheduled jobs into a specific position of a partial sequence until the final schedule is completed. NEH (Nawaz et al. 1983) is a typical phase II heuristic for an n-job-m-machine TFS problem to minimize C_{max}. The NEH heuristic is described as follows:

Step 1 For each job i ($i = 1, 2, \ldots n$), NEH sums the processing times on all of the m machines, and then arranges these sums in a non-ascending order.

Step 2 The NEH schedules the first two jobs to get a partial sequence with minimum C_{max}, and then inserts the 3rd job into three possible positions, i.e. before the two jobs, between the two jobs, and after the two jobs, to get another partial sequence of three jobs with minimum C_{max}. Following this insertion manner, the NEH inserts the 4th job into the partial sequence of three jobs in four possible positions, the fifth job into the partial sequence of four jobs in five possible positions, and so on, until the last job into the partial sequence of $(n - 1)$ jobs in n possible positions.

Step 3 Finally, the NEH selects the sequence of n jobs with minimum C_{max}.

In phase III, solution improvement, there are two main characteristics: (1) there must be an initial schedule, and (2) after using artificial intelligence techniques, the quality of solution is better than the initial schedule. For the development of better heuristics, Framinan et al. (2004) clearly stated the importance of heuristic development in phase I, index development, should not be underestimated, as it is required for the other two phases.

In a case study including 19 constructive heuristics for $Fm/prmu/C_{max}$ problems, Ruiz and Maroto (2005) concludes that the NEH heuristic is the best, the CDS heuristic is 8th, and two PDRs (LPT and SPT rules) are the worst. However, the CDS heuristic has the simplest computational complexity among the first 8 heuristics, i.e. $O(m^2n + mn \log n)$. Moreover, King and Spachis (1980) did case studies of five PDRs and the CDS heuristic for two different TFS problems, $Fm/prmu/C_{max}$ and $Fm/nwt/C_{max}$. From the studies, Spachis concluded that the CDS heuristic and LWBJD (least weighted between jobs delay) PDR (priority dispatching rule) work the best for $Fm/prmu/C_{max}$ problems and MLSS (maximum left shift savings) PDR works the best for $Fm/nwt/C_{max}$ problems, but no single method works the best for both $Fm/prmu/C_{max}$ and $Fm/nwt/C_{max}$ problems.

Compared with TFS scheduling problems, the literature on HFS scheduling problems is still scarce (Linn and Zhang 1999; Wang 2005). According to Botta-Genoulaz (2000), the CDS heuristic is the best among the six heuristics, including the NEH heuristic, for HFS problems. The problem that Botta-Genoulaz solved is an n-job S-work-centre HFS problem to minimize the maximum lateness, i.e. min (L_{max}), which was cleverly converted to an n-job $S + 1$-work-centre HFS problem to minimize C_{max}. The processing time of job i in work centre $s + 1$ is calculated by $p_{iS+1} = D_{max} - d_i$, $i = 1\ldots n$, where $D_{max} = \max\ [d_k]$, and d_k is the due date for job k $(k = 1\ldots n)$. Although the CDS heuristic was constructed for TFS problems to minimize C_{max}, when applying the CDS heuristic to HFS problems, Botta-Genoulaz also converted the processing times, $p'_{is} = p_{is}/OPTR_s$, $s = 1\ldots S + 1$, where p_{is} is the original processing time of job i in work centre s, and $OPTR_s$ is the number of operators/machines assigned to work centre s. Therefore, it is reasonable to compare our SS (state space) heuristic with the CDS heuristic for $FFs/FCFS/C_{max}$ problems, which is computationally efficient and it is based-on Johnson's algorithm, because the basis of Botta-Genoulaz's conclusion is to minimize C_{max} for n-job $S + 1$-work-centre HFS problems.

For HFS problems with identical parallel machines and with a no wait constraint, i.e. $FFs/nwt/C_{max}$ problems, Thornton and Hunsucker (2004) proposed a no intermediate storage heuristic, i.e. the NIS heuristic, which works the best among the CDS heuristic, LPT and SPT PDRs, and a heuristic of random sequence generation. Like the CDS heuristic, the NIS heuristic also applies Johnson's algorithm and uses a filter concept to convert an $FFs/nwt/C_{\max}$ problem to a two-machine problem. The work centres before the filter are regarded as virtual machine 1, the work centres after the filter are regarded as virtual machine 2, and the work centres that are covered by the filter are omitted. The filter goes from work centre 2 to work centre $S - 1$ (the work centre before the last work centre or Work Centre S), and the width of filter changes from 1 to $S - 2$, which means filtering out processing times in 1 work centre or $S - 2$ work centres at a time. In total, there are $1 + (S - 1) \times (S - 2)/2$ sequences generated by the NIS heuristic and the one with the minimum make-span is the final schedule, from which we can see that the NIS is not very computational efficient. Again, because the NIS is based-on Johnson's algorithm, comparing our SS heuristic to NIS is reasonable.

To evaluate the suitability and efficiency of heuristics in OKP scheduling and control, there are three main evaluation criteria: optimality, computational complexity, and flexibility. Usually the optimality is used to evaluate the performance of a heuristic for offline production scheduling. However, when adaptive production control is taken into consideration, the computational complexity becomes critical. That is why the artificial intelligence based heuristics are not suitable for adaptive control, although they can get better solutions. Another criterion is the flexibility of a heuristic, i.e. whether a heuristic can deal with disturbances. Of course, different situations have different requirements for optimality, computational complexity, and flexibility of a heuristic. There is inevitably a trade-off among those three criteria, and the selection of heuristic(s) for adaptive control depends on the specifics of different situations, e.g. the value of optimality as compared to a rational result that is close-to-optimal scheduling solution, the type and volume of disturbances that underlies the requirements of response time, etc.

For example, given two heuristics, H_1 and H_2, two types of disturbances, D_1 and D_2, and a maximum response time for online rescheduling t, some characteristics of heuristics are shown in Table 7.1 for a given number of jobs.

When disturbance D_1 happens to production, the selection of a heuristic between H_1 and H_2 depends on optimality and computational complexity, because both heuristics are flexible enough to deal with D_1. If $T_{11} \leq t$ and $T_{21} \leq t$, the selection of a heuristic depends only on optimality. Otherwise, the selection depends on computational complexity. When disturbance D_2 happens, we can only choose heuristic H_2 if $T_{22} \leq t$. However, if $T_{22} > t$, and the maximum response time t cannot be relaxed, neither heuristic H_2 nor heuristic H_1 can be selected.

Table 7.1 Characteristics of heuristics H_1 and H_2 for disturbances D_1 and D_2

Heuristics	Disturbance		Computation time	
	D_1	D_2	D_1	D_2
H_1	Yes	No	T_{11}	∞
H_2	Yes	Yes	T_{21}	T_{22}

In general, the flexibility of a heuristic must meet the requirement of specific scheduling problems. After that, the computational complexity is a critical selection criterion for online rescheduling (i.e. adaptive production control), whereas the optimality is a dominant selection criterion for offline scheduling.

7.2 State Space (SS) Heuristic

The SS heuristic minimizes the maximum completion time (or make-span) for HFS problems, in which the *FCFS* discipline is applied and the capacity of WIP (work-in-process) inventories is limited. Because there are multiple operators in each work centre of a flow line, SS heuristic also maximizes the utilization of the flow line, i.e. max (*Util*).

There are two main concepts applied in our SS heuristic, which we introduce in Sect. 7.2.1. Then the main steps of the SS heuristic are introduced in Sect. 7.2.2. Finally, the computational complexity of the SS heuristic is analyzed in Sect. 7.2.3.

7.2.1 Two Concepts in the SS Heuristic

There are two concepts in the SS Heuristic:

1. State Space Concept
2. Lever Concept

7.2.1.1 State Space Concept

Consider a hybrid flow line with three work centres and two operators in each work centre (Fig. 7.2). The operators in each work centre follow a *FCFS* rule, that is, the operator who is available first for the next job should serve immediately in the work centre. Then there is a next available time of each work centre, A_s, where $A_s = \min(a_{sk})$, for $k = 1 \ldots OPTR_s$, in which a_{sk} is the next available time of operator k in work centre s, and $OPTR_s$ is the number of operators allocated to work centre s. For an S-work-centre flow line, there are $S - 1$ time differences

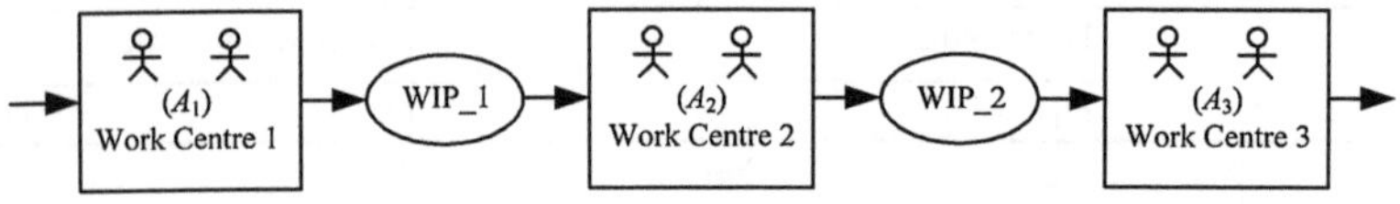

Fig. 7.2 A three-work-centre flow line with two operators in each work centre

between work centre available times. In the example above, there are two differences of the next available times, $A_2 - A_1$, and $A_3 - A_2$. If we regard such a difference of the next available times as a space, $SPACE_s = A_{s+1} - A_s$ for $s = 1...S - 1$, then $SPACE_s$ is a time period available for work centre s to finish a job without causing an operator in work centre $s + 1$ to be idle. If the completion time of job i in work centre s is larger than the next available time of work centre $s + 1$, such a job causes work centre $s + 1$ to be idle for $IDLE_{is} = c_{is} - A_{s+1}$ time units, where c_{is} is the completion time of job i in work centre s, $c_{is} = \max(A_s, c_{is-1}) + p_{is}$, where p_{is} is the processing time of job i in work centre s. The completion time c_{is} depends on the maximum of job i's completion time from the previous work centre $s - 1$, c_{is-1}, and the next available time of current work centre s, A_s. If the completion time of job i in work centre s is smaller than the next available time of work centre $s + 1$, then there are two possibilities depending on whether WIP is full. If the WIP inventory, WIP_s, is full, a block happens to operator k who processed job i in work centre s, $DELAY_{is} = A_{s+1} - c_{is}$. The block means that, after finishing job i, operator k in work centre s has to hold it in hand for $DELAY_{is}$ time units until there is a vacancy in WIP_s. Therefore, the next available time of operator k in work centre s is delayed. Alternatively, if WIP_s is not full, job i goes into inventory, and there is no *IDLE* and no *DELAY*.

The main idea of SS is to find a job that fits $S - 1$ spaces, without causing *IDLE* or *DELAY* time. After a job i is processed on a line, the state of the line, i.e. the next available time of work centres, is changed, and the space is changed accordingly. It is clear that greater *IDLE* and *DELAY* are not good for production if the objectives are to minimize C_{max} and maximize *Util*, while greater *SPACE* is good for production to some extent. Therefore, in SS, job i is chosen according to its performance on *IDLE*, *DELAY*, and *SPACE*.

From the foregoing description of SS, we can see that *IDLE* and *DELAY* are evaluated according to job i and work centre s, but *SPACE* is only evaluated by work centre s. To make *SPACE* both job and work centre dependant, there could be two ways to model *SPACE*. One model is $SPACE_{is} = c_{is+1} - A_s$, for $s = 1...S - 1$, which means using the difference between the completion time of job i in work centre $s + 1$ and the next available time of work centre s as the *SPACE* created by job i for work centre s to process the next job. The other model is $SPACE_{is} = p_{is+1}$, for $s = 1...S - 1$, which means using the processing time of job i in work centre $s + 1$ as *SPACE*. In our current version of the SS heuristic, we use the latter model of *SPACE* because there is one less calculation for each iteration, increasing the computation speed for adaptive control. However, we illustrate the alternative model of *SPACE*, $SPACE_{is} = c_{is+1} - A_s$, in a case study in Sect. 7.4 to show the flexibility of the SS concept.

7.2.1.2 Lever Concept

From our previous research on TFS problems, we found that the lever concept was suitable for flow shop production (Li et al. 2007), which means that *IDLE* (or *DELAY*) in an earlier work centre is worse for min ($C_{\max}$) objective than in a later work centre. Consider a lever where force F takes effect and causes a torque of $F \times L$, where F is the unit of force and L is the length of force arm. We model an S work centre flow line as a lever, and $IDLE_{is}$ caused by job i, or $DELAY_{is}$, has a torque effect manifested as $IDLE_{is} \times LVR\_IDLE_s$ or $DELAY_{is} \times LVR\_DELAY_s$, where LVR_IDLE_s or LVR_DELAY_s is the length of arm for $IDLE_{is}$ or $DELAY_{is}$ respectively.

The lever concept for *IDLE* (LVR_IDLE_s) in SS is shown in Fig. 7.3. For an S-work centre flow line, a job could cause at most $S - 1$ times of *IDLE*, since no *IDLE* is caused in work centre 1 and an *IDLE* takes effect in the next work centre. Therefore, the fulcrum of a lever for *IDLE* is set between work centre $s - 1$ and work centre s, and the length of arm for an *IDLE* caused by work centre s in work centre $s + 1$ is $LVR\_IDLE_s = S - s$, for $s = 1 \ldots S - 1$.

A lever concept for *DELAY* (LVR_DELAY_s) in SS is shown in Fig. 7.4. Like the number of possible *IDLE*s, there could be $S - 1$ *DELAY*s, since no *DELAY* is caused in work centre S (the last work centre). But a *DELAY* takes effect in current work centre s, whereas *IDLE* is for the next work centre. Therefore, one unit of *DELAY* in work centre s should be worse for min ($C_{\max}$) objective than one unit of *IDLE* in work centre s. So the length of arm for a *DELAY* is $LVR\_DELAY_s = S - s + 1$, for $s = 1 \ldots S - 1$. The fulcrum of a lever for *DELAY* is set in work centre S.

There is also a lever concept for *SPACE* (LVR_SPACE_s) in SS, shown in Fig. 7.5. The length of force arm for a space is $LVR\_SPACE_s = s$, for $s = 1 \ldots S - 1$. The fulcrum of a lever for *SPACE* is set between work centre 1 and work centre 2, which means *SPACE* in a later work centre is better for min ($C_{\max}$) than in an earlier work centre.

Therefore, all *SPACE*s, *IDLE*s, and *DELAY*s are converted to torques, that is, $SPACE'_{is} = p_{is+1} \times LVR\_SPACE_s$, $IDLE'_{is} = IDLE_{is} \times LVR\_IDLE_s$, and $DELAY'_{is} = DELAY_{is} \times LVR\_DELAY_s$. The job selection scheme is then

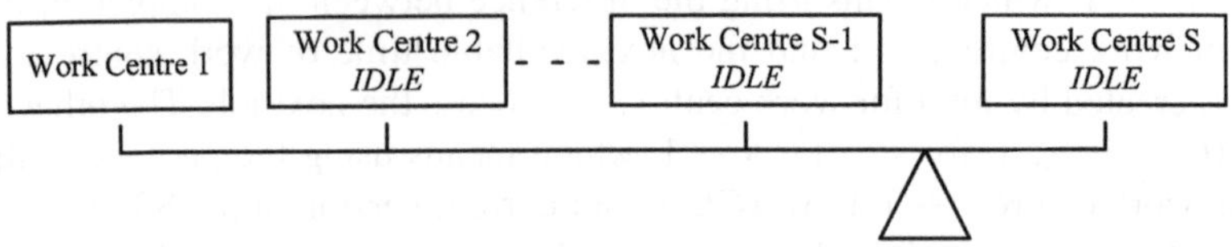

Fig. 7.3 A lever concept for *IDLE* in SS

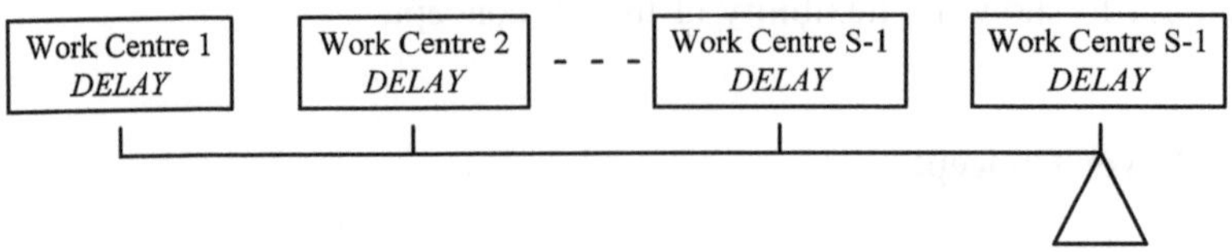

Fig. 7.4 A lever concept for *DELAY* in SS

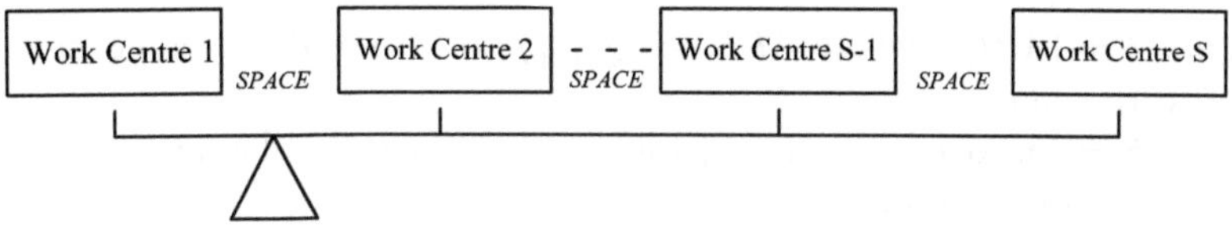

Fig. 7.5 A lever concept for *SPACE* in SS

$$\max_{1\leq i\leq n}\left[\sum_{s=1}^{S-1} SPACE'_{is} - \left(\sum_{s=1}^{S-1} IDLE'_{is} + \sum_{s=1}^{S-1} DELAY'_{is}\right)\right] \tag{7.1}$$

that is, to select one out of n unscheduled jobs with the maximum torque difference between $SPACE'_{s}$ and $IDLE'_{s} + DELAY'_{s}$.

Take the process of job i on a 4-work centre flow line as an example. The operation of job i in work centre 1 would cause a 3-time unit idle to work centre 2. The operation of such job in work centre 2 would cause 0-time unit idle to work centre 3, but a 2-time unit delay in work centre 2, and the operation of such job in work centre 3 would cause a 1-time unit idle to work centre 4. Accordingly, such job would create a 2-time unit space between work centres 1 and 2, a 6-time unit space between work centres 2 and 3, and a 3-time unit space between work centres 3 and 4. Thus, $\sum_{s=1}^{3} IDLE'_{is}$ created by job i equals to $3 \times 3 + 0 \times 2 + 1 \times 1 = 10$, $\sum_{s=1}^{3} DELAY'_{is}$ equals to $0 \times 4 + 2 \times 3 + 0 \times 2 = 6$, and $\sum_{s=1}^{3} SPACE'_{s}$ equals to $2 \times 1 + 6 \times 2 + 3 \times 3 = 23$.

7.2.2 The Steps to Achieve the SS Heuristic

The main idea of SS is to select a successive job according to the state space generated by previous jobs in a sequence. Therefore, it is important to select initial jobs to create an initial state space, which affects the selection of successive jobs and ultimately the schedule performance.

7.2.2.1 Initial Job Selection

Two items should be taken into consideration for initial job selection in SS. One is the number of initial jobs—which is straightforward, and the other is the initial job selection scheme. The number of initial jobs is set as min ($OPTR_s$, for $s = 1 \ldots S$). The reason is that if the number of initial jobs is smaller than min ($OPTR_s$), the state (the first available time of a work centre) is zero since all operators in all work centres are available at time zero; if the number is greater than min ($OPTR_s$), the number of jobs (initial job number—min ($OPTR_s$)) are not selected by the state space concept.

For initial job selection scheme, five $1 \times S$ vectors are introduced as follows: $Vector\_1 = [0]$;

Vector_3 = [APT_s], where $APT_s = \left(\sum_{i=1}^{N} p_{i,s}\right)/n$, i.e. the average processing time of work centre *s*;

Vector_5 = [max(p_{is}), $i = 1 \ldots n$] for $s = 1 \ldots S$, i.e. the maximum processing time in each work centre;

Vector_2 = (*Vector_3*)/2;

Vector_4 = *Vector_3* + (*Vector_5* − *Vector_3*)/2 = (*Vector_3* + *Vector_5*)/2

Take an *n*-job four-work-centre production problem as an example. *Vector_1* should be [0, 0, 0, 0], if *Vector_3* = [4, 6, 8, 10], and *Vector_5* = [10, 12, 16, 20], then *Vector_2* = [2, 3, 4, 5] and *Vector_4* = [7, 9, 12, 15].

The initial number of jobs are selected according to min $\left(\sum_{s=1}^{S} (|p_{is} - Vector\_v(s)|)\right)$ for $i = 1 \ldots n$, which means the minimum absolute difference between one job's processing times and the vector.

7.2.2.2 Steps

The following steps represent the generic programming logic of the SS heuristic.

Step 1 Determine the number of operators in each work centre, i.e. $OPTR_s$. (1a): Calculate *n* and *S*. (1b): Set an expected throughput rate, *r*, which means in every *r* time units a job is expected to be finished in a work centre. (1c): $OPTR_s$ = Roundup (APT_s/r). (1d): Set the start time of every operator to 0. (1e): Put all of *n* jobs into a candidate pool. (1f): Set an output sequence to be $1 \times n$ zero vector, *Sequence_v*.

Step 2 Set the capacity of each of $S - 1$ WIP inventories. The capacity of each WIP inventory could be different.

Step 3 Calculate five vectors for initial job selection.

Step 4 FOR $v = 1{:}5$. This is an iteration loop to select initial jobs according to one of five vectors, i.e. *Vector_v*.

Step 5 Select min ($OPTR_s$, for $s = 1 \ldots S$) number of jobs according to *Vector_v* by the equation $\min\left(\sum\limits_{s=1}^{S} (|p_{is} - Vector\_v(s)|)\right)$ for $i = 1 \ldots n$. Then put selected jobs into a *Sequence_v* and eliminate them from the candidate pool. Calculate the next available time of each operator, and then the next available time of each work centre, namely *STATE*, and WIP inventory status, namely *WIP_Status*, which is now $1 \times (S - 1)$ zero vector.

Step 6 FOR $i = \min(OPTR_s) + 1{:}\ n$. This is an iteration loop to sequence the rest of $n - \min(OPTR_s)$ jobs in the pool.

Step 7 According to *STATE* and *WIP_Status*, calculate $IDLE'_{is}$, $DELAY'_{is}$, and $SPACE'_{is}$.

Step 8 Select job *i* according to $\max\limits_{i}\left[\sum_{s=1}^{S-1} SPACE'_{is} - \left(\sum_{s=1}^{S-1} IDLE'_{is} + \sum_{s=1}^{S-1} DELAY'_{is}\right)\right]$, and then put such job number into *Sequence_v* and eliminate it from the candidate pool.

Step 9 Calculate intermediate completion time of a partial schedule *Sequence_v*, update *WIP_Status*, and update *STATE*.

Step 10 END *i*. Calculate the utilization of a line. (10a): Calculate utilization of each work centre first, $Util_s = \left(\sum_{i=1}^{n} p_{is}/OPTR_s\right)/(c_{nks} - c_{1k's-1})$, $c_{1k0} = 0$, $s = 1\ldots S$, in which c_{nks} is the completion time of the last finished job in work centre s, and $c_{1k's-1}$ is the start time of the first job in work centre s, i.e. the completion time of the first finished job in work centre $s - 1$. (10b): Calculate the average utilization of each work centre, i.e. the utilization of a line, $Util$ = average ($Util_s$), $s = 1\ldots S$.

Step 11 END v. Output each of five sequences and related make-span and utilization, and the minimum make-span and the maximum utilization are regarded as the final performance of SS.

7.2.3 The Computational Complexity of the SS Heuristic

The computational complexity of a heuristic is the highest order of operations for a specific problem. For an n-job S-work-centre HFS problem the computational complexity of the SS heuristic consists of two parts: job selection and the make-span calculation. The main scheme of SS heuristic to construct a job sequence is to select the next successive job according to the current state of the flow line. This scheme is for both offline scheduling and online adaptive control. Make-span calculation is to calculate the current state of the flow line, but it is carried out five times in the SS heuristic for offline scheduling, since there are five sequences generated by the SS heuristic, although only one of them will be chosen as the final production schedule.

7.2.3.1 Computational Complexity of Job Selection

If the state of the flow line is known, to select one out of n unscheduled-jobs takes $S \times n$ computational operations. Secondly, to choose one out of $(n - 1)$ jobs takes $S \times (n - 1)$ computational operations. Thirdly, to choose one out of $(n - 2)$ jobs takes $S \times (n - 1)$ computational operations. If we continue this job sequencing and computational operation counting until the last job left, the total number of computational operations of n job sequencing is:

$$(n + n - 1 + n - 2 + n - 3 + \cdots + 3 + 2 + 1) \times S = S \times (n + 1) \times n/2.$$

There are five sequences generated by the SS heuristic, the total number of computational operations of job sequencing of SS heuristic is:

$$5 \times S \times (n+1) \times n/2.$$

Because only the highest order of operation computation equation is counted in computational complexity, the computation complexity of SS heuristic is $O(Sn^2)$, which is considered as a very simple computational complexity among the existing heuristics.

7.2.3.2 Computational Complexity of the Make-Span Calculation

An n-job and S-work-centre HFS problem can be modelled by a two-dimension matrix, where the row dimension is for n jobs, and the column dimension is for S work centres. The make-span calculation could be carried out along the column dimension. It means that, if the input sequence of n jobs of work centre 1 is known, the output sequence of n jobs of work centre 1, which is also the input sequence of work centre 2, can be calculated; the output sequence is in a non-descending order of completion times of n jobs; and then the output sequence of work centre 2 can be calculated, and so on, finally the output sequence of the final work centre (i.e. work centre S) can be calculated. However, the capacities of WIP inventories are limited, which means the completion times of jobs in work centre s are constrained by the next available times of operators in work centre $s+1$. For example, when calculating the output sequence of work centre s, if a job i's completion time in work centre s causes the WIP_s inventory to be overloaded, which means at that time the WIP_s inventory is full and there is no operator available in work centre $s+1$ to process a job in the WIP_s inventory, then a *DELAY* happens to such job i. This *DELAY* means the job i's completion time is delayed to a later time, and so is the next available time of operator k, who processes the job i in work centre s. Consequently, the *DELAY* affects the completion times of all jobs following job i in work centre s, and the completion times in the previous work centre need to be checked because of the limitation on WIP inventories. In an extreme situation, when a *DELAY* happens in work centre $S-1$ (the second last work centre), the completion times of jobs in all previous work centres have to be recalculated. Because of the many recalculations, it is time consuming to calculate make-span along the column dimension when the capacity of each WIP inventory is small. Therefore, the make-span calculation should be done along the row dimension, i.e. calculated by jobs and not by work centres.

7.3 Adaptive Production Scheduling and Control

As mentioned at the outset of this chapter, in OKP, the following uncertainties need to be coped with by adaptive production scheduling and control: job insertion or cancellation, change operation routines, operator absence, machine breakdown, and variations in processing times. The change of operation routines can be handled by setting the processing time of a job at the breakdown or skipped work centre as zero. The variation in processing times can be handled by the robustness of the SS heuristic. Through a large number of randomly generated cases which vary the processing times from −5 to 20%, it has validated that the job sequence generated by using SS heuristic is very robust to the variation of job processing times (Li et al. 2011a). This means that we use a fixed job sequence is much better than we change job sequence according to the changes of processing times.

Therefore only two types of disturbances in OKP need to be coped with the adaptive production scheduling and control system. These two types of disturbances and the coping strategies are described as follows:

1. Job insertion/cancelation. The system needs to insert or cancel the job from the job sequence and employs the SS heuristic to re-schedule all the jobs which are out of production schedule time fence, a.k.a. planning time fence (Krajewski and Ritzman 1999). The production schedule time fence is a period in which it is not allowed to change the production schedule in order to keep the stability of the production system.
2. Operator absence or machine breakdown. For the work centre in which the operator is absence or a machine is broken down, the processing times of all the jobs at this work centre need to be prolonged proportionally to cover the absence of the operator or the machine breakdown. Then the system uses the SS heuristic to re-schedule the jobs which are out of production schedule time fence.

To adaptively schedule and control OKP, a close loop control system structure, which is common in adaptive control theory, is suggested in Fig. 7.6. This system structure consists of the SS heuristic and a simulation module that is called temporized hierarchical object-oriented coloured Petri nets with changeable structure (THOCPN-CS) (Li 2006). High customization and dynamic uncertainties in OKP demand for a great effort on a simulation module. Simultaneously, adaptive

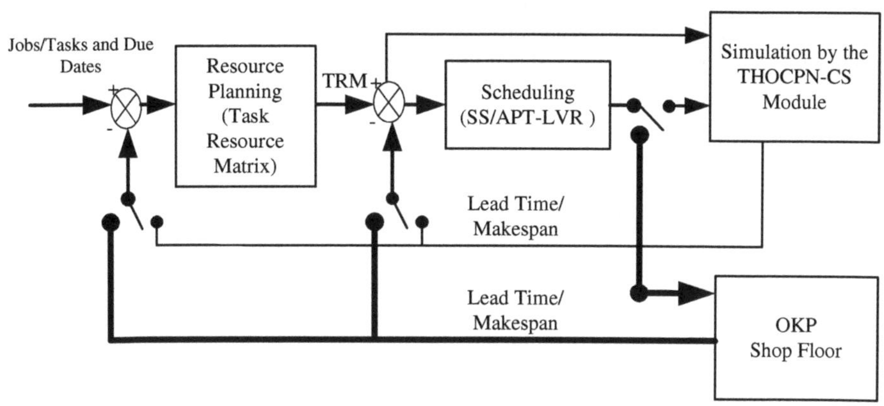

Fig. 7.6 A close loop production scheduling and control system structure

production control demands for prompt solutions in time. Therefore, the unique feature of the THOCPN-CS simulation module, changeable structure, makes it easy and flexible to simulate frequent changes in OKP for adaptive production control. Steps to achieve adaptive production scheduling and control in OKP are summarized as follows.

Step 1 IF Disturbance 1 (i.e. job insertion/cancelation) occurs, THEN Switch the control loop from OKP Shop Floor control mode to Simulation Mode and Go to Step 3; IF Disturbance 2 (operator absence) occurs, THEN Change the processing times of the jobs at the work centre in which the operator is absence, Switch the control loop from OKP Shop Floor control mode to Simulation Mode and Go to Step 3; Go to Step 6.

Step 2 Manually assign possible manufacturing resources (e.g. operators, portable tools and devices, etc.) to the work centres, form a task-resource matrix (TRM) with jobs, and set up the processing times of jobs in the changed work centres.

Step 3 Schedule the jobs by the SS heuristic.

Step 4 Use the THOCPN-CS module to simulate the job sequence (or the new production schedule) generated from Step 2. IF the new schedule meets the due date, THEN Go to Step 5, ELSE re-schedule the production under senior (1) add more (if late than the due date) or remove some (if early than the due date) resources to/from the work centres, or (2) change operation routines, Go to Step 2.

Step 5 Deliver the production schedule to the shop floor and switch the control loop from the Simulation mode to the OKP Shop Floor Control mode.

Step 6 Control the production running on the shop floor, Check the occurrence of the disturbance, and Go to Step 1.

Through the steps as mentioned above, the production of an OKP shop floor can be adaptively scheduled and controlled to cope with disturbances. However, to avoid the system instability, this adaptive control system should be applied according to the different situations in manufacturing companies. In general, a production schedule time fence should be added when we are using this close loop control system. As mentioned before, within this time fence, the system is not allowed to change the production schedule of the jobs. The manufacturing companies normally wish their production schedule time fence as shorter as possible so that they can response to market demands and unexpected disturbances promptly. However, a shorter production schedule time fence may result in instability of the production system. Therefore, the production schedule time fence is limited by the flexibility of the production system of a company. The higher flexibility of the production system, the shorter production schedule time fence is obtained. For a custom windows and doors manufacturing company in Calgary, Alberta, Canada, with its manual production schedule and control system, its production schedule time fence is three working days. By adapting this computer aided adaptive production scheduling and control system, its production schedule time fence could be reduced to 2 h.

The second solution for achieving adaptive production scheduling and control is using the distributed intelligence strategy. This strategy suggests that an OKP company plans its production within a planning time bucket. A planning time bucket could be a shift (i.e. 2 h or 4 h) or a day. Within a planning time bucket, the planner makes sure that the production capacity meets the demand. The shop floor manager can release all the jobs or operations within the planning time bucket to the shop floor. Through using an agent-based distributed control module, the manager can assign operations to the operators on-line or the operators can select operations by themselves once they are free. By this distributed shop floor control module and on-line operation assignment, the production on the shop floor can be adaptively scheduled and controlled by the shop floor manager and operators. The shop floor production control module in the OKP management and control software system© employs an agent-based distributed control structure. Please refer to Sect. 8.4 for detailed descriptions of this shop floor production scheduling and control module.

7.4 Heuristics Comparison Through Case Studies

To validate the efficiency of the SS heuristic, we carry out three kinds of case studies: SS performance as compared to other heuristics, SS performance with disturbances, and an industrial case study from a local customized window and door manufacturer. For the performance of SS compared with other heuristics, we carry out four separate studies on different flow shop configurations: $Fm/prmu/C_{\max}$, $Fm/nwt/C_{\max}$, $FFs/FCFS/C_{\max}$, and $FFs/nwt/C_{\max}$. Taillard's (1993) benchmarks are used in case studies, because, to our best knowledge, such benchmarks are well designed and the most accepted for flow shop scheduling problems. In Taillard's benchmarks, the machine number ranges from 5 to 20, the job number ranges from 20 to 500, the processing time follows a uniform distribution from 1 to 99, and there are 120 instances in total. Although all these case studies are carried out based on Taillard's benchmarks, the result of $Fm/prmu/C_{\max}$ is a comparison with the best known upper bounds. For the remaining case studies on SS performance, the results are mutual comparisons, that is, the improvement of one heuristic over another.

To examine the performance of SS in the presence of disturbances, we examine the performance of SS with operator absence/machine breakdown under two definitions of SPACE, and using a simple optimization method together with SS heuristic.

In the third kind of case study we compare the SS heuristic with data provided by a local customized window and door manufacturer, and measure the improvement due to SS. Finally, we show with a simple two-machine TFS example that the State Space concept can provide more solutions than Johnson's algorithm, and thus provide more opportunities as the core of a more elaborate heuristic.

7.4.1 Case Studies on Taillard's Benchmarks

Four separate studies on different flow shop problems where conducted.

7.4.1.1 $Fm/prmu/C_{max}$

For traditional permutation (or no pre-emption) flow shop scheduling problems, the deviation (*DEV*) from the best known upper bounds is used to evaluate the performance of a heuristic, where *DEV* = (C_{max} *of a heuristic* − *The upper bound*) ÷ (*The upper bound*) in percentage. The results of the deviation studies for CDS, NIS, SS, and a version of SS without the lever concept (SSnoLVR) heuristics are shown in Table 7.2.

In Table 7.2, the column "Scale" means the size of problems. For example, 20*5 means 20-job five-machine problems. The column "Instance" means the number of instances in each scale. Columns 3, 4, 5, and 6 represent the average deviation of each of the CDS, NIS, SS, and SSnoLVR heuristics respectively. For the total average deviation, the SS heuristic has the smallest deviation from Taillard's benchmarks, 8.11%, SSnoLVR heuristic ranks second at 8.80%, the NIS heuristic ranks third at 9.01%, and the CDS heuristic ranks last at 11.28%. These results show the advantage of the SS heuristic and the effectiveness of the lever concept.

Table 7.2 Average deviation from Taillard's benchmarks (%)

Scale	Instance	CDS	NIS	SS	SSnoLVR
20*5	10	9.05	7.41	9.14	7.80
20*10	10	13.48	9.46	10.18	13.13
20*20	10	11.07	7.30	10.64	14.02
50*5	10	7.15	4.96	3.60	3.38
50*10	10	14.46	11.57	9.67	9.24
50*20	10	18.13	14.50	16.15	16.12
100*5	10	5.25	4.70	1.60	1.75
100*10	10	9.51	8.27	6.71	6.05
100*20	10	16.45	13.50	11.83	15.71
200*10	10	7.55	6.61	3.09	2.48
200*20	10	13.75	11.33	9.10	11.31
500*20	10	9.56	8.44	5.63	4.60
Total average		11.28	9.01	8.11	8.80

7.4.1.2 Fm/nwt/C_{max}

For traditional no wait flow shop scheduling problems, an improvement (*IMPR*) over the NIS heuristic is used to evaluate the performance of the CDS and SS heuristics based on Taillard's benchmarks, where $IMPR = (C_{max}$ *of NIS* $- C_{max}$ *of CDS or SS*$) \div (C_{max}$ *of NIS*$)$ in percentage is shown in Table 7.3.

For $Fm/nwt/C_{max}$ problems, the CDS heuristic performs 1.04% worse than the NIS heuristic. In contrast, the SS heuristic performs on average 2.27% better than the NIS heuristic. However, we recognize that for HFS problems the improvements of SS over NIS will shrink as the number of operators/machines in each work centre increases. For example, if the number of resources in each work centre is the same as the number of jobs, the C_{max} is fixed as max $\left(\sum_{s=1}^{S} p_{is}\right)$ for $i = 1\ldots n$, even for no wait or no pre-emption flow shop problems.

7.4.1.3 FFs/nwt/C_{max}

For hybrid no wait flow shop problems with identical parallel operators/machines in each work centre, two operators/machines are assigned to each work centre. The improvement of the CDS or SS heuristics over the NIS heuristic is shown in Table 7.4. For such hybrid no wait flow shop problems with two operators/machines in each work centre, the SS heuristic has a small improvement over the NIS heuristic, 0.39%, and the CDS heuristic still performs slightly worse, −1.33%.

7.4.1.4 FFs/FCFS/C_{max}

For HFS problems with the *FCFS* discipline applied to jobs in WIP inventories, we set two variables. One is a throughput rate $r = 31$, used to calculate the number of resources in each work centre, where $OPTR_s$ = Roundup (APT_s/r). The average

Table 7.3 Improvement over NIS heuristic for $F_m/nwt/C_{max}$ problems (%)

Scale	Instance	CDS	SS
20*5	10	−0.32	2.01
20*10	10	−2.59	−2.86
20*20	10	−3.50	−2.71
50*5	10	0.29	8.29
50*10	10	−1.29	0.49
50*20	10	−2.42	−1.67
100*5	10	−0.27	9.20
100*10	10	−0.61	3.78
100*20	10	−1.00	−0.02
200*10	10	−0.22	5.59
200*20	10	−0.41	1.69
500*20	10	−0.10	3.46
Total average		−1.04	2.27

Table 7.4 Improvement over NIS heuristic for *FFs/nwt/*$C_{\max}$ problems (%)

Scale	Instance	CDS	SS
20*5	10	−1.71	−2.66
20*10	10	−2.72	−2.02
20*20	10	−3.06	−2.88
50*5	10	−0.77	3.34
50*10	10	−1.50	−2.18
50*20	10	−3.48	−2.04
100*5	10	0.21	7.15
100*10	10	−0.55	0.60
100*20	10	−1.75	−1.13
200*10	10	−0.15	3.54
200*20	10	−0.50	0.97
500*20	10	0.01	2.00
Total average		−1.33	0.39

processing time of each work centre ranges from 30.75 to 64.40 for all of 120 instances in Taillard's benchmarks, therefore, $OPTR_s$ varies from 1 to 3 for each work centre. Another variable is the capacity of WIP inventories. Different configurations of WIP inventories have different impacts on production (Vergara and Kim 2009). For the ease of case study, we simply set each WIP inventory at the same $WIP_s = 5$, even though in theory each could be set to a different value. The calculation of processing times in CDS is $p'_{is} = p_{is}/OPTR_s$, $s = 1\ldots S$ (Botta-Genoulaz 2000). The improvement (*IMPR*) of the SS heuristic over the CDS heuristic is used to evaluate performance for such scheduling problems for both min ($C_{\max}$) and max (*Util*) objectives, where $IMPR_1 = (C_{\max}$ *of CDS* $- C_{\max}$ *of SS*$) \div (C_{\max}$ *of CDS*) in percentage, and $IMPR_2 =$ (*Util of SS* $-$ *Util of CDS*) $\div$ (*Util of CDS*). The results are shown in Table 7.5. Over the 120 instances in Taillard's benchmarks, the SS heuristic has an average 1.16% improvement over the CDS heuristic on the min ($C_{\max}$) objective and 3.96% on the max (*Util*) objective.

7.4.1.5 Short Summary

In all four separate studies on different flow shop problems, on average, the SS heuristic performs better than the CDS and NIS heuristics, although worse for a few instances. However, the SS heuristic performs much better than the CDS and NIS heuristics for each of 30 large sized instances, in which the job number $\geq$ 200.

7.4.2 Case Studies on Operator Absence or Machine Breakdown

To test the suitability of the SS heuristic to adaptive control, a case study of operator absence or machine breakdown is carried out based on the data from

Table 7.5 Improvement over CDS heuristic for *FFs/FCFS/*C_{max} problems (%)

Scale	Instance	Min (C_{max})	Max (*Util*)
20*5	10	−2.39	7.33
20*10	10	0.27	5.66
20*20	10	−2.65	−0.02
50*5	10	2.87	4.90
50*10	10	2.47	6.17
50*20	10	0.08	1.45
100*5	10	2.42	3.47
100*10	10	1.34	4.69
100*20	10	1.54	2.43
200*10	10	3.14	3.96
200*20	10	2.03	4.41
500*20	10	2.79	3.14
Total average		1.16	3.96

Taillard's benchmarks. For this case study, we assume that, when a half of jobs are finished, one operator is absent or a machine is broken down in a middle work centre of a flow line, specifically in work centre 3, 6, or 11 according to the scale of instances in Taillard's benchmarks. For the remaining half of the jobs, if the production is carried on according to the original schedule when such a disturbance happens to the shop floor, the completion time is recorded as *Original.* If adaptive control is applied, i.e. using the SS heuristic to re-schedule the remaining jobs, the completion time is recorded as *Adaptive*. The improvement of adaptive control over no adaptive control is used to evaluate the performance, which means (*Original* − *Adaptive*) ÷ (*Original*) in percentage.

To show the potential of the SS heuristic as a core heuristic, case studies on operator absence or machine breakdown are carried out under the two definitions of *SPACE*, i.e. $SPACE_{is} = p_{is+1}$ and $SPACE_{is} = c_{is+1} - A_s$. As a third case study in this section, we integrate a simple optimization method with the SS heuristic.

7.4.2.1 $\mathbf{SPACE_{is} = p_{is+1}}$

The results are given in Table 7.6. Table 7.6 shows that adaptive control is slightly better than no adaptive control with a 0.10% improvement if $SPACE_{is} = p_{is+1}$.

7.4.2.2 $\mathbf{SPACE_{is} = c_{is+1} - A_s}$

The results are given in Table 7.7. We can see that adaptive control has a 2.02% improvement over no adaptive control.

Table 7.6 Improvement adaptive control over no adaptive control, where $SPACE_{is} = p_{is+1}$

Scale	Instance	Min (C_{max})
20*5	10	2.46
20*10	10	1.81
20*20	10	3.01
50*5	10	0.88
50*10	10	2.17
50*20	10	−2.80
100*5	10	0.39
100*10	10	0.29
100*20	10	−4.18
200*10	10	−0.41
200*20	10	−1.25
500*20	10	−1.21
Total average		0.10

Table 7.7 Improvement adaptive control over no adaptive control, where $SPACE_{is} = c_{is+1} - A_s$

Scale	Instance	Min (C_{max})
20*5	10	7.75
20*10	10	6.62
20*20	10	−8.49
50*5	10	1.23
50*10	10	3.10
50*20	10	2.99
100*5	10	0.22
100*10	10	3.26
100*20	10	4.50
200*10	10	0.55
200*20	10	1.64
500*20	10	0.86
Total average		2.02

7.4.2.3 Integration with an Optimization Method

It is easy to integrate other optimization techniques into the SS (state space) concept. For example, in the SS heuristic, there are two effects affecting the final production performance. Increased *SPACE* improves production, whereas *IDLE* and *DELAY* make production worse. If we introduce a weighting factor, α, into the SS heuristic that allows different weights for *SPACE* and for *IDLE* or *DELAY*, we could sequence the jobs according to

$$\max_i \left[(1-\alpha) \times \sum_{s=1}^{S-1} SPACE'_{is} - \alpha \times \left(\sum_{s=1}^{S-1} IDLE'_{is} + \sum_{s=1}^{S-1} DELAY'_{is} \right) \right] \quad (7.2)$$

As α changes from 0 to 1 with increment of 0.1, the performance of the SS heuristic, in which $SPACE_{is} = p_{is+1}$, is shown in Table 7.8. The columns represent the performance of each α integrated with the SS heuristic. A weight $\alpha = 0.0$

Table 7.8 Improvement of adaptive control over no adaptive control with integration of SS heuristic with a weighted factor, where $SPACE_{is} = p_{is+1}$

Scale	Instance	0.0	0.1	0.2	0.3	0.4	0.5	0.6	0.7	0.8	0.9	1.0
20*5	10	−0.59	0.75	2.17	3.12	1.97	2.46	−2.31	−7.08	−7.68	−9.25	−2.30
20*10	10	0.67	0.77	0.56	5.01	3.68	1.81	−2.86	−4.62	−2.11	−1.59	−4.27
20*20	10	7.50	10.00	8.69	6.74	6.15	3.01	0.00	−9.46	−9.12	−8.99	−9.79
50*5	10	0.22	1.11	1.17	0.97	1.55	0.88	0.15	−1.66	−1.33	−1.24	−0.64
50*10	10	5.50	4.97	4.06	2.80	2.75	2.17	−4.35	−7.10	−6.20	−8.76	−8.89
50*20	10	1.65	3.67	1.94	2.90	−5.41	−2.80	−5.53	−5.22	−7.50	−0.22	−1.07
100*5	10	0.43	0.88	0.36	0.49	0.58	0.39	0.17	−0.98	−1.09	−1.16	−2.24
100*10	10	3.05	3.02	2.20	2.42	1.05	0.29	−2.35	−1.97	−2.76	−1.42	−2.50
100*20	10	5.94	5.24	5.79	4.76	−0.39	−4.18	−5.26	−5.76	−6.61	−5.73	−7.49
200*10	10	0.47	0.47	0.05	0.27	0.15	−0.41	−0.75	−1.26	−1.10	−1.58	−1.15
200*20	10	1.53	2.24	2.39	0.82	−0.15	−1.25	−1.44	−1.48	−2.34	−2.36	−1.98
500*20	10	0.17	0.18	0.16	−0.02	−0.64	−1.21	−1.63	−1.69	−1.66	−1.59	−1.54
Average		2.21	2.77	2.46	2.52	0.94	0.10	−2.18	−4.02	−4.96	−5.33	−6.15

means no *IDLE* or *DELAY* is taken into consideration to sequence the jobs, and $\alpha = 1.0$ means no *SPACE* is taken into consideration. The weighting factor α reveals the relationship between *SPACE*, *IDLE* and *DELAY* in adaptive control with Taillard's benchmarks: *SPACE* affects the production performance more than *IDLE* or *DELAY*, where $\alpha = 0.1$ has the greatest improvement, 2.77%.

7.4.3 Industrial Case Study

To validate the SS heuristic in a real setting, an industrial case study has been carried out in a custom windows and doors manufacturing company in Calgary. This case consists of 1,396 jobs on a flow line with five work centres for 1-day production. These jobs are delivered to customers in 28 batches. Each batch of products is destined for customers in a given geographic area. Using data provided by the company and using the SS heuristic in its basic form, the SS heuristic produced the results shown in Table 7.9.

As shown in Table 7.9, the company used 42,300 time units to finish 1,396 jobs. The production of 1,396 jobs in 42,300 time units was achieved by the company's original schedule, which was generated by a production scheduler who has worked in the company for many years. The SS heuristic can generate a new schedule, reducing 42,300 time units to 41,771, a 1.25% improvement in productivity. Such an improvement on daily production translates into the production of 17 additional products daily, or more than $5,000 revenue per day.

The SS heuristic was programmed in C++ language, and run on an Intel Pentium IV personal computer with a 3.20 GHz CPU and 1.00 GB RAM. The computation time of the SS heuristic to schedule 1,396 jobs on five machines was 70.67 s. Those 70.67 s are for offline production scheduling, generating and evaluating five sequences for all 1,396 jobs. However, for online adaptive

Table 7.9 An industrial case study

	The company	SS	Improvement		The company	SS	Improvement
1	1,795	1,711	84	16	1,489	1,489	0
2	1,458	1,444	14	17	1,477	1,477	0
3	1,698	1,697	1	18	1,743	1,712	31
4	2,292	2,261	31	19	1,751	1,745	6
5	1,570	1,556	14	20	1,434	1,430	4
6	1,798	1,753	45	21	1,587	1,570	17
7	1,420	1,420	0	22	1,587	1,393	194
8	1,573	1,567	6	23	1,196	1,165	31
9	1,828	1,805	23	24	1,094	1,083	11
10	1,676	1,676	0	25	1,362	1,362	0
11	1,568	1,568	0	26	1,281	1,281	0
12	1,691	1,691	0	27	923	923	0
13	1,465	1,465	0	28	857	851	6
14	1,364	1,353	11	Total	42,300	41,771	529
15	1,323	1,323	0	Percent			1.25%

production control, the computational complexity of SS heuristic is $O(Sn)$ (please refer to Sect. 7.2.3). Therefore, if there are 1,396 jobs left unproduced when a disturbance occurs, it takes only 10.12 ms ("the overall computation time" ÷ "5 sequences" ÷ "1,396 jobs") for SS heuristic to select the next successive job, which is fast enough to respond to a disturbance on the shop floor.

Originally, it took a long time for the experienced production scheduler to off-line schedule 1-day production of customer orders in the company. This manual scheduling process made it difficult for the company to respond to changes of customer orders in time and at a low cost. The company also used a fixed 3-day production scheduling policy to avoid complicated re-scheduling activities. This policy means that customers are not able to change or place an urgent order 3 days before production. With the SS scheduling heuristic, the company could change the production scheduling policy from 3 days to 2 h, taking into consideration of other constraints, e.g. raw material handling speed and capacity in the company, convenience of operators to switch between workstations or production lines.

7.4.4 An Extension of SS Heuristic

To further illustrate the SS (state space) concept, we compare a scaled down version of SS heuristic with Johnson's algorithm (Johnson 1954) for a two-machine TFS scheduling problem, $F2/prmu/C_{\max}$. For $F2/prmu/C_{\max}$ problems, the lever concept that is part of the SS heuristic has no effect on the job selection. This is because for such problems, there is no limit to the WIP inventory between machines 1 and 2, then no *DELAY* is taken into consideration. In addition, the

Table 7.10 A two-machine flow shop example

	Machine 1	Machine 2
Job 1	5	20
Job 2	20	10
Job 3	10	15
Job 4	15	12

length of force arm for *SPACE* or *IDLE* is equal to one for two-machine TFS scheduling problems. However, the SS concept can yield different job sequences than Johnson's algorithm. A numerical example is provided in Table 7.10.

Johnson's algorithm sequences the jobs according to the following scheme. If min $(p_{a1}, p_{b2}) \le$ min (p_{a2}, p_{b1}), job a should be processed earlier than job b. Therefore, for the example in Table 7.10, Johnson's algorithm generates a sequence of [Job 1, 3, 4, 2] with $C_{\max} = 62$.

According to the SS concept (again, not exactly the SS heuristic), and using Johnson's algorithm for initial job selection, we obtain two additional sequences, both of which have $C_{\max} = 62$, and are different from the one generated by Johnson's algorithm: [Job 1, 2, 3, 4] and [Job 1, 4, 3, 2]. Thus, we can see that the SS concept can yield different sequences than Johnson's algorithm with the same level of optimality, and hence can provide greater opportunities for improvement as the core of a more elaborate heuristic.

7.5 eGantt Chart Module for OKP Scheduling and Control

In OKP, some companies produce large and complicated customer ordered products, e.g. a compressor for oil and gas production. This type of OKP products can take a few weeks or few months to complete its production. To manage and control the production of this type of large OKP products somehow likes the management of projects. For adaptively scheduling and control the production of these large OKP products, a computer module, named electronic Gantt chart module or eGantt module is presented in this section. By using this eGantt chart module, an OKP company can schedule the production under the resource constraints or adaptively re-schedule and control the production to cope with the disturbances in OKP.

The resource as referred here mean a kind of critical resource in an OKP company, which often means operators/workers or machining centres. In project management, the resource planning aims at achieving the two objectives, i.e. resource levelling and resource allocation under the constraints of limited available resources. Resource levelling aims at minimizing resource assignment fluctuation along the period of a project. The resource allocation under limited available resources is looking for a resource allocation schedule that will not exceed the maximum available number of resources but has the minimum completion period of the project. In fact, the resource planning is the major activity in adaptive OKP scheduling and control for large OKP products. This is because an OKP company often produces products under the various resource constraints. The eGantt chart module is particularly developed to help OKP company to quickly and accurately generate a production schedule under the resource constraints.

For the resource planning in project management, Burgess and Kilebrew (1962) developed a heuristic for resource levelling for a single type of resource. For resource allocation under the limited available resources, two heuristics are commonly used, i.e. longest duration heuristic and least slack heuristic. These three heuristics for single type of resource planning are widely used in project management today. Some advanced resource planning heuristics or approaches for multi-types of resource planning have been developed, such as Petrinet and memetic approach (Raja and Kumanan 2007), computer simulation approach (Kumar and Rajendran 1993), etc. In general, the resource planning is a NP hard problem. These advanced techniques normally demand a relative long computation time and under some assumptions that may not be realistic for all kinds of OKP companies. Readers can use these advanced techniques to get a better resource allocation plan and the eGantt chart module does support and facilitate the application of these advanced resource planning techniques.

In this section, we first slightly modify the Burgess and Kilebrew (1962) heuristic, longest duration heuristic and least slack heuristic so that these simple and widely used resource planning heuristics can be applied in eGantt Chart module. By case studies, we demonstrate that a rational or nearly optimal production schedule under various resource constraints can be achieved by using these three simple resource planning heuristics and eGantt chart module. We also empirically validate the flexibility, applicability and efficiency of the eGantt chart module for adaptively scheduling and control OKP of large products or managing complicated engineering projects. To speed up the production to meet a specific due date, we also include a subsection to describe a cost-effective heuristic for crashing a project, which is also a widely applied heuristic for crashing projects in project management.

7.5.1 Resource Levelling

The resource levelling problem can be modelled as follows:

$$\text{Min}\, Z = \sum_{k=1}^{N} (D_k - \overline{D})^2 \; \triangleright \; \text{Min}\, Z = \sum_{k=1}^{N} (D_k)^2 \qquad (7.3)$$

$$\text{S.t.}: \text{The project duration} \leq N$$

Where, D_k is the resource demand at time k ($k = 1, 2, \ldots, N$), N is the project duration, and $\overline{D}$ is mean resource demand.

Burgess and Kilebrew (1962) proposed a simple heuristic for a single type of resource levelling. To apply this heuristic to the OKP scheduling by using eGantt chart, this heuristic is slightly modified as follows:

Step 1 Convert the PPS of a product which is realised to be produced in an OKP company into the project network diagram, i.e. simply delete all the boxes from the PPS and link the circles/ovals directly in the PPS.
Step 2 Draw all the operations of the network diagram on a Gantt chart and start an operation as early as possible.
Step 3 Select the operation with the maximum slack among the operations that start latest among all the others and move the operation within its slack times to obtain $\text{Min}\, Z = \sum_{k=1}^{N} (D_k)^2$. This implies a backward schedule process.
Step 4 Repeat Step 3 until the operations that start first.

Example 7.1 Assume that the network diagram for producing an OKP product is converted from the PPS as shown in Fig. 7.7.

According to Step 2 of the modified Burgess and Kilebrew heuristic, we have the Gantt chart as shown in Fig. 7.8.

In Fig. 7.8, D means duration of an operation. ES means early start time of an operation and EF means early finish time of an operation. Sum of Squares means $Z = \sum_{k=1}^{N} (D_k)^2$. The operation in red color means these operations with the minimum slacks, i.e. zero (or 0) slacks for this example. These operations will not be rescheduled.

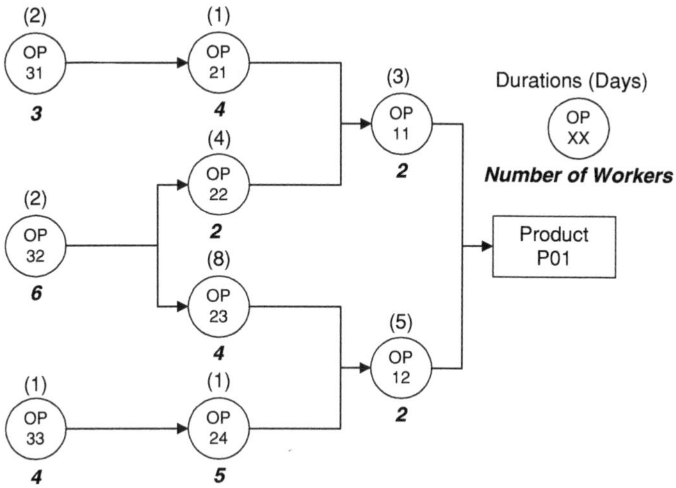

Fig. 7.7 The project network diagram for producing an OKP product

Operation	D	ES	EF	Worker	0	1	2	3	4	5	6	7	8	9	10	11	12	13	14
OP31	2	0	2	3	3	3													
OP32	2	0	2	6	6	6													
OP33	1	0	1	4	4														
OP21	1	2	3	4			4												
OP22	4	2	6	4			2	2	2	2									
OP23	8	2	10	4			4	4	4	4	4	4	4	4					
OP24	1	1	2	5		5													
OP11	3	6	9	2							2	2	2						
OP12	5	10	15	2											2	2	2	2	2
	Workers				13	14	10	6	6	6	6	6	6	4	2	2	2	2	2
	Workers^2				169	196	100	36	36	36	36	36	36	16	4	4	4	4	4
	Sum of Suqares				717														

Fig. 7.8 Gantt chart of the Example 7.1

Repeat Step 3: (1) Moving OP11; (2) Moving OP21 and then OP22; (3) Moving OP24; Moving OP31 and then OP33. We finally have the schedule as shown in Fig. 7.9.

It ends with the minimum Sum of Squares of 539.

If there are a number of products to be produced in an OKP company, we suggest that: (1) using the Burgess and Kilebrew heuristic to schedule the project as the example shown in Fig. 7.9; and then (2) using the eGantt chart (see Sect. 7.7) to schedule all the projects. This suggestion does not guaranty the optimal solution but it does provide a rational solution since each of the projects has been levelled before it lays on an overall schedule Gantt chart.

7.5.2 Resource Allocation Under Limited Number of Resource

The resource allocation problem under limited number of resource can be expressed by the following model:

Operation	D	ES	EF	Worker	0	1	2	3	4	5	6	7	8	9	10	11	12	13	14
OP31	2	4	6	3					3	3									
OP32	2	0	2	6	6	6													
OP33	1	2	3	4			4												
OP21	1	11	12	4												4			
OP22	4	7	11	4								2	2	2	2				
OP23	8	2	10	4			4	4	4	4	4	4	4	4					
OP24	1	6	7	5							5								
OP11	3	12	15	2													2	2	2
OP12	5	10	15	2											2	2	2	2	2
	Workers				6	6	8	4	7	7	9	6	6	6	4	6	4	4	4
	Workers^2				36	36	64	16	49	49	81	36	36	36	16	36	16	16	16
	Sum of Squares				539														

Fig. 7.9 The final schedule of Example 7.1

$$\begin{aligned} &\text{Min}\, Z = N \\ &\text{S.t.}: \; D_k \leq L; \quad k = 1, 2, \ldots, N \end{aligned} \tag{7.4}$$

Where, D_k is the resource demand at time k ($k = 1, 2, \ldots, N$), N is the project duration which is a variable to be determined through a resource allocation plan, and L is the maximum number of available resources.

To apply the longest duration (LD) heuristic in OKP scheduling by using eGantt chart, the LD heuristic is modified as follows:

Step 1 Convert the PPSs of all the products which are realised to be produced in an OKP company into the project network diagram, i.e. simply delete all the boxes from the PPSs and link the circles/ovals directly in the PPSs.

Step 2 Draw all the operations of the network diagram on a Gantt chart and start an operation as early as possible.

Step 3 At time t when the resource demand exceeds the maximum capacity, i.e. $D_t > L$, Assign the resource to the operation with the longest duration among the operations that compete for the same resources and delay the rest of competing operations to time $t + 1$.

Step 4 Repeat Step 3 until the minimum duration of project, i.e. N.

Example 7.2 Assume that two OKP products are released to an OKP shop floor. The project network diagram of one of the products is shown in Fig. 7.7 and the other one is shown in Fig. 7.10. The maximum number of available workers is 14.

The two products are placed on the Gantt chart as shown in Fig. 7.11.

Both products P01 and P02 has been levelled using Burgess and Kilebrew heuristic. The operations in red colour constitute the critical paths of the two case product projects. Figure 7.11 shows that at days 2, 4 and 5 (Please note that the completion day of an operation equals the Gantt chart reading plus 1), the resource demand exceeds the limit of 14 workers. The competing operations are

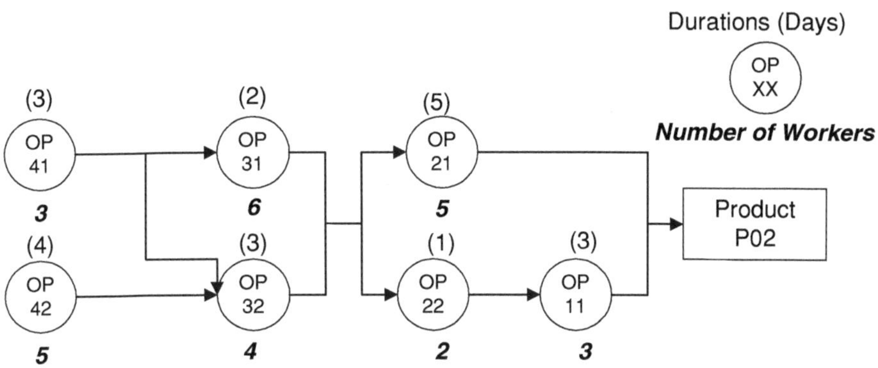

Fig. 7.10 The project network diagram of the second case product

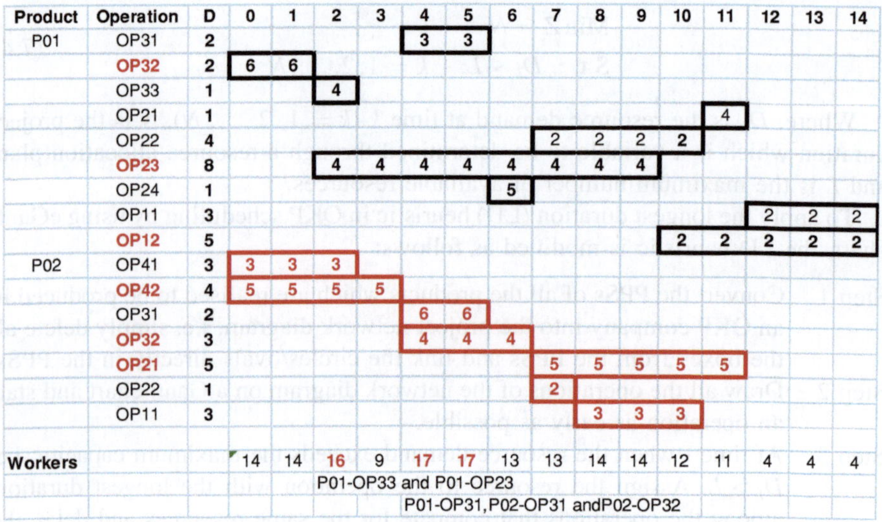

Fig. 7.11 The initial Gantt chart for scheduling the two case products

also illustrated on Fig. 7.11, i.e. at day 2: P01-OP33 and P01-OP23; at day 4: P01-OP31, P02-OP31 and P02-OP32. Table 7.11 illustrates the detailed steps or trails to apply LD heuristic.

We have the final resource allocation plan as shown in Fig. 7.12.

As shown in Fig. 7.12, the minimum duration for completing the two case products under 14 workers is 15 days. The resource allocation plan as shown in Fig. 7.12 can be further levelled using Burgess and Kilebrew heuristic. This maybe an interested exercise by readers through using the eGantt chart as described in Sect. 7.5.4.

Finally, we modify and present the modified least slack (LS) heuristic as follows:

Step 1 Convert the PPSs of all the products which are realised to be produced in an OKP company into the project network diagram, i.e. simply delete all the boxes from the PPSs and link the circles/ovals directly in the PPSs.

Step 2 Draw all the operations of the network diagram on a Gantt chart and start an operation as early as possible.

Table 7.11 Application of LD heuristic to the case in Fig. 7.9

Day	2	5	7
Competing resources	P01-OP33, P01-OP32	P01-OP31, P02-OP31, P02-OP32	P01-OP31, P01-OP22, P02-OP21, P02-OP22
Decision	Move P01-OP33 to Day 3	Move P01-OP31 to Day 7	Move P02-OP22 and P02-OP11 to Day 9

Product	Operation	D	0	1	2	3	4	5	6	7	8	9	10	11	12	13	14
P01	OP31	2								3	3						
	OP32	2	6	6													
	OP33	1				4											
	OP21	1												4			
	OP22	4								2	2	2	2				
	OP23	8			4	4	4	4	4	4	4	4					
	OP24	1							5								
	OP11	3													2	2	2
	OP12	5											2	2	2	2	2
P02	OP41	3	3	3	3												
	OP42	4	5	5	5	5											
	OP31	2					6	6									
	OP32	3					4	4	4								
	OP21	5								5	5	5	5	5			
	OP22	1										2					
	OP11	3											3	3	3		
Workers			14	14	12	13	14	14	13	14	14	13	12	14	7	4	4
					P01-OP33 and P01-OP23												
							P01-OP31,P02-OP31 andP02-OP32										

Fig. 7.12 The final Gantt chart for scheduling the two case products

Step 3 At time t when the resource demand exceeds the maximum capacity, i.e. $D_t > L$, Assign the resource to the operation with the least total slack among the operations that compete for the same resources and delay the rest of competing operations to time $t + 1$.

Step 4 Repeat Step 3 until the minimum duration of project, i.e. N.

Product	Operation	Slack	0	1	2	3	4	5	6	7	8	9	10	11	12	13	14
P01	OP31	9										3	3				
	OP32	0	6	6													
	OP33	8				4											
	OP21	9												4			
	OP22	6								2	2	2	2				
	OP23	0			4	4	4	4	4	4	4	4					
	OP24	8							5								
	OP11	6													2	2	2
	OP12	0											2	2	2	2	2
P02	OP41	2	3	3	3												
	OP42	0	5	5	5	5											
	OP31	2					6	6									
	OP32	0					4	4	4								
	OP21	0								5	5	5	5	5			
	OP22	1												2			
	OP11	1													3	3	3
Workers			14	14	12	13	14	14	13	11	11	14	12	13	7	7	7
					P01-OP33 and P01-OP23												
							P01-OP31,P02-OP31 andP02-OP32										

Fig. 7.13 The final Gantt chart for producing the two case products using LS heuristic

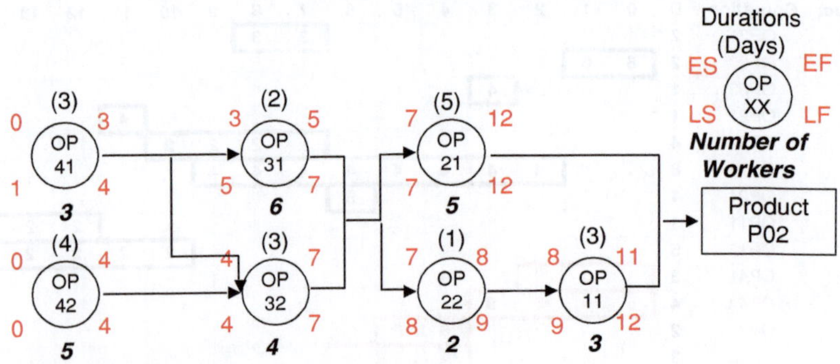

Fig. 7.14 Determination of LF, EF, LS and ES

To apply LS heuristic to solve the problem as described in Example 7.2, we have the resource allocation plan as shown in Fig. 7.13.

In Fig. 7.13, Slack means the total slack of an operation in the project network diagram. It can be calculated by:

$$\text{Total Slack of an Operation} = \text{LF} - \text{EF or LS} - \text{ES} \tag{7.5}$$

where, LF means the latest finish time of an operation, EF means earliest finish time of the operation, LS means the latest start time of the operation, and ES means earliest start time of the operation. The LF, EF, LS and ES for the case product as shown in Fig. 7.10 can be determined on the project network diagram as demonstrated in Fig. 7.14.

7.5.3 Minimum Cost Project Crashing Heuristic

Sometimes, we need to speed up the progress of the production of an OKP product or a project. This is refereed as project crashing in project management. The project crashing can be realized by adding more resources into the production, e.g. assign more workers to an OKP flow line or workshop. This will cost more for the production or the project. To minimize the extra cost to crash a project, a minimum cost project crashing heuristic is commonly used in project management. This heuristic is realized by the following steps:

Step 1 Determine the project's critical path(s).
Step 2 Find the cheapest operation(s) on the critical path(s) to crash.

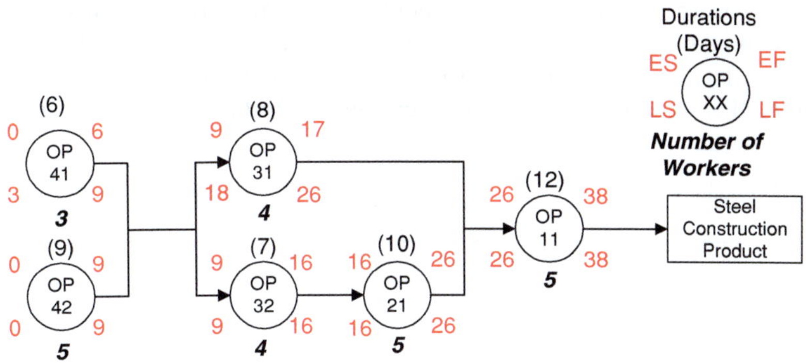

Fig. 7.15 The project network diagram for a steel construction product

Step 3 Reduce the time for the operation(s) until (a) it/they cannot be further reduced, (b) another path or the other paths becomes critical, or (c) the increase in direct costs exceeds the savings that results from shortening the project. If more than one path is critical, the time for an operation on each path may have to be reduced simultaneously.

Step 4 Repeat Steps 1, 2 and 3 until the increase in direct costs is less than the savings generated by crashing the project.

Through the following simple example, we demonstrate this heuristic.

Example 7.3 A steel construction product has been ordered by a customer to an OKP company which produces various customized steel construction products. The project network diagram for this product has been derived from the PPS of the product and it is shown in Fig. 7.15. Suppose that by allocating additional resources to an activity, the company can reduce the duration of any activity by up to 5 days. The cost per day of reducing the duration of an operation is listed in Table 7.12. The customer requires the company to produce the product within 25 days. Beyond 25 days, the customer will reduce its payment to the company by $5000 per day.

Table 7.12 The crash cost of the project for case product as shown in Fig. 7.15

OP41	OP42	OP31	OP32	OP21	OP11
$1000	$2000	$300	$3000	$4000	$5000

As we can see from Fig. 7.15, without crashing the project, it needs 38 days for the company to produce the product. Therefore, we need to find a minimum cost schedule to crash the project to 25 days. To find a minimum cost crashing schedule, the minimum cost project crashing schedule method is applied as follows:

Trail 0 Do nothing

Crash cost	Saving	Balance
0	0	0

After crashing:
Critical path(s): OP42-OP32-OP21-OP11
Cheapest and crushable operation(s) along the critical path(s): OP42 ($2000/day)
Project duration: 38 days

Trail 1 Crashing OP42 by 3 days, i.e. OP42 becomes 6 days

Crash cost	Saving	Balance
3 * ($2000) = $6000	3 * ($5000) = $15,000	−6000 + 15000 = $9000

After crashing:
Critical path(s): OP42/OP41-OP32-OP21-OP11
Cheapest and crushable operation(s) along the critical path(s): OP42&OP41 ($3000/day), OP32 ($3000/day)
Project duration: 38 − 3 = 35 days

Trail 2 Crashing OP42&41 by 2 days, i.e. OP42&41 become 4 days; crashing OP32 by 5 days, i.e. OP32 becomes 2 days

Crash cost	Saving	Balance
7 * ($3000) = $21,000	7 * ($5000) = $35,000	−21,000 + 35,000 = $14,000

After crashing:
Critical path(s): OP42/OP41-OP32-OP21-OP11
Cheapest and crushable operation(s) along the critical path(s): OP21 ($4000/day)
Project duration: 35 − 7 = 28 days

Trail 3 Crashing OP21 by 3 days, i.e. OP21 become 7 days

Crash cost	Saving	Balance
3 * ($4000) = $12,000	3 * ($5000) = $15,000	−12,000 + 15,000 = $3000

After crashing:
Critical path(s): OP42/OP41-OP32-OP21-OP11
Cheapest and crushable operation(s) along the critical path(s): OP21$OP31 ($4300/day)
Project duration: 28 − 3 = 25 days

If one more day to be crashed, it will cost $4000, i.e. crash OP21 by one more day. However, this cost cannot be paid by the saving since the customer only reduces the payment for any day beyond 25 days. To complete the project early, the company will get nothing. Therefore, the heuristic stops at the end of Trail 3. The total cost for crashing the project is 6000 + 21000 + 12000 = $39,000, which will save the company 13*(5000) = $65,000, and the total balance is −39000 + 65000 = $26,000 surplus.

7.5.4 eGantt Chart Module

As mentioned at the outset of Sect. 7.5, the electronic Gantt chart or eGantt chart module is a computer module for one-of-a-kind production (OKP) scheduling and control or project planning in project management. Before the production of products starts, a user can use it to schedule the production so that the production capacity meets the demand and the customers' due dates are met. During the production, the user can use it to adjust the production schedule adaptively to cope with changes of production progress and unexpected disturbances. The overflow of a work-in-process (WIP) buffer (or inventory) can also be checked by the eGantt chart module. An eGantt chart module has been developed as a module in the OKP management and control software system©. In this section, through a case study, we demonstrate the principles and functionalities of the eGantt chart module.

To use an eGantt chart module to schedule the production of a product or a project, we need to define a few concepts as follows.

Place means a work centre, section, inventory, WIP buffer, etc., in which a production material can stay. The maximum capacity of a Place needs to be specified in the database so that the eGantt chart module (or the scheduling module) can access it. The maximum capacity of a Place is defined by a number which means the number of operations can be held in the place within a given time period (or scheduling time bucket), e.g. an hour, a shift or a day.

Resource means a worker or operator in an OKP company. The capacity of resource (simply called Overall Resource Capacity or ORC) is the maximum number of workers/operators for the production or projects within a business unit. A business unit for a small company may mean the whole company, whereas for a medium size or large size company could mean a work shop or a branch of the company. The ORC needs to be specified in the database so that the eGantt chart module can access it.

Resource type means a skill type or technical trade of a worker/operator. The capacity of a resource type (simply called Resource Type Capacity or RTC) means the maximum number of workers who have the similar skill type or technical trade under this resource type. The RTC of each skill type or technical trade needs to be specified in the database so that the eGantt chart module can access it.

Since one worker may have multi-skills, the sum of RTCs maybe more than the ORC. When a user uses the eGantt chart module to plan the production or projects, the eGantt chart module will check both the sum of RTC and ORC to make sure that there is no capacity overflow problem for RTCs and ORC.

By the concepts as defined above, we can use the eGantt chart module in the OKP management and control software system© to plan Example 7.2. In fact the two projects as referred in Example 7.2 are the projects or production of two customized steel-frame structures in a structural steel construction company. These two products were produced by a team of 14 workers (i.e. OTC = 14) which consists of six steel plate cutting workers and eight welding workers. Therefore, for this case study, there are two Resource Types:

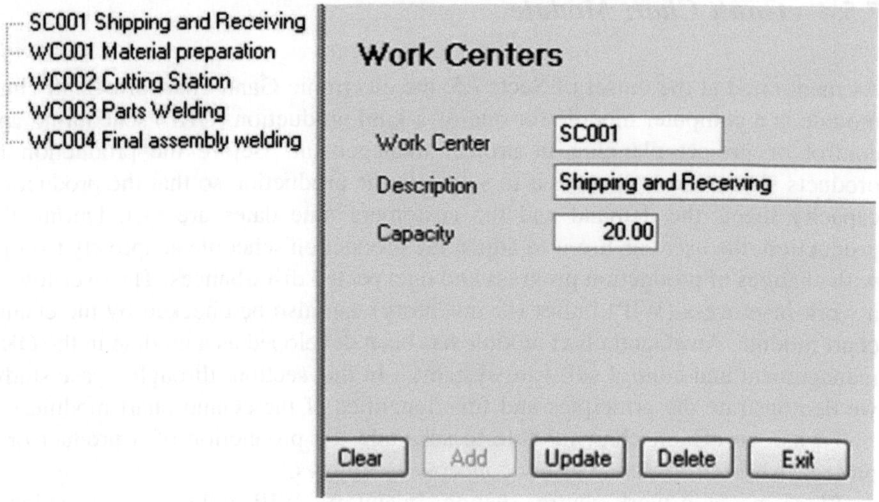

Fig. 7.16 The work centres definition interface

1. SK001: Cutting
2. SK003: Welding

Therefore, there are six RTCs of SK001 and eight RTCs of SK002.

The team performs the production in 5 work centres which are defined in the OKP Management and Control software system as shown in Fig. 7.16.

The capacity of a work centre, which is input by the interface as shown in Fig. 7.16, often means the capacity of the WIP buffer (or inventory) associated or in front of the work centre. By using this interface, the various working Places in a company can be defined in the OKP management and control software system©, which formulate a part of enterprise production data (EPD).

The workers and their skills (or technical trades) are defined by using the User Definition Interface of the system. This interface is shown in Fig. 7.17. By using this interface, the Resources and Resource Types are defined in the OKP management and control software system©, which formulate the other part of the EPD.

After the Places, Resources and Resources Types of this team are defined in the system, the eGantt chart module shows the constraints for production scheduling automatically in Fig. 7.18.

By default, the eGantt chart module turns off the display (as indicated by red cross in Fig. 7.18) of Skills and Work Centres on the eGantt chart in order to simplify the display of the eGantt Chart. However, if a user is going to schedule the production of products or projects under the limitation of critical Skills and bottle-neck Work Centres, he/she can click on these Skills and Work Centres, and check the box as an example shown in Fig. 7.19.

For this case study, we are interested in the constraints or limitations of SK001 (Cutting) and SK002 (Welding) since these two trades cannot be replaced between

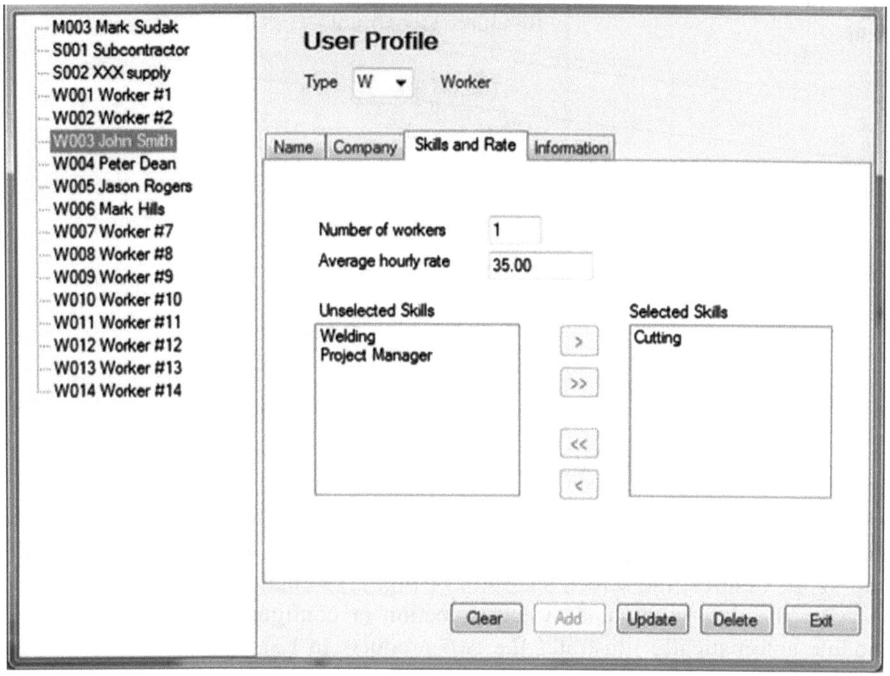

Fig. 7.17 The user definition interface

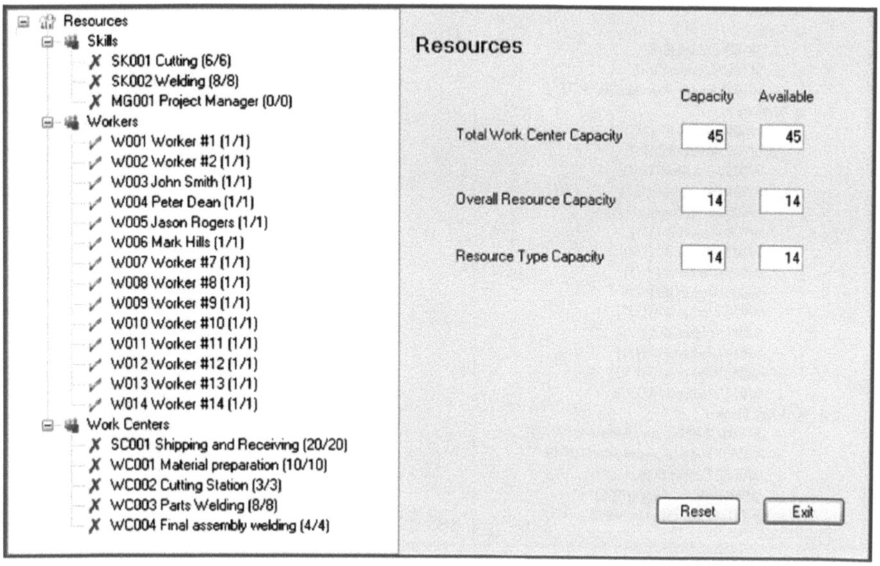

Fig. 7.18 The production scheduling constraints

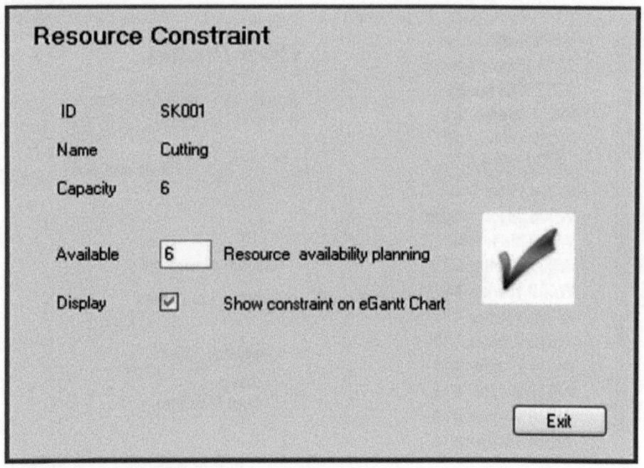

Fig. 7.19 Select a critical Skill

each other as well as the work centres WC002 and WC004 since they have the much less capacity compared with the other work centres. By clicking these Skills and Work Centres, the screen snapshot of Fig. 7.18 changes into Fig. 7.20.

After the data inputs and system selection or configuration, the eGantt chart module automatically illustrates the two products in Example 7.2 on the eGantt chart as shown in Fig. 7.21.

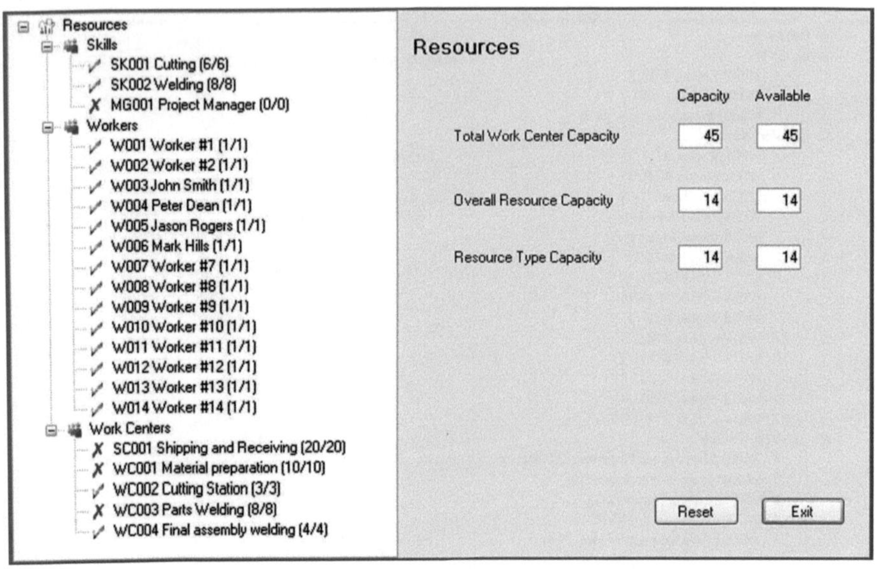

Fig. 7.20 The changed production scheduling constraints screen snapshot

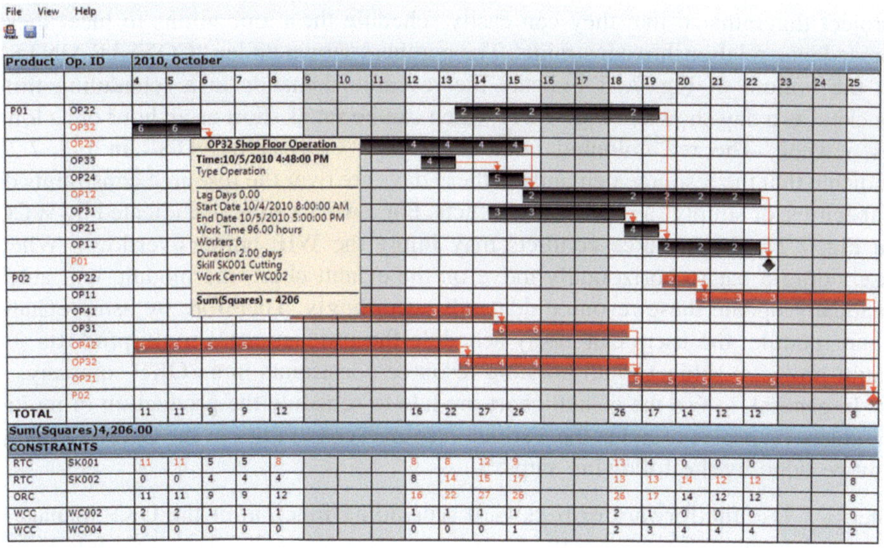

Fig. 7.21 The screen snapshot of the eGantt chart

As shown in Fig. 7.21, the filled columns by gray colour indicate the weekends or public holidays, e.g. October 9 and 10, 2010 are Saturday and Sunday and October 11 is Thanksgiving Day. These are non-working days. The non-working days are specified by a user on the work calendar through the enterprise production data interface of the OKP management and control software system© (see Sects. 8.2.2). The horizontal bars on the chart are the operations of the production of products P01 and P02. The eGantt chart module is able to draw an operation bar automatically to skip the non-working days, i.e. not to count these non-working days for the completion time the production. When a user places the mouse cursor on an operation bar, the module pops up a window to show the details of the operation, e.g. P01-OP32 (i.e. operation OP32 for making product P01) is shown in Fig. 7.21. At the bottom of this popup window, it shows "Sum(Squares) = 4206" which is the sum of the squares as defined by model (7.3) in Sect. 7.5.1. As suggested by Burgess and Kilebrew (1962) heuristic, when we schedule the production by using the mouse to drag a bar (or operation) horizontally to move on the eGantt chart within the given completion time of the production, we should make this number as small as possible. The red coloured operation identification codes, e.g. P01-OP32, P01-OP23, P01-OP12, P02-OP42, P02-OP32 and P02-OP21, indicate the critical operations that formulate the critical paths in the project diagram for producing products P01 and P02. The red arrow lines between operation bars to indicate the precedence relationships between the operations. When users drag a bar to horizontally move on the eGantt chart, these red colour lines conveniently indicate the precedence relationships between the operations in the project diagrams so that they can easily schedule these operations to meet these precedence relationship constraints. The numbers shown under "CONSTRAINTS" at the bottom of Fig. 7.21 indicate the resource demands in a scheduling time bucket (as a day shown in Fig. 7.21) which can be set as short as an hour or as long as a week. The red coloured numbers under "CONSTRAINTS" in Fig. 7.21 indicate that the resource demands in these days are over the resource constraints or capacities, or simply say resource conflicts. For a work centre as indicated by WCC in Fig. 7.21, the resource conflicts may imply the WIP buffer overflows. When users drag a bar to horizontally move on the eGantt chart, the module will automatically update these resource demands accordingly. Therefore by using eGantt chart module, the users can easily schedule the production of multi-products (or multi-projects) subject to all possible resource constraints in an OKP company.

In general, to use the eGantt chart module to schedule the production of multi-products or projects under the various resource constraints in an OKP company can be done by the following steps:

Step 1 Use the forward or backward scheduling function of the OKP management and control software system© (see Sects. 8.2.4) to schedule the production of all the products (or projects). This will lead to automatically place the production (or projects) on the eGantt chart. As shown in Fig. 7.21, both products P01 and P02 are scheduled by forward schedule function start at the same day, i.e. October 4, 2010.

Step 2 By using the modified LD or LS heuristic (see Sect. 7.5.2), drag the operation bars to move horizontally on the eGantt chart to avoid the resource conflicts, or red coloured numbers under "CONSTRAINTS" as shown in Fig. 7.21.

Step 3 By using the modified Burgess and Kilebrew heuristic (see Sect. 7.5.1), level the resource through moving the operation bars not to violate the resource constraints but making the value of "Sum(Square)" as small as possible.

By using the three steps as mentioned above, we have a final production schedule for making products P01 and P02 as shown in Fig. 7.22.

As shown in Fig. 7.22, the total completion time is 18 working days from October 4, 2010 to October 28, 2010. All the resource conflicts as the red numbers under "CONSTRAINTS" as shown in Fig. 7.21 have been solved by the new schedule as shown in Fig. 7.22. The value of "Sum(Squares)" as shown in Fig. 7.22 is 1800 which is much less than 4206 as shown in Fig. 7.21. This means this new schedule has a much balanced or levelled resource demands the days during the production of products P01 and P02 (or projects for producing P01 and P02). To get this new production schedule, a user can try it through following the three steps as mentioned above within a short time, e.g. we used about 10 min to get this schedule. Therefore, by using the eGantt chart it should be not difficult to plan the various resources to achieve the optimal or nearly optimal production schedule under the limitations or constraints of the resources. This type of

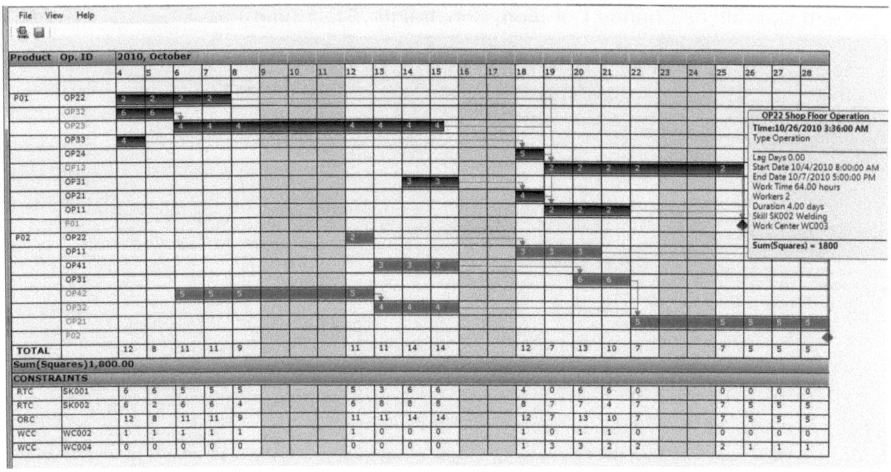

Fig. 7.22 The production schedule for producing case products P01 and P02

production scheduling work is often considered as time-consuming and difficult tasks in OKP companies or project management.

If we need to speed up the production of products P01 and P02, we can use the minimum cost project crashing heuristic (see Sect. 7.5.3) to add more resources into the critical operations, i.e. these operations along the longest path of the production schedule of products P01 and P02. As shown in Fig. 7.22, operations P01-OP32, P02-OP42, P02-OP32, P01-OP34 and P01-OP21, P01-OP11, P02-OP21, form the longest path. If we are trying to get the operation bar of P02-OP21 shorter, we can right-click P02-OP21 bar and the eGantt chart module automatically pops up a window as shown in Fig. 7.23.

The duration of P02-OP21 can be crashed in two ways: (1) add more workers and the eGantt chart module will automatically calculate the "Duration" of P02-OP21, or (2) directly change "Work Effort" by clicking "Task" button. To change "Work Effort" may mean reducing work contents, improving production efficiency or adding advanced production facilities or tools.

After changing the operation duration, click the "Update" button as shown in Fig. 7.23. The operation bar of P02-OP21 is shortened accordingly by the eGantt Chart module. By meaning of the operation specifying function of the eGantt chart and minimum cost project crashing heuristic, we can speed up or slowdown the production of OKP products to meet a specific due date.

By the operation specifying function in the eGantt chart module, we can also update the production schedule according to value changes of the "Work Efforts" of the operations. These changes possibly occur when errors of initial estimate of the "Work Efforts" are identified or production efficiency is improved during the production progress as well as unexpected disturbances change the "Work Efforts" of some operations.

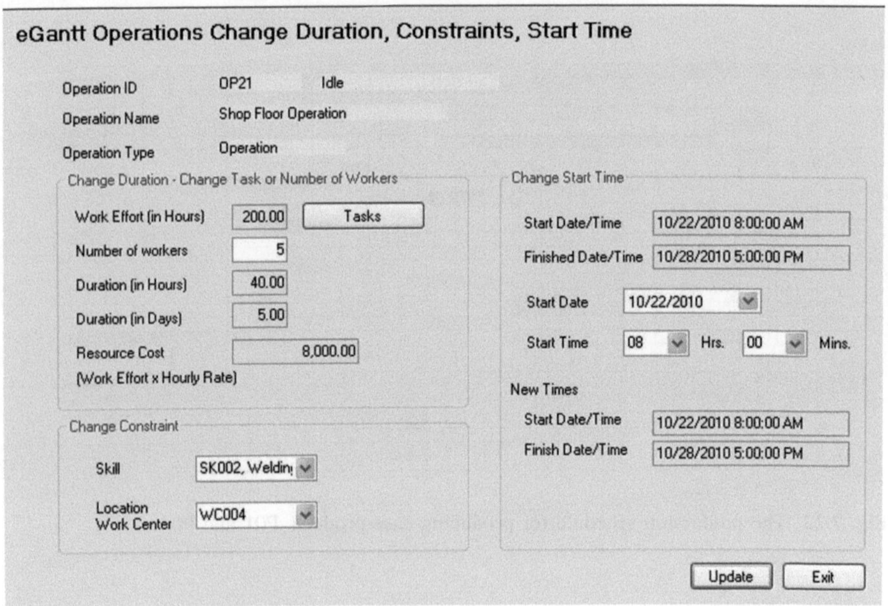

Fig. 7.23 The screen snapshot for specifying operation P02-OP21

Besides the project crashing technique, we can also shorten the production completion time through dynamic allocation of the resources to those longer operation bars. In this case, there are six cutting workers and 8 welding workers. As shown in Fig. 7.22, in most of days, we do not use the maximum capacities of SK001 (cutting workers) and SK002 (welding workers). Therefore, we can assign these un-allocated workers to the longer operation bars to shorten the operation bars and hence to shorten the overall completion time of the production. This dynamic resource allocation strategy will help an OKP company to further improve the resource utilization and production efficiency.

7.6 Chapter Summary

In this chapter, the concepts, methods and a computer module for adaptive one-of-a-kind production (OKP) scheduling and control are presented. For OKP companies that produce a large number of customized products per day, e.g. a windows and doors manufacturing company in Calgary produces about 2000 custom windows and doors daily, the production scheduling means to determine a job sequence so that the make-span of the 2000 windows or doors can be minimized. This is because all the customer orders to be produced in the same day have the same due date. In the company, for example, the customer orders that are produced in a day will be transported to the customers in the next day. Therefore, the OKP scheduling problem becomes a hybrid flow shop scheduling problem. However, due to the frequent disturbances in OKP, the production schedule needs to be adjusted to cope with these disturbances, i.e. adaptive OKP scheduling and control. The adaptive OKP scheduling and control requires a suitable and efficient scheduling heuristic. To evaluate the suitability and efficiency of a heuristic, we suggest the three criteria, i.e. optimality, computational complexity, and flexibility. The state space (SS) heuristic has been developed under these three evaluation criteria. Through the Taillard's benchmarks and an industrial case from the custom windows and doors manufacturing company in Calgary, we validate the suitability and efficiency of the SS heuristic for adaptive OKP scheduling and control. By using the SS heuristic and the close loop control system, the OKP for a large number of customer orders on in an OKP company can be adaptively scheduled and controlled.

For OKP companies that produce large and complicated customized products which lead-times are often counted in weeks or even months, e.g. a compressor for oil/gas production, a computer aided production scheduling and control technique has been developed, namely eGantt chart module. By a case study, we demonstrate that the eGantt chart module can efficiently handle the production scheduling and control problems of multi-products under various resource constraints. Since the OKP schedule and control prefers management by project, the eGantt chart module can be applied to project management. However in comparison with existing project management software systems, e.g. Microsoft Project 2007, the eGantt chart module has the following advantages or novel functionalities:

1. It employs the PPS (product production structure) to model the data structure of a project to avoid the logic error of separating the BOM (bill of material) and BOO (bill of operation). This integrated data structure makes easily to change the product design or workflow to cope with unexpected disturbances during the production progress. It also provides a cost tracking and control function (see Chap. 4) during the production progress or project progress to control the overall budget or cost of a project that includes all possible costs associated with the project. The existing project management software systems do not have this function and require the project managers to record and track some costs of a project manually, i.e. material, outsourcing and subcontracting costs.
2. It provides a novel computer aided scheduling platform (i.e. eGantt chart module) by which a scheduler can easily schedule the production of multi-products or planning multi-projects under the constraints of various resources in a company. The existing project management software systems do not have this platform and hence make the resource planning as a difficult task for the schedulers.

It is worth to note that all OKP products, regardless of whether the products are simple or complicated, are managed individually and modeled by PPSs (product production structures) in the system so that the production efficiency and customer satisfaction can be achieved. If an OKP company needs to mass-produce a large number of OKP products daily, e.g. a windows and doors manufacturing company in Calgary produces about 2000 custom windows and doors daily, the PPS of a product is automatically generated by the system according to the PPS of a template product. For a complicated OKP product, its PPS is normally generated manually through modifying the PPS of a similar product that was produced in the past. From this point of view, the OKP is referred in this book as project-based production.

References

Blecker T, Friedrich G (eds) (2006) Mass customization: challenges and solutions. Springer, New York

Botta-Genoulaz V (2000) Hybrid flow shop scheduling with precedence constraints and time lags to minimize maximum lateness. Int J Prod Econ 64(1):101–111

Brucker P, Drexl A, Mohring R, Neumann K, Pesch E (1999) Resource-constrained project scheduling: notation, classification, models, and methods. Eur J Oper Res 112(1):3–41

Burgess AR, Kilebrew JR (1962) Variation in activity level on a cyclical arrow diagram. J Ind Eng 13:76–83

Campbell HG, Dudek RA, Smith ML (1970) A heuristic algorithm for the n-job, m-machine scheduling problem. Manage Sci 16(10):630–637

Framinan JM, Gupta JND, Leisten R (2004) A review and classification of heuristics for permutation flow-shop scheduling with makespan objective. J Oper Res Soc 55(12):1243–1255

Goyal SK, Mehta K, Kodali R, Deshmukh SG (1995) Simulation for analysis of scheduling rules for a flexible manufacturing system. Integr Manuf Syst 6(5):21–26

Gupta JND, Stafford EF Jr (2006) Flowshop research after five decades. Eur J Oper Res 169(3):699–711

Herroelen W, De Reyck B, Demeulemeester E (1998) Resource-constrained project scheduling: a survey of recent developments. Comput Oper Res 25(4):279–302

Johnson DS (1954) Optimal two- and three-stage production schedules with set-up times included. Naval Res Logist Q 1:61–68

King JR, Spachis AS (1980) Heuristics for flow-shop scheduling. Int J Prod Res 18(3):345–357

Krajewski LJ, Ritzman LP (1999) Master production scheduling: operations management, 5th edn. Addison, Wesley, p 725

Kumar VKA, Rajendran CR (1993) Manpower resource levelling in the maintenance of the drill shop computer simulation approach. Int J Model Simul 13:152–155

Li W (2006) Adaptive production scheduling and control in one-of-a-kind production. Thesis (M.Sc.), University of Calgary, Canada

Li W, Luo XG, Tu YL, Xue D (2007) Adaptive production scheduling for one-of-a-kind production with mass customization. Trans N Am Manuf Res Inst SME 35:41–48

Li W, Chang G, Nault RN, Tu YL (2011a) Variation in processing times in one-of-a-kind production: case studies using state space heuristic. In: Proceedings of 39th North American manufacturing research conference, June 13–17, 2011, Corvallis, Oregon, USA, CD-ROM

Li W, Nault RN, Xue D, Tu YL (2011b) An efficient heuristic for adaptive production scheduling and control in one-of-a-kind production. Comput Oper Res 38:267–276

Linn R, Zhang W (1999) Hybrid flowshop scheduling: a survey. Comput Ind Eng 37(1):57–61

Nawaz M, Enscore EE Jr, Ham I (1983) A heuristic algorithm for the m-machine, n-job flow-shop sequencing problem. OMEGA Int J Manage Sci 11(1):91–95

Ovacik IM, Uzsoy R (1997) Decomposition methods for complex factory scheduling problems. Kluwer Academic Publishers, London

Park SC, Raman N, Shaw MJ (1997) Adaptive scheduling in dynamic flexible manufacturing systems: a dynamic rule selection approach. IEEE Trans Robot Autom 13(4):486–502

Pinedo M (2002) Scheduling theory, algorithms and systems. Prentice Hall, Englewood Cliffs

Raja K, Kumanan S (2007) Resource levelling using Petrinet and memetic approach. Am J Appl Sci 4(5):317–322

Ruiz R, Maroto C (2005) A comprehensive review and evaluation of permutation flowshop heuristics. Eur J Oper Res 165(2):479–494

Taillard E (1993) Benchmarks for basic scheduling problems. Eur J Oper Res 64(2):278–285

Thornton HW, Hunsucker JL (2004) A new heuristic for minimal makespan in flow shops with multiple processors and no intermediate storage. Eur J Oper Res 152(1):96–114

Tu YL (1996a) A framework for production planning and control in a virtual OKP company. Technical papers of North American Manufacturing Research Institution of SME, pp 121–126

Tu YL (1996b) Automatic scheduling and control of a ship welding assembly line. Comput Ind 29(3):169–177

Vergara HA, Kim DS (2009) A new method for the placement of buffers in serial production lines. Int J Prod Res 47(16):4437–4456

Wang H (2005) Flexible flow shop scheduling: optimum, heuristic and artificial intelligence solutions. Expert Syst 22(2):78–85

Wortmann JC, Muntslag DR, Timmermans PJM (1997) Customer-driven manufacturing. Chapman & Hall, London

Chapter 8
One-of-a-Kind Production Management and Control Software System©

In Canada, most manufacturing companies are small or medium sized enterprises (SMEs) that produce one-of-a-kind products. They are typical one-of-a-kind production (OKP) companies. As discussed in Sect. 3.4.1, the logic error of separating bill of material (BOM) and bill of operation (BOO) in current MRP II (manufacturing resource planning) or ERP (enterprise resource planning) systems makes these systems difficultly implemented in these SMEs. These SMEs commonly believe that both MRP II and ERP systems have very little benefit to them. Therefore, they are implementing Lean production techniques to manage and control their production, and aim to improve the production efficiency and hence the competitive ability against keen competitions from overseas.

Lean production has been developed from JIT (just-in-time) production philosophy. It mainly focuses on reducing wastes resulted from various inventories (e.g. work-in-process inventory, pipeline inventory, etc.) and idle/waiting times (or none value added processes) in production. A Lean production system is normally characterized by a pull system with one piece flow technique. However, most Lean production systems have been implemented mainly by managerial methods and are lack of technology support, particular the computer aided production management and control technology. Kanban (a visual managerial method) and one piece flow technique are commonly used techniques in Lean production systems to control the production and inventory. As indicated by Krajewski et al. (2007), the Lean production system is normally applied in large batch or mass production modes. When customer orders are diversified, it is difficult to use Lean production principles and techniques to balance a production line and hence it is hard to achieve a high production efficiency. Therefore, a computer aided production management and control technology or software system is needed for those SMEs or OKP companies.

Mass customization is understood as an idea to achieve both customization and scale of economy (Pine 1999). To realize the goal of mass customization, a lot of manufacturing companies, particularly those SMEs or OKP companies in Canada, are trying to make customized products at the nearly efficiency of large batch or

Y. Tu and P. Dean, *One-of-a-Kind Production*,
DOI: 10.1007/978-1-84996-531-6_8, © Springer-Verlag London Limited 2011

even mass production within a short delivery time. This also demands a computer aided production management and control software system to manage and control the frequently changeable data flows in the production of customized products and the supply chain.

Through the research and development (R&D) in the past nearly two decades, we have developed such a software system, viz. OKP Management and Control Software system© (OKPMCS system) which copyright has been registered with Canadian Intellectual Property Office. This system has been mentioned in the previous chapters. In this chapter, we will give a brief description of the OKPMCS system. For the detailed description of this system, please refer to 'One-of-a-Kind Production Management and Control Software System—User Manual' which can be downloaded from our webpage: http://www.cimsys.org/. Please refer to Appendix A for guidelines on how to download the manuals and the OKP Management and Control Software system©.

8.1 Introduction

The OKPMCS system has been designed to support OKP companies to realize Lean production and mass customization. It has been developed based on the PPS (product production structure, see Sect. 3.4.1) which is an integrated data structure of bill of material (BOM) and bill of operation (BOO). By the PPS data structure, the design, process plan and work flow through a whole supply chain of a customer specifically ordered product can be jointly modelled, i.e. integrated. The OKPMCS system provides a graphic interface for users to easily create or modify the PPS. Therefore, the OKPMCS system can cope with frequent changes of product design, process planning and work flow in OKP. It also suggests OKP companies to manage and control the production of a customized product as a project so that the company can effectively fulfil the customer order at a low cost and within a short lead-time. By using the OKPMCS system and the relevant methods and concepts as presented in previous chapters, the OKP companies are moving toward to project-based production for achieving both high customization and high production efficiency, i.e. mass customization. If we view the management of OKP as the management of multiple projects, this software system can be easily applied in project management.

In addition to the novel concept of PPS, the OKPMCS system also has the following features which are best fit with the OKP management and control:

1. It has a PPS graphic modelling interface which can be employed to rapidly and flexibly model a production of an OKP product through the whole supply chain.
2. It employs the customer order oriented production constraints and capacity planning and control function to avoid resource conflicts or operation chaos in OKP shop floors and the supply chain.

3. It can be linked with a real-time monitoring system, e.g. barcode reading system or digital camera, to monitor production progress in a nearly real-time manner.
4. It has a PPS-based electronic Kanban that shows the real status of shop floor operations, subcontracting operations and outsourcing operations. By this electronic Kanban, the factory managers of OKP companies can easily view and monitor the progress of customer orders.
5. It has a project (or customer order) oriented cost estimating, tracing and controlling function.
6. It has the rapid quotation and lead-time estimate function.
7. It has the prototype-based incremental and concurrent product development function.
8. It has an electronic platform for automatic inquiry and bargain with subcontractors and suppliers on the specifically ordered parts.
9. It has the electronic Gantt chart (or eGantt chart) for production and resource planning under constraints and capacity limitations of resources and critical facilities in an OKP company.
10. It can be facilitated by two different databases according to the scale or size of an OKP company, i.e. Microsoft Access and SQL Server.
11. It can be easily integrated with the market available MRP II or ERP systems.
12. It has a bilingual function which can easily switch between English and Chinese languages.
13. It has a simple program structure and user friendly interface to let it be easily learned and installed in an OKP company.

Most of these features as listed above are unique and novel.

The OKPMCS system has four modules, namely PPSMaint, PPSManager, PPSShopFloor, and Web Application, which are described in the following sections.

8.2 PPSMaint Module

This is the main module of the OKPMCS system. The main functions of PPSMaint module are summarized below:

8.2.1 PPS Graphic Interface

Generally speaking, the production of an OKP product can be viewed as a material flow process through a whole supply chain, which includes various outsourcing operations (i.e. purchasing raw materials or components), subcontracting operations (i.e. subcontracting some parts or manufacturing processes to subcontractors), and shop floor operations (i.e. manufacturing component or parts, storing and

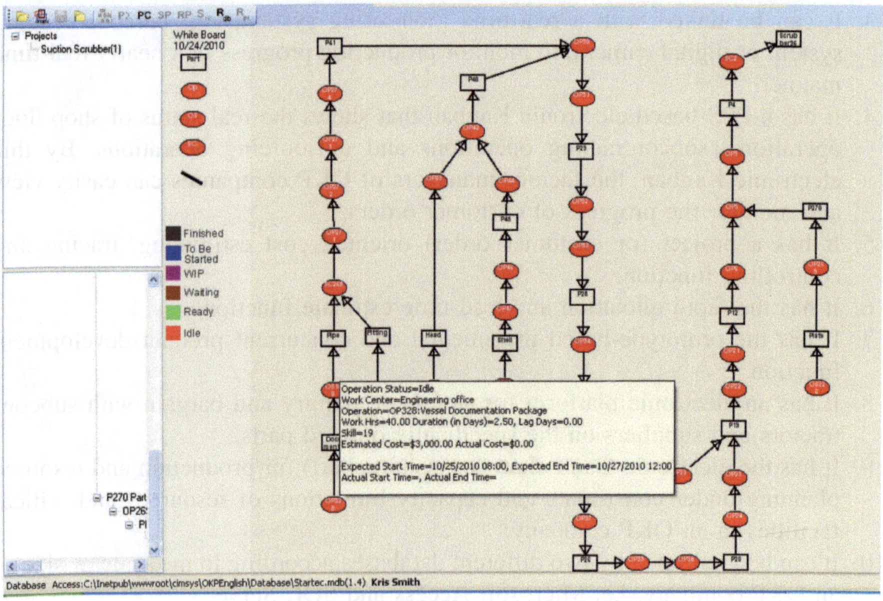

Fig. 8.1 The PPS of a customized high-pressure vessel

handling work-in-process parts or components, assembling the product and finally delivering the product to the customer). The delivering operation could be a subcontracting operation if the company would like to hire a delivery subcontractor to do so. Along with the material flow through this whole supply chain, the value will be added to the product and consequently the cost of the product is accumulated. The PPS data structure likes an integrated data media for describing this process. Figure 8.1 shows the PPS for the production of a customized high-pressure vessel in a manufacturing company in Canada.

As shown in Fig. 8.1, the OKPMCS system provides a graphic interface for users to easily create (or draw) and modify a PPS in terms of work-flow technique. A user can use a computer mouse to click a graph legend (i.e. Part, OP, SC, OS or Link) and drop it at anywhere on the screen or whiteboard to quickly draw a PPS to model a global work-flow of manufacturing an OKP product. The OKPMCS system will build the data relationships in the database automatically according to the PPS drawing. The colour of an oval and the pop-up window associated with an operation show the production progress and the detailed information of a selected operation. In this way, the OKPMCS system shows an electronic Kanban in the production.

The PPS records and displays the bill of material (BOM), bill of operation (BOO) and the work-flow of the production of a customized product. It also records the consumption of time, resource and hence the cost of the parts and operations. Through the achieving and restoring function (see Fig. 8.2) of the OKPMCS system, a user can create a PPS through modifying an old PPS in the

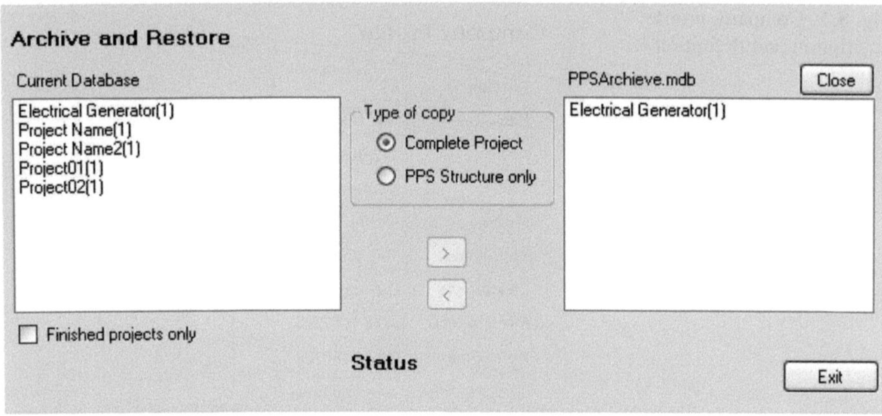

Fig. 8.2 Archiving and restoring function of the OKPMCS system

archiving database and store a PPS of a complete product in the archiving database. This achieving and restoring function is designed to facilitate the practice of the prototype-based incremental and concurrent product development method as described in Sects. 3.2 and 3.3. This function also provides an OKP company an effective way to management its knowledge of product design and production. Since the PPS is an integrated data structure for BOM, BOO and BOI (bill of information), the PPS graphic modelling interface can be also viewed as an integrated product data management (PDM) module.

8.2.2 Enterprise Production Data (EPD) Maintenance and Management

The enterprise production data (EPD) include the description and definition of company profile (see Fig. 8.3), users (including production manpower, outsourcing suppliers and subcontractors, see Fig. 8.4), work centres, worker skills (or trades), hourly rates of resources, work times (or shift times), and production calendar. These EPD are used for production management and control. The OKPMCS system provides an easily and simple user interface for users to manage and maintain the EPD in an OKP company (Please refer to the User Manual at www.cimsys.org).

8.2.3 Cost Tracing and Monitoring

As described in Chap. 4, the OKPMCS system supports the cost estimate methods, i.e. generative and variant cost estimate methods, and cost index structure (see Fig. 4.2). Also as demonstrated in Chap. 4 by a case study, the OKPMCS system

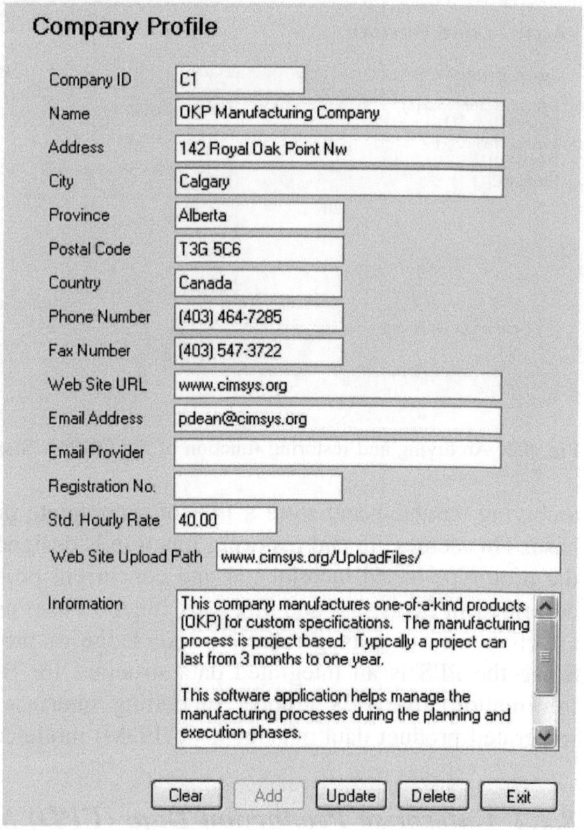

Fig. 8.3 Company profile description and definition

can automatically calculate: PB (Planned Budget), UPDE (up-to-date expenditure), CPI (cost performance index) and CV (cost variance). One example of these cost calculations for an on-going project is shown in Fig. 8.5.

Based on these cost tracing and calculation values, a project manager can properly control the budget and cost of the production. These cost data are also recorded in the PPS. When a project completes, the PPS will be stored in the archiving database. These recorded PPSs in the archiving database can be used as knowledge base for cost estimate and quick quotation for a new customer order.

8.2.4 Production Scheduling and Control

The OKPMCS system supports the adaptive production scheduling and control techniques as described in Chap. 7. By the SS heuristic, the sequence for producing different customer orders can be determined in order to minimize

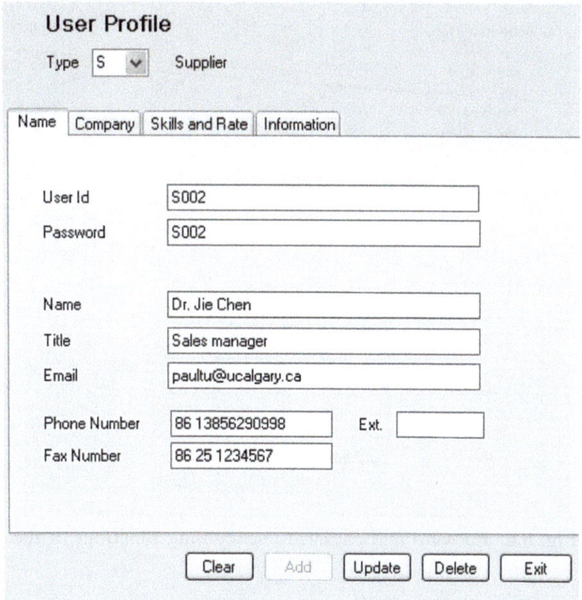

Fig. 8.4 Example of user definition of the OKPMCS system

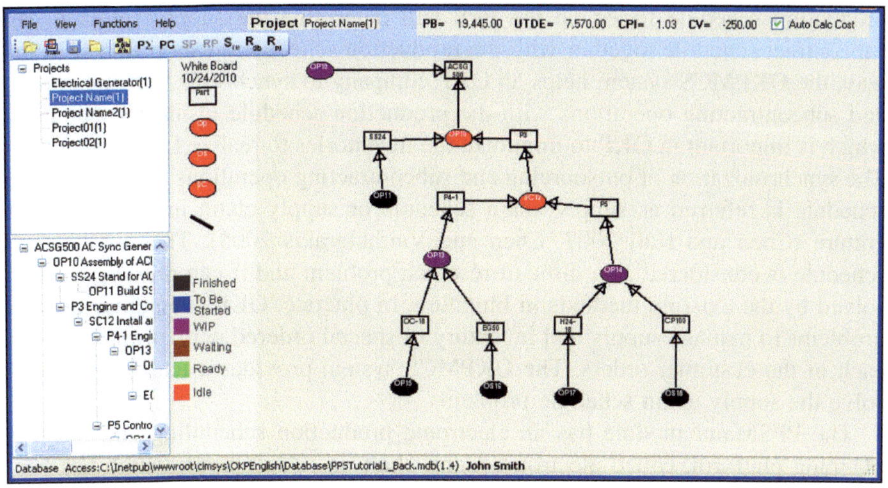

Fig. 8.5 An example of cost tracing and calculation of an on-going project

make-span. By the time estimates of operations recorded in the PPS and production calendar in the EPD, a product production can be scheduled by either Forward Scheduling or Backward Scheduling function (see Fig. 8.6) and hence the starting time and end time of each operation in the PPS can be determined automatically by the OKPMCS system.

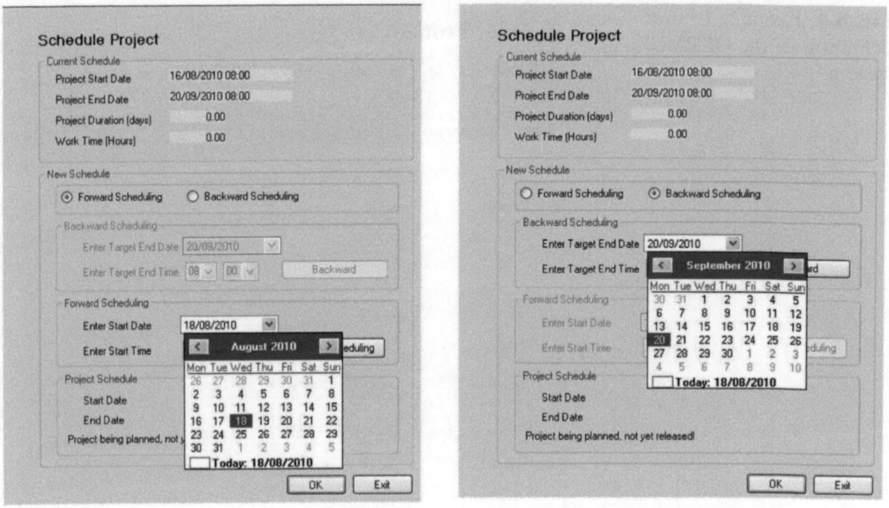

Fig. 8.6 Forward and backward scheduling functions of the OKPMCS system

Since the PPS models a whole material flow process through a supply chain, the production schedule based on the PPS will determine the supply schedule and subcontract schedule together with the production schedule on shop floors. In this way, the OKPMCS system helps an OKP company to synchronize its outsourcing and subcontracting operations with the production schedule inside the company, which is important in OKP to minimize the inventories to realize Lean production. The synchronization of outsourcing and subcontracting operations with production schedule is referred as supply chain schedule or supply chain integration in literature (Chen and Hall 2007, Chen and Vairaktarakis 2005). The supply chain schedule is considered as a difficult research problem and it can only be partially solved by the existing methods in literature. In practice, OKP companies do have problems to manage supply and inventory of special ordered parts or materials for each of the customer orders. The OKPMCS system provides a feasible means to solve the supply chain schedule problems.

The PPSMaint module has an electronic production scheduling and resource planning platform, called electronic Gantt chart or eGantt chart, which has been demonstrated in Chap. 7. With this platform, a user can schedule multi-projects (or products production) under consideration of resource constraints and/or limited capacities. The eGantt chart module can be also used as an electronic platform to facilitate the project planning and scheduling techniques as presented in Chap. 7, i.e. Burgess and Kilebrew resource levelling heuristic (Sect. 7.5.1), longest duration heuristic (Sect. 7.5.2), least slack heuristic (Sect. 7.5.2), and minimum cost project crash technique (Sect. 7.5.3). The eGantt chart module is a novel and unique feature of the OKPMCS system compared with existing production scheduling software systems.

8.2.5 On-line Production Progress Monitoring and Control

The PPS as shown on the PPS graphic interface of the PPSMaint module is also a real-time production progress monitor or indicator. As shown in Fig. 8.7, the colours of various operations (or ovals) indicate the different status of the operations, i.e. black means "Finished", blue means "To Be Started", purple means "WIP (work-in-process)", brown means "Waiting", and red means "Idle". To be further added by pop-up window function, e.g. as shown in Fig. 8.7 OP11 is displayed on the pop-up window, a production manager can easily view the production progress on-line.

Therefore, the PPS is viewed not only as an electronic Kanban, but also as a real-time monitor or indicator of the production progress of a customized product. With this monitor or indicator, the production manager has a full and updated picture of progress and status of outsourcing, subcontracting and shop floor operations. This function helps OKP companies to synchronize or integrate various in-house operations (i.e. shop floor operation as indicated by OP in the PPS as shown in Fig. 8.7) with outsourcing (as OS in the PPS) and subcontracting (as SC in the PPS) operations, which is important and critical for OKP companies to improve their production efficiency.

8.2.6 Miscellaneous Functions

The PPSMaint module also provides the following miscellaneous functions:

1. Project summary: When a user clicks the project summary button, the PPSMaint module will summarize the project on a display window as shown in Fig. 8.8.

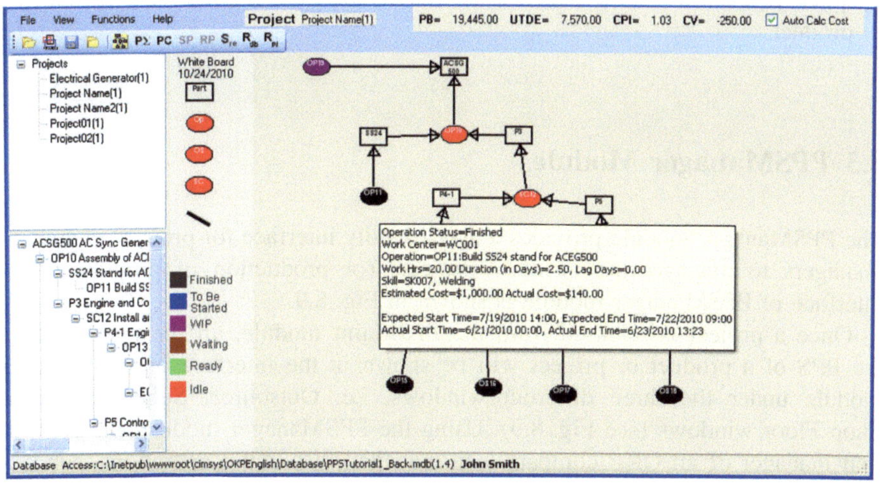

Fig. 8.7 PPS is used as a real-time production progress monitor or indicator

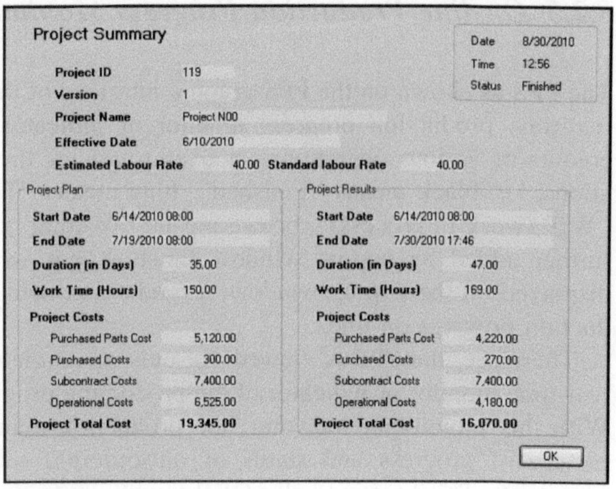

Fig. 8.8 Screen snapshot of a project summary

2. Importing Bill of Material (BOM): The PPSMaint module can import a BOM which is prepared in Microsoft Excel worksheet.
3. On-line language switch: The PPSMaint module supports both Chinese and English languages. These two languages can be switched on-line between each other.
4. Tree-view: The PPSMaint module can automatically generate a tree view of PPS (see lower left part of Fig. 8.7).
5. PPS saving: In addition to "Archive or Restore" function as described in Sect. 8.2.1, the PPSMaint module also provides a "Save as (copy)" function which can save a PPS under a different project name. By this function, a user can quickly generate a PPS which is similar to the PPS of a currently producing product.

8.3 PPSManager Module

The PPSManager module provides a user-friendly interface for project/production managers to manage one or more projects (or production of products). The interface of PPSManager module is shown in Fig. 8.9.

Once a project is released from the PPSMaint module, all the operations in the PPS of a product or project will be shown in the interface of PPSManager module under the three different windows, i.e. Outsource, Subcontract, and Shop Floor windows (see Fig. 8.9). Using the PPSManager module, the production manager of an OKP company can manage outsourcing operations, subcontracting operations and shop floor operations, which are briefly described as follows:

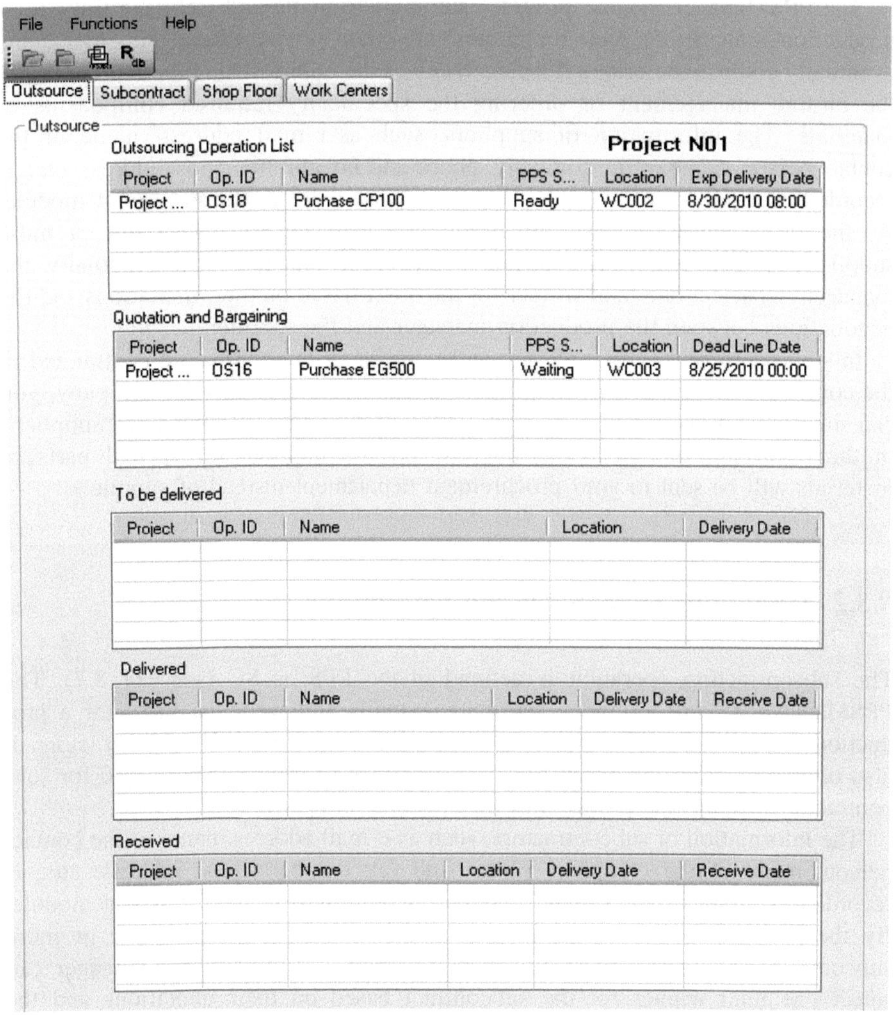

Fig. 8.9 The interface of PPSManager module

8.3.1 Managing Outsourcing Operations

The outsourcing operation, which is defined in the PPS as OS, is used for ordering special component or material that is specifically required by an individual customized product. For those commonly used components or materials, we would suggest users to use the other way to order and specify these common components or materials in the PPS as the items in stock. In this way, an OKP company can improve its material purchasing efficiency and reduce the ordering and purchasing costs of these common components and materials.

The PPSManager module provides automatic e-mails and webpage links for a production manager to seek for quotes, negotiate prices, chase the orders, and eventually confirm the orders if he/she receives the orders. This function facilitates the on-line management of ordering the specifically required components or materials. The information of suppliers, such as e-mail address, name of the contact person, name of the company, phone and fax numbers, post address, etc., is recorded as a part of EPD (enterprise production data) in the PPSMaint module. By the PPSManager module, the production manager can select one or more suppliers to deal with for ordering a specific component and eventually the manager can select one final winner for the order based on their quotations and the negotiations between the production manager and the suppliers.

In some companies, the ordering and purchasing operations are centralized in the company's procurement department. If this is the case in your company, you can simply define your company's procurement department as the sole "supplier" in the PPSMaint module. Consequently, all the inquires for special parts or materials will be sent to your procurement department instead of suppliers.

8.3.2 Managing Subcontracting Operations

The subcontracting operation is defined in the PPS as SC (see Fig. 8.7). The PPSManager module provides automatic e-mails and webpage links for a production manager to seek for quotes and negotiate prices with subcontractors. It also offers the Webpage-based reporting interface as shown in Fig. 8.10 for subcontractor to report the progress of the subcontracting work.

The information of subcontractors, such as e-mail address, name of the contact person, name of the company, phone and fax numbers, post address, etc., is recorded as a part of EPD (enterprise production data) in the PPSMaint module. By the PPSManager module, the production manager can select one or more subcontractors to get quotes and negotiate prices. Eventually the manager can select one final winner for the subcontract based on their quotations and the negotiations between the production manager and the subcontractors.

8.3.3 Managing Shop Floor Operations

The shop floor operation is defined in the PPS as OP. As defined in Sect. 3.4.1, 'OP' means any operations to be carried by an OKP company in house. It could be a manufacturing operation (e.g. machining, welding, cutting, assembly, surface processing, etc.), a material handling operation, delivery, or a storage operation. The PPSManager module provides a user friendly interface, i.e. Shop Floor window of the PPSManager module (see Fig. 8.9), to list all the shop floor operations under three status, i.e. Ready, WIP (work-in-process) and Finished. The

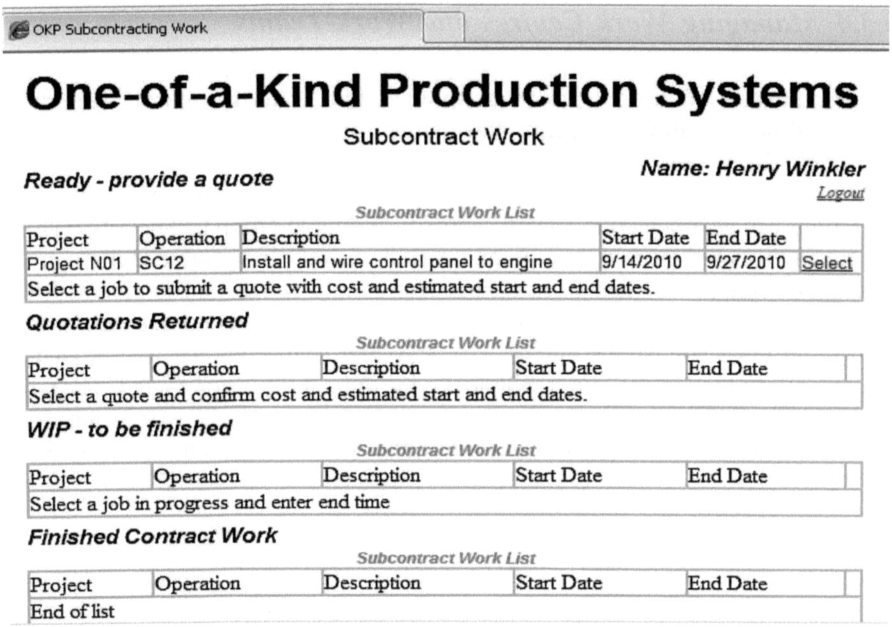

Fig. 8.10 Webpage-based reporting interface for subcontractors

production manager can use this interface (or window) to assign workers or work-teams to Ready operations, monitor the progress of WIP operations and finally record the Finished operations as references for deciding payments or incentives to workers or work-teams. It should note that the shop floor operations have no status, like "To Be Started" or "Waiting", which is applied to outsourcing and sub-contracting operations only. The shop floor operations with "Idle" status will not be listed in the Shop Floor window of PPSManager module until these Idle operations become Ready. This feature has the following advantages in OKP shop floor control:

1. It will let the production manager focus on Ready operations and avoid assigning a worker or work-team to a shop floor operation on "Idle" status that is not ready to be carried out due to the precedence relationship between the operations in the PPS. This will help reducing idle or waiting time of workers in an OKP shop floor.
2. All the Ready and WIP operations give a real-time total demand of production resources at present time to the production manager. Based on this total demand, the production manager can properly assign production resources to those critical operations.
3. When the production manager assigns workers or work-team to an operation, he/she will change this operation status from Ready to WIP immediately regardless its scheduled start time. This facilitates the on-line production scheduling which is highly demanded by OKP shop floor control.

8.3.4 Managing Work Centres and Work-Teams

OKP companies often organize or group operations in terms of work centres so that a real or logic flow production line can be formulated to improve production efficiency. In a real production line, the work centres are physical places, machines, or work places in which a group of operations can be carried out. The operations can be assigned to a work centre according to group technology (GT) or a company's experience. Then a product will be flowed and produced through the first work centre to the last work centre. Normally a work-team is assigned to a work centre to carry out the operations. In a logic flow production line that is often used in an OKP company to produce a large and heavy product, the product is actually located in a place and it will not be moved until it is completed. The work centres in a logic flow line are actually work-teams. These work-teams will sequentially work on the product according to a production plan or schedule. To

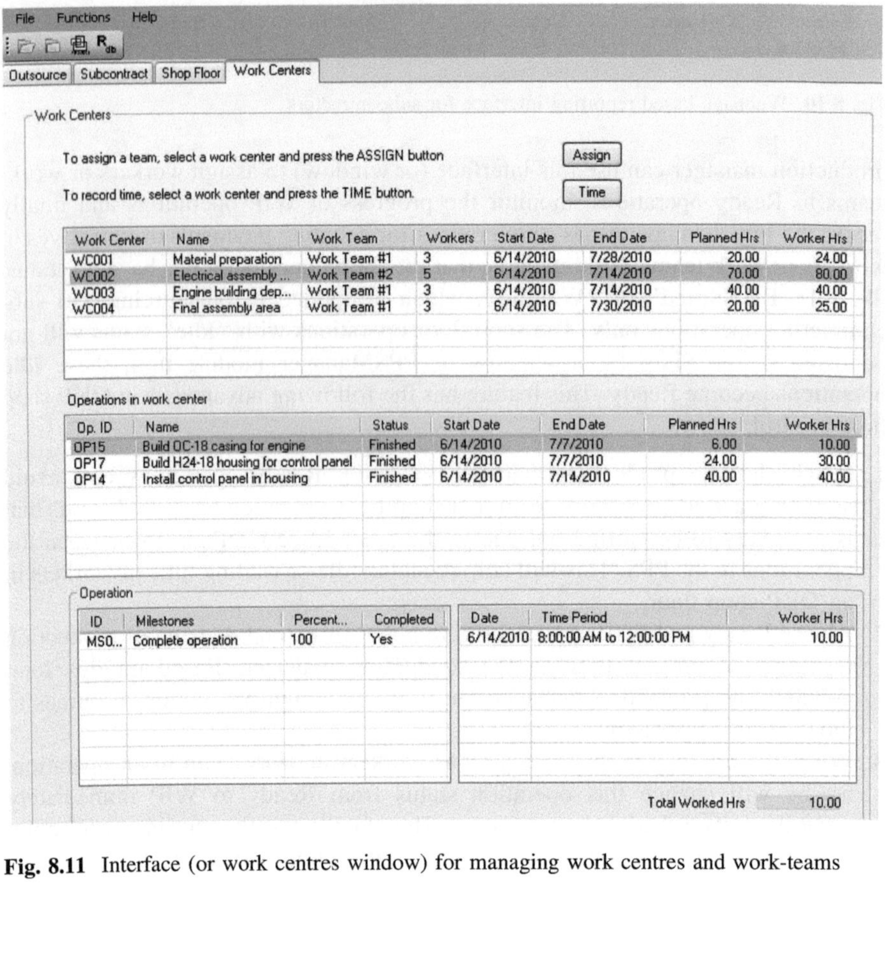

Fig. 8.11 Interface (or work centres window) for managing work centres and work-teams

facilitate this production management method in OKP companies, the PPSManager module has an interface (or work centres window) as shown in Fig. 8.11.

Every time, work centres window shows only a selected project. Using the interface, a production manager can assign work-teams to different work centres. The operations in each work centre are assigned when the PPS of a product or project is created. The production routine is defined in the PPS. With this interface, the production manager can also report the actual time of a work-team to work on a specific operation.

To facilitate the production management of a real or logic flow production line, the work centre and operation have been defined in the PPSManager module as follows:

A *work centre* is defined as a group of operations that produces either a complete product or a component/part of a product. Usually either the same person or same work-team works in this work centre to produce or assemble a component/part of a product or the final product. The operations in a work centre have a pre-defined sequence of performance in order to produce the product. Each operation has a known or calculable cycle or production time and therefore can be scheduled during the planning phase.

An *operation* is usually comprised of one or more tasks. Each task has a known or calculable production time. The sum of these task times equals the time of the operation time. The tasks can have a pre-define sequence or the sequence can be left to the determination of the worker or work-team in the work centre.

In the PPSManager module, a worker is assigned to a single operation as they become available to work on. A team of workers is assigned to all the operations in a work centre and they work on the operations in a scheduled sequence.

8.4 PPSShopFloor Module

This module is designed for shop floor workers or work-teams to check the operations that are available to be assigned by themselves, have been assigned by the production manager, being carried out by them (i.e. WIP) or have been completed by them. Figure 8.12 shows a screen snapshot when Work-team #1 logins into the PPSShopFloor module.

The operations under "Available Operations" window are the operations that Work-team #1 can select and assign to them. The operations under "WIP Operations" window are the ones on which Work-team #1 are working. These operations may be either selected by Work-team #1 or assigned by the production manager. This means that once an operation is selected or assigned to Work-team #1, the PPSShopFloor module will move it to the "WIP Operations" window. The operations under "Completed Operations" window are the operations which have been completed by Work-team #1. Since in the PPS each operation has been specified to consist of a number of tasks and milestones, Work-team #1 can click on the WIP operations to see these pre-specified tasks and milestones. Once a milestone is achieved, they can check the "Completion" box after the milestone. When all the milestones of an operation are achieved, the PPSShopFloor module will move the operation to the "Completed Operations" window. The progress and status of the operations are immediately shown on the PPS in the PPSMaint module, i.e. the colours of ovals in the PPS immediately change. The production manager can view the progress of a project or production immediately. Consequently, the shop floor production is monitored and controlled in time.

In addition to the real-time shop floor monitoring and control, the PPSShopFloor module also has the following functions:

1. The interface of the PPSShopFloor module as shown in Fig. 8.12 can be used as a time sheet for a worker or work-team in OKP companies.
2. The interface as shown in Fig. 8.12 is in fact an electronic Kanban in a shop floor. By this electronic Kanban, the shop floor production management is visible. This is helpful to reduce WIP inventory, smooth production flow in the shop floor, and hence avoid production chaos which often occur in OKP shop floors.
3. If the company reward system is linked with the recorded completion operations in the database, it will motivate workers on the shop floors to actively improve the production efficiency.

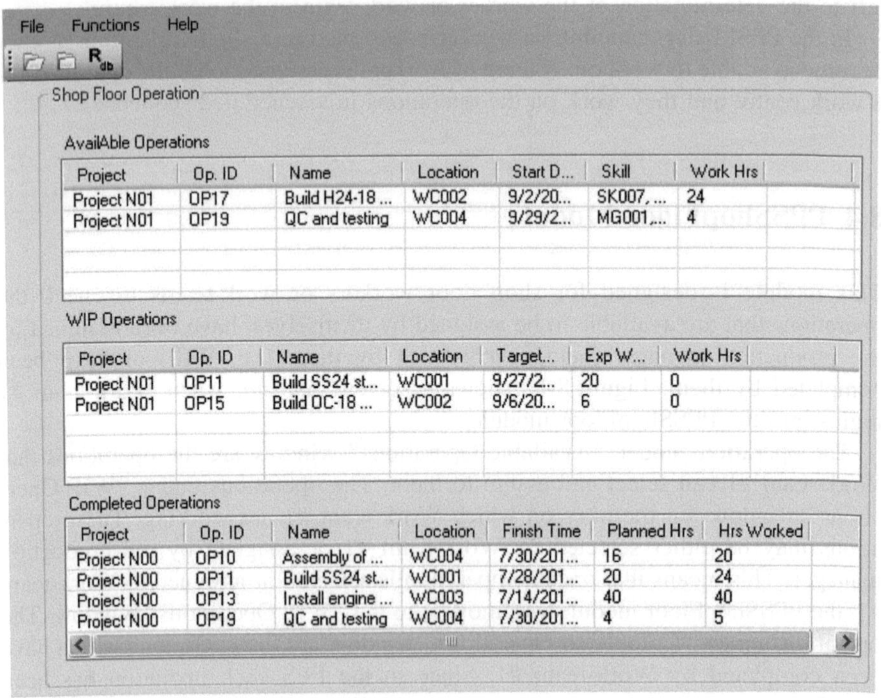

Fig. 8.12 Interface of PPSShopFloor module

8.5 Managing Complicated Projects or Products

Some OKP companies produce complicated products, e.g. steel ships, airplanes, electrical generators, customized high pressure vessels, etc., which result in a large and complicated PPS for an overall product production. If a user is trying to create such a complicated PPS using the PPSMaint module, it can be done, but it will be awkward and the PPS is too big to be viewed on one screen display.

For a complicated project or production of a complicated product, a hierarchical decomposition of the overall PPS into a number of simple sub-PPSs is suggested. In fact, Fig. 3.3 in Sect. 3.4.1 shows a ship building production. On the bottom level as shown in Fig. 3.3, the PPSs are production projects for making web-panels. The PPSs on intermediate levels in Fig. 3.3 are production projects for making different sizes of blocks. On the top level, the PPS is the project for assembling blocks into the ship at the yard area. Figure 3.3 shows a typical ship building process in a shipyard. Following this production decomposition in a shipyard, the hierarchical decomposition of a PPS is illustrated by a case as shown in Fig. 8.13.

From Fig. 8.13, we can see that by a coding system it is clearly show the hierarchical relationship between a PPS on the top level and the sub-PPSs on low levels. Sub-PPS "Project N01 Part 21" is Part 21 and OP21 on PPS "Project N01". Sub-PPS "Project N01 Part 21–31" is Part 21–31 and OP21–31 on sub-PPS "Project N01 Part 21". These two sub-PPSs are jointly shown in Fig. 8.14.

Obviously, PPS "Project N01" has been decomposed into three hierarchical levels. PPS "Project N01" is on the top level, sub-PPS "Project N01 Part 21" is on the second level, and sub-PPS "Project Part 21–31" is on the third (or bottom)

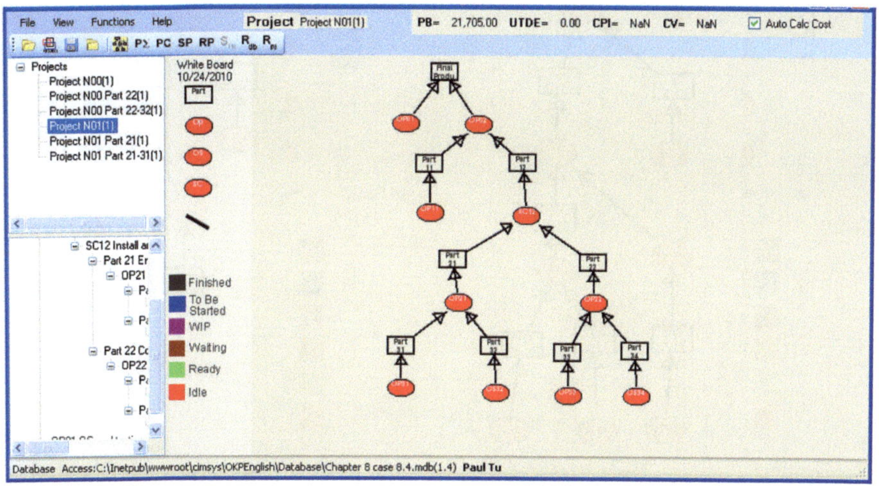

Fig. 8.13 The PPS "Project N01" consists of two sub-PPSs, Project N01 Part 21 and Project N01 Part 21–3

level. For the production of a more complicated product, further decomposition levels can be hierarchically added to the bottom of the hierarchy, and consequently more sub-PPSs can be decomposed on these levels. Even if in this case, more sub-PPSs can be decomposed on the middle level and bottom level. To simplify the description, we only decompose one sub-PPS on each of the hierarchical levels. By a proper coding system, a user should be easy to identify a PPS and its sub-PPSs on different hierarchical levels. The "Tree View" function of the system also helps users to identify the hierarchical structure of the PPS and its sub-PPSs.

However, the OKP management and control software (OKPMCS) system treats the PPS and its sub-PPSs as individual projects and it will not automatically establish links or relationships between a PPS and its sub-PPSs. Users should maintain the relationships between the PPS and its sub-PPSs by the ways which are suitable and convenient for them, e.g., a coding system, a flowchart to record the overall production (or project) plan that includes the PPS and its decomposed sub-PPSs at the different production (or project) stages in the different shop floors or work areas of an OKP company. The following concerns let us not design the OKPMCS system to automatically build the links or relationships between a PPS and its sub-PPSs.

1. Keep the production flexibility in the user's company. In some OKP companies, they would like to pool all the common parts to be produced at a shop floor or

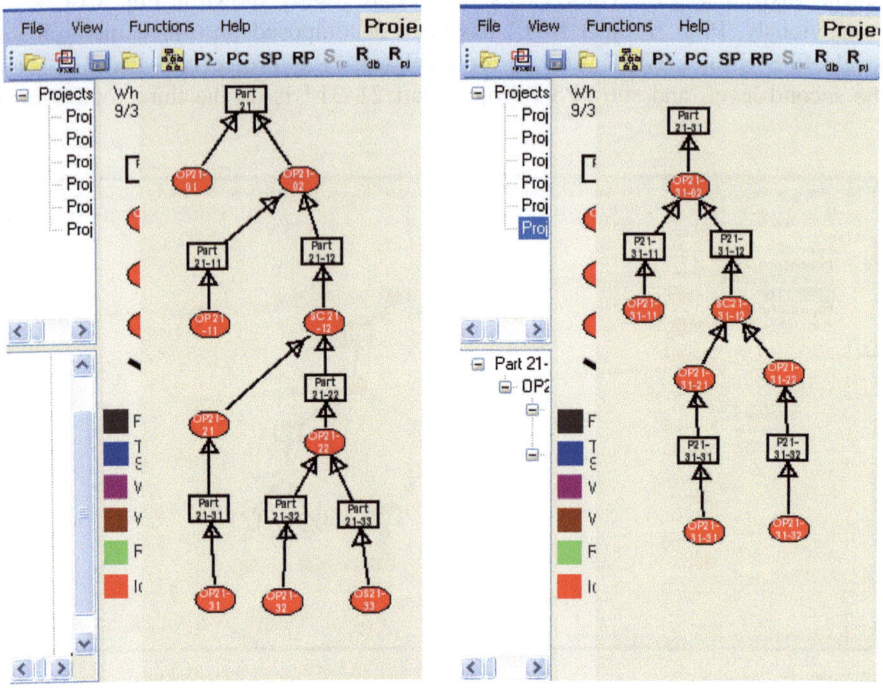

Fig. 8.14 Sub-PPS "Project N01 Part 21" and sub-PPS "Project Part 21–31"

by a production line to achieve a local mass production in order to gain high production efficiency. This practice means that the companies will set up the starting date and end date of producing these common parts separately from the overall production of the final products (or the overall project).
2. Keep the software system structure simple and clear. This is important for implementing the OKPMCS system in manufacturing companies. It overcomes one of the major drawbacks of existing ERP systems which often have a complicated software system structure and operating system which result in a long learning period.
3. Change the structure of a PPS easily. Since there is no automatic links between a PPS and its sub-PPSs in the OKPMCS system, a change made on any part of the PPS or its sub-PPSs will not automatically transferred to the other parts of the PPS and its sub-PPSs until the user makes the necessary change (or changes) to the other parts of the PPS or its sub-PPSs. Therefore a production manager can freely change a sub-PPS in his shop floor according to his needs or production changes. This separation or autonomous feature is critical for OKP shop floor control since frequent changes need to be made in OKP due to the customization and unexpected uncertainties. If the automatic links exist, any change in a sub-PPS will be automatically transferred to the other sub-PPSs which may be used to control the production in different shop floors. These frequent automatic updating will impact the stability of the production systems in these shop floors.

To schedule the production of a complicated product, which PPS has sub-PPSs, the following methods are suggested, i.e. (1) forward scheduling—bottom-up scheduling, and (2) backward scheduling—top-down scheduling.

8.5.1 Forward Scheduling—Bottom-Up Scheduling

This method will use the "Forward Scheduling" function in the PPSMaint module as shown in Fig. 8.15.

This scheduling method is very straight forward. It is described by the following steps:

Step 1: Start scheduling the sub-PPSs on the bottom level by a desired start time T_0 (e.g. T_0 = September 6, 8:00 as shown in Fig. 8.15) of the production. Using the "Forward Scheduling" function in the PPSMaint module, a user can get the completion time of each of the sub-PPSs on the bottom level, e.g. T_{21-31} = September 14, 2010, 9:00 as shown in Fig. 8.15. By subtracting start time (T_0) from end time (T_{21-31}), the user can get the total time. According to Fig. 8.15, for example, Total time = 9/14/2010 9:00 − 9/6/2010 8:00 = 6 working days × 8 h per working day + 9:00–8:00 = 49 h. It is worth to note that this total time may be different from the 'Work Time' in project summary, e.g. 55 h as shown in Fig. 8.16, due to the concurrently carrying out some operations during the production process.

Step 2: Input this total time into the PPS or a sub-PPS on the higher level, e.g. input 49 h into OP21–31 of Project N01 Part 21 as shown in Fig. 8.17, and schedule the PPS or sub-PPSs on the higher level using "Forward Scheduling" function in the PPSMaint module by setting up start date as T_0.

Figure 8.17 also shows that the resource cost (i.e. 4000 + 2675 = $6675 as shown in Fig. 8.16) of Project N01 Part 21–31 has been added into OP21–31 in the sub-PPS of Project N01 Part 21. The material cost (i.e. $620 as shown in Fig. 8.16) is added into Part 21 in the sub-PPS of Project N01 Part 21. The display of Part 21 of Project N01 Part 21 omits here.

Step 3: Repeat Steps 1 and 2 until to the PPS on the top level.

Step 4: Check the PPS or sub-PPSs from the top level down to the bottom level, find the start date for each of the sub-PPS on the lower level, and re-schedule the sub-PPSs on the lower level using these start dates.

Step 5: Repeat Step 4 until to all the sub-PPSs on the bottom level.

Through the five steps as described above, the overall production of a complicated product can be scheduled. Finally, using eGantt Chart, the user can check the production schedule or adjust the production schedule under consideration of resource constraints and production capacity limitations.

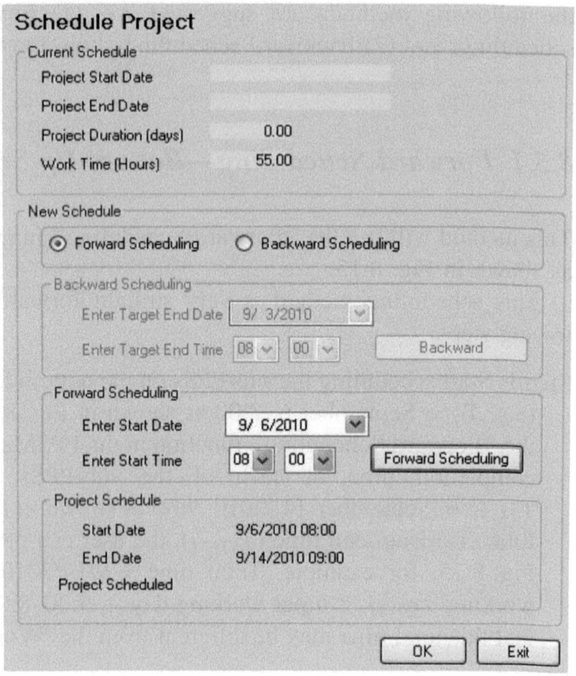

Fig. 8.15 Scheduling production of Part 21–31

8.5.2 Backward Scheduling—Top-Down Scheduling

This method will help an OKP company to schedule the production of a complicated product by a desired completion time T_c. If we assume that the desired completion time of Project N01 as shown in Fig. 8.13 is October 18, 2010, at 11:00 am, i.e. $T_c = 10/18/2010$ 11:00. This method is described as follows by using the case as shown in Fig. 8.13:

Step 1: Use Steps 1 and 2 as described in Sect. 8.5.1 to estimate the total times for all the decomposing operations, e.g. OP21–31 = 49 h and OP21 = 99 h.

Step 2: Set the desired production completion time (e.g. $T_c = 10/18/2010$ 11:00) and backward schedule the PPS on the top level. Read the completion times for the decomposing operations (e.g. OP21, $T_{c21} = 9/30/2010$ 11:00 as shown in Fig. 8.18).

Step 3: Use the completion times of decomposing operations and backward schedule the sub-PPSs of the decomposing operations on a lower level. Repeat this process down to the sub-PPSs on the bottom level. For example, use $T_{c21} = 9/30/2010$ 11:00 to backward schedule "Project N01 Part 21" on second level, and $T_{c21-31} = 9/22/2010$ 9:00 to backward schedule "Project N01 Part 21–31" on the bottom level (or third level).

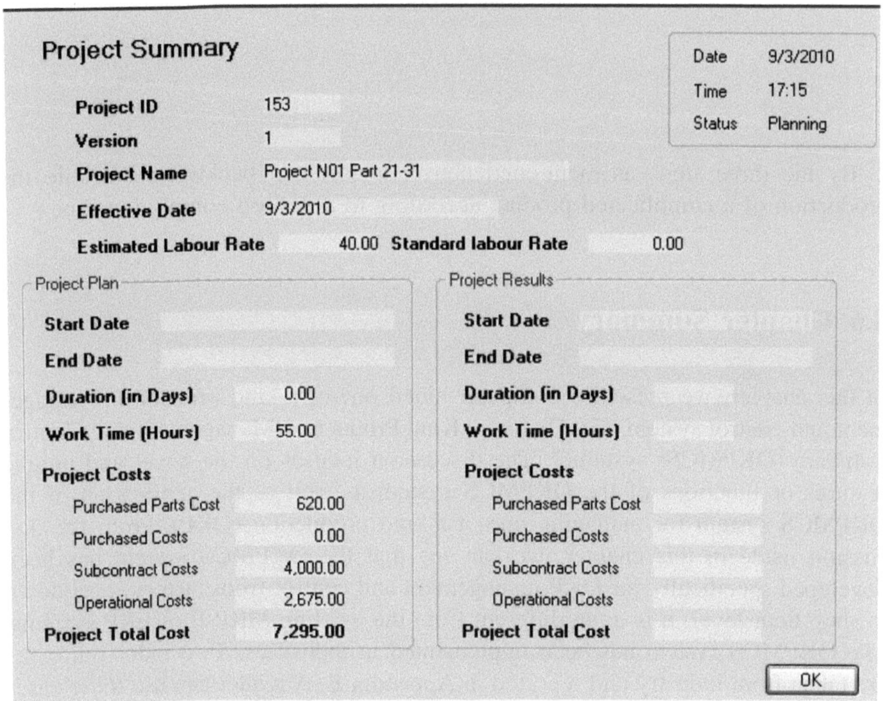

Fig. 8.16 Project summary of Project N01 Part 21–31

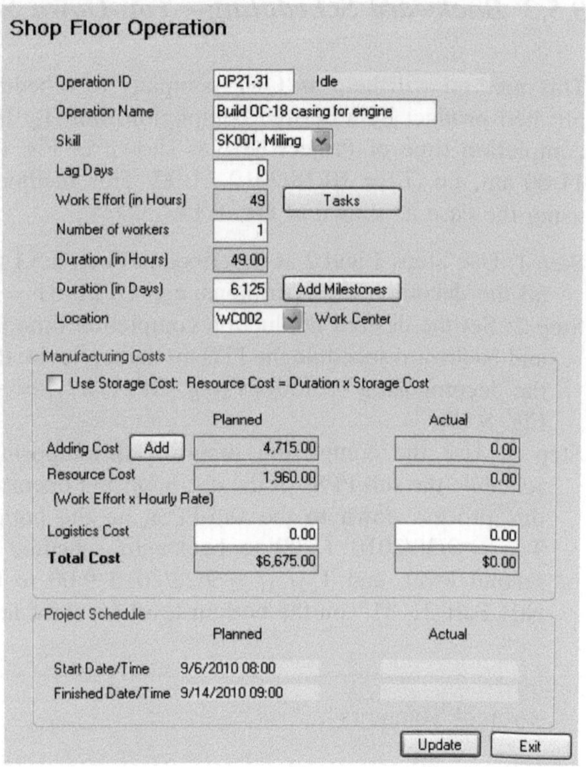

Fig. 8.17 Inputting the total time of 49 h into OP21–31 on the PPS of Project Part 21

By the three steps as mentioned above, a user can backward schedule the production of a complicated product according to a desired completion time.

8.6 Chapter Summary

In this chapter, we present a computer aided one-of-a-kind production management and control system, viz. One-of-a-Kind Production Management and Control Software (OKPMCS) system©. The discussion focuses on the novel and unique features or functions of the OKPMCS system as well as the application of the OKPMCS system for managing one-of-a-kind production (OKP). From the discussion made in this chapter, we can see that the OKPMCS system has been developed specifically for OKP management and control from factory level down to shop floor level. It is quite different from the existing MRP II or ERP systems. The OKPMCS system has been implemented in industries. Two industrial cases are taken from industry and reported in Appendix E. A reader can use these cases to learn how to use the OKPMCS system to manage and control one-of-a-kind production.

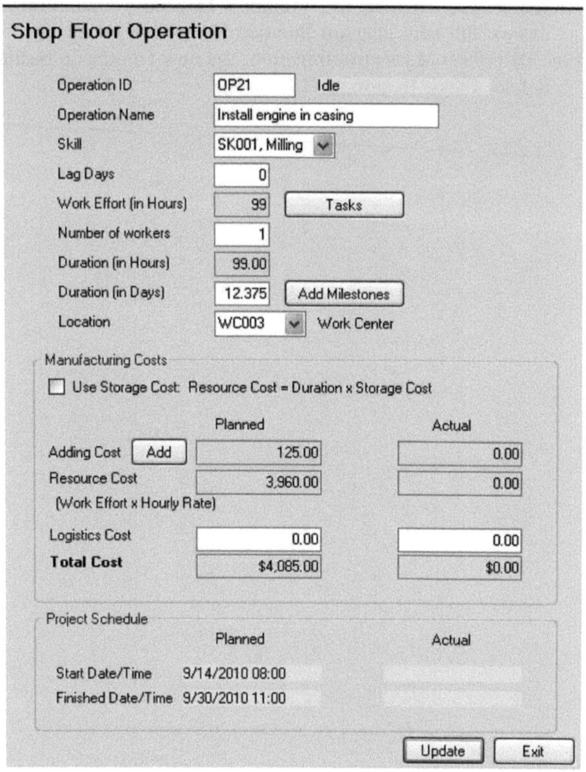

Fig. 8.18 Display window of OP21 of "Project N01" in backward scheduling

In this chapter, we focus our discussions on the novel features or functions of the OKPMCS system as well as the suitability, feasibility and effectiveness of applying the OKPMCS system to solve the production management and control problems in OKP. Therefore, the functionalities of the OKPMCS system are not fully exposed in this chapter. Following the guidelines in Appendix A, users can refer to the user manual of the OKPMCS system from our webpage: www.cimsys.org. From the same webpage, users can also download the installation manual of the OKPMCS system. The installation manual describes the detailed requirements of a computer system and its software and operating system configuration, as well as the procedure to install the OKPMCS system.

References

Chen ZL, Hall NG (2007) Supply chain scheduling: conflict and cooperation in assembly systems. Oper Res 55(6):1072–1089

Chen ZL, Vairaktarakis GL (2005) Integrated scheduling of production and distribution operations. Manage Sci 51:614–628

Krajewski LJ, Ritzman LP, Malhotra M (2007) Operations management: processes and value chains, 8th edn. Pearson Prentice Hall, New Jersey
Pine BJ (1999) Mass customization: the new frontier in business competition. Harvard Business School Press, Boston

Appendix A
Accessing the CIM Web Site to Download the OKP Management and Control Software System©

When you buy this book, you are entitled to free access to the CIM web site and freely use the One-of-a-Kind Production Management and Control Software (OKPMCS) System© for a year for either your business or private learning. The user manual and installation manual of the OKPMCS system are available on the CIM web site. Use the following steps to download the OKPMCS system and these manuals from the CIM web site.

Step 1: Use the following link to login the CIM web site:
URL: http://www.cimsys.org/Login/Login.asp

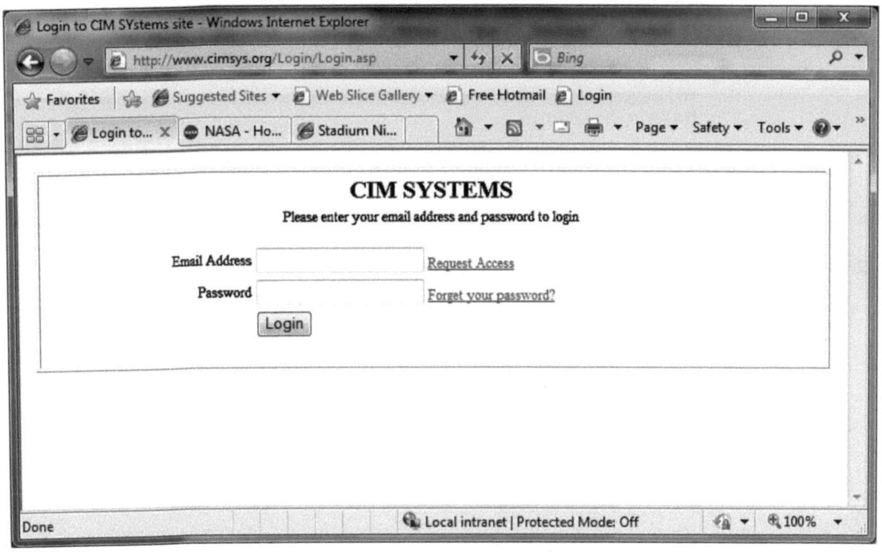

Fig. A.1 CIM systems login screen

Y. Tu and P. Dean, *One-of-a-Kind Production*,
DOI: 10.1007/978-1-84996-531-6, © Springer-Verlag London Limited 2011

Step 2: Press the Request Access link. Complete the form as shown in Fig. A.2 using your information and the Access Code of ***OKP2011***. You will receive an e-mail confirming your access using the e-mail address and password you entered in the form above.

Step 3: Return to the Login page and enter your e-mail address and password and then press the *Submit* button. You will see a web page similar to the one shown in Fig. A.3.

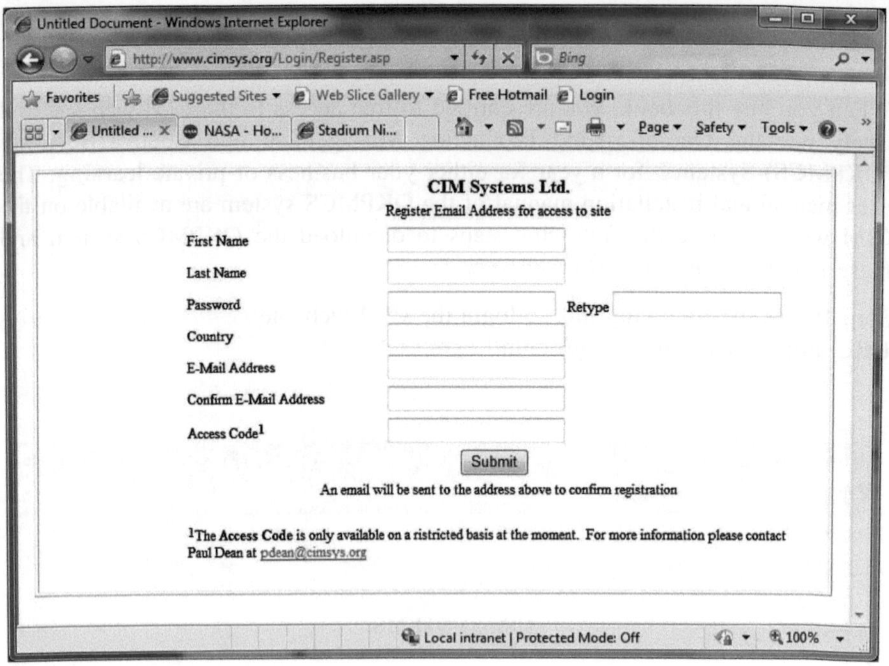

Fig. A.2 Request access to CIM system download site

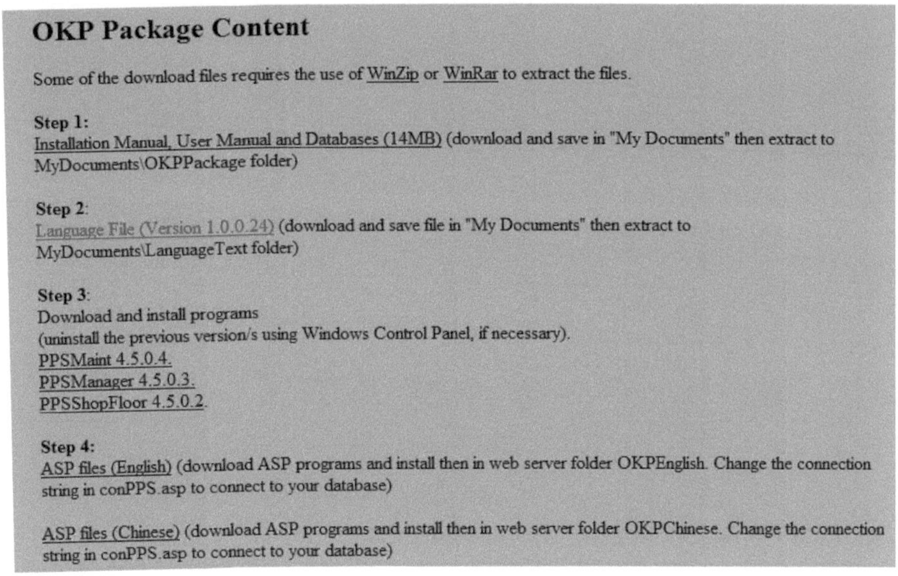

OKP Package Content

Some of the download files requires the use of WinZip or WinRar to extract the files.

Step 1:
Installation Manual, User Manual and Databases (14MB) (download and save in "My Documents" then extract to MyDocuments\OKPPackage folder)

Step 2:
Language File (Version 1.0.0.24) (download and save file in "My Documents" then extract to MyDocuments\LanguageText folder)

Step 3:
Download and install programs
(uninstall the previous version/s using Windows Control Panel, if necessary).
PPSMaint 4.5.0.4.
PPSManager 4.5.0.3.
PPSShopFloor 4.5.0.2.

Step 4:
ASP files (English) (download ASP programs and install then in web server folder OKPEnglish. Change the connection string in conPPS.asp to connect to your database)

ASP files (Chinese) (download ASP programs and install then in web server folder OKPChinese. Change the connection string in conPPS.asp to connect to your database)

Fig. A.3 Download screen

Appendix B
Relational Database Specification

This definition is a simplified version or sub-set of the full definition of the OKPMC software system. The purpose of the definition is to explain the use of a database in the real-time process algebra (RTPA) definition of the framework. A relational database is used for storage and retrieval of data. The database contains one or more tables. A table is made up of rows and column sets.

Table B.1 Table matrix of row and columns

Row	Column 1	Column 2	Column 3	…	Column N
Row 1	Value $_{1,\ 1}$	Value $_{1,\ 2}$	Value $_{1,\ 3}$		Value $_{1,\ n}$
Row 2	Value $_{2,\ 1}$	Value $_{2,\ 2}$	Value $_{2,\ 3}$		Value $_{2,\ n}$
Row 3	Value $_{2,\ 1}$	Value $_{3,\ 2}$	Value $_{3,\ 3}$		Value $_{3,\ n}$
……..					
Row M	Value $_{m,\ 1}$	Value $_{m,\ 2}$	Value $_{m,\ 3}$		Value $_{m,\ n}$

A table is an M $\times$ N matrix of data values. Tables can have relationships based on a connection between one or more of the same columns. For example, a table with a set of customers can be related to a table of orders by having the customer account number in both tables. As a result, a list of orders for a specific customer can easily be found. Table rows can be accessed directly through their key values, which are stored in one or more columns. In the example above, the customer account number column is the primary key of the customer table while it is a secondary or foreign key in the order table.

Tables can be created and deleted and they can be updated by having columns added, deleted or changed. Also, tables can have any of their matrix values changed, with rows added, deleted or updated with new column values. A table has a number of behaviours. For the majority of relational databases available for application development, most of these behaviours are executed by sending a command in simple query language (SQL) format.

In the framework specification, it is assumed that a database system will be used by all clients and that the database management system will provide the

Y. Tu and P. Dean, *One-of-a-Kind Production*,
DOI: 10.1007/978-1-84996-531-6, © Springer-Verlag London Limited 2011

administrative functions of the database. This framework confines itself to making a connection to (and disconnection from) the database, executing SQL commands and receiving a status indicator. The main SQL commands that the framework uses are the "update table", "insert into table" and "delete from table". These three change the content of a database table in some way, according to the processing logic of the application. The fourth SQL command that the framework uses is the "Select from" command. This command causes the database to return a "record set" to the application which is used by process logic in the application.

A record set ($\mathbf{RS}$) is defined as

$$\mathbf{RS} = \underset{i=1}{\overset{NofRows}{R}} \ \text{Index}(i\mathbf{N}){:}\mathbf{N},\ \text{NofColumns}(i\mathbf{N}){:}\mathbf{N} \ \underset{j=1}{\overset{NofColumns}{R}} \ \text{Name}(j\mathbf{N}){:}\mathbf{S},\ \text{Type}(j\mathbf{N}){:}\mathbf{S},\ \text{Value}(j\mathbf{N}){:}\ \mathbf{RT}$$

Where $\mathbf{RT} \in \{\mathbf{N}, \mathbf{Z}, \mathbf{B}, \mathbf{S}, \mathbf{D}, \mathbf{DT}, \mathbf{DI}\}$

The relational database can be described by the following unified modelling diagram

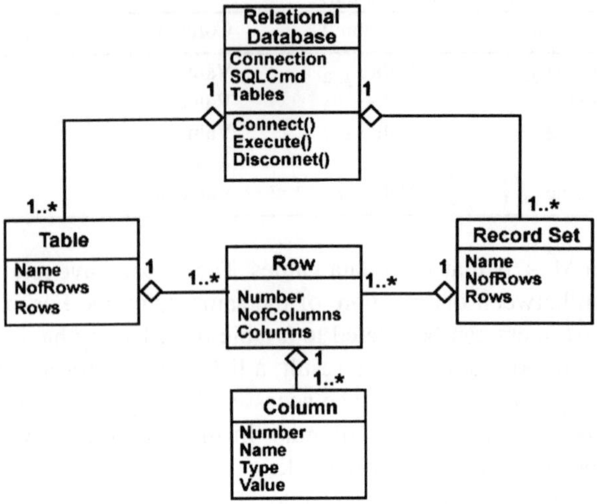

Fig. B.1 Relational database

Method: Select a specific record set from the database

The Item class sets the Connect and SQL strings to required values and then attempts to connect to the database. If a connection is made, the required record set is retrieved from the database by executing the SQL command statement. If the execution of the statement fails, the Status message notifies the application.

RTPA specifications for the relational database used in the Framework are now defined on the following pages.

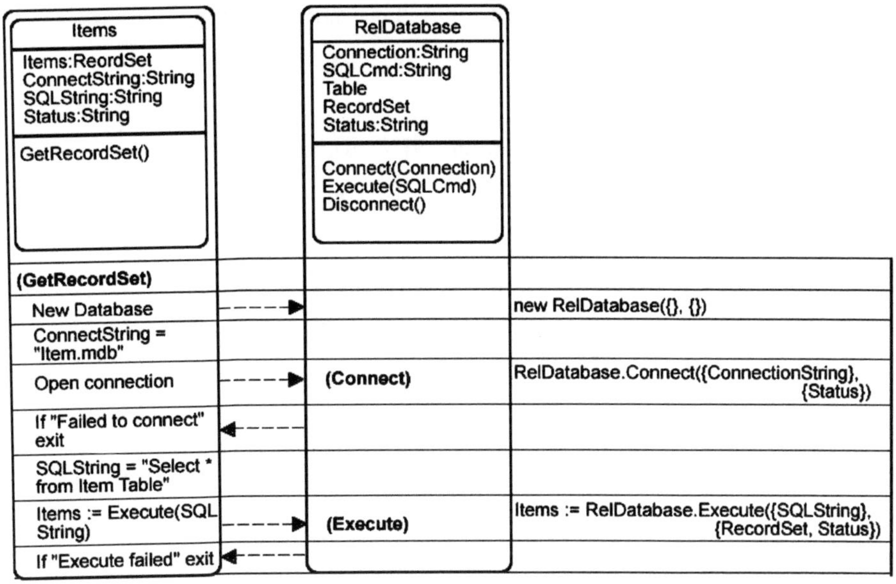

Fig. B.2 Get record set from database

B.1 Relational Database Specification

RelDb ≙ RelDb.Architecture
|| RelDb.StaticBehaviors
|| RelDb.DynamicBehaviors

B.1.1 Relational Database System Architecture

1RelDb.Architecture ≙ CLM : $\mathbf{ST}$ // Relational Database
|| AccessModel : $\mathbf{ST}$
|| Events : $\mathbf{ST}$
|| Status : $\mathbf{ST}$

RelDb.Architecture.CLM ≙ RelDB ID$\mathbf{S}$:
(<RelDbName : $\mathbf{S}$>, <NofTables : $\mathbf{N}$ | $0 \leq$ NofTables$\mathbf{N} \leq$ MaxSize$\mathbf{N}$>,
< $\overset{NofTables}{\underset{i=1}{R}}$ RelTbl(i$\mathbf{N}$) : $\mathbf{ST}$ >)

RelDb.Architecture.CLM.RelTbl$\mathbf{ST}$ ≙ RelTblID$\mathbf{S}$: (<RelTableName:$\mathbf{S}$ >,
<NofRows : $\mathbf{N}$ |$0 \leq$ NofRows$\mathbf{N} \leq$ MaxSize$\mathbf{N}$ >,

< $\underset{i=1}{\overset{NofRows}{R}}$ (RelRow(i**N**): **ST** >)

RelDb.Architecture.CLM.RelRow**ST** ≜ RelRowID**S** : (
<NofColumns : **N** |0 ≤ NofColumns**N**≤ MaxSize**N** >,
< $\underset{i=1}{\overset{NofColumns}{R}}$ (<RelColumnName(i**N**) : **S** >,
<RelColumnType(iN) : **S**>,
<RelColumnValue(i**N**) : **RT** >, where RT ∈ {N, Z, B, S, D, DT, TI}
<Key : **BL**=Y/N>, <ForeignKey : **BL**=Y/N >) >)

RelDb.Architecture.CLM.RelRecordSet**ST** ≜ RelRecordSetID**S** : (
<NofRows : **N** |0 ≤ NofRows**N**≤ MaxSize**N** >,
<NofColumns : **N** |0 ≤ NofColumns**N**≤ MaxSize**N** >,
< $\underset{i=1}{\overset{NofRows}{R}}$ (RelRowNo(i**N**): **ST** $\underset{i=1}{\overset{NofColumns}{R}}$ (<RelColumnName(i**N**) : **S** >,
<RelColumnType(iN) : **S**>,
<RelColumnValue(i**N**) : **RT** >) where **RT** ∈ {**N, Z, B, S, D, DT,TI**}
|| Connection : **S**
|| SQLCmd : **S**

RelDb.Architecture.AccessModel**ST** ≜ RelDbID**ST**(Connection**S**)

RelDb.Architecture.Events**ST** ≜ @Connect**S**
|| @Disconnect**S**
|| @Execute**S**

RelDb.Architecture.Status**ST** ≜ ⓢRelDbIDConnected**BL**
|| ⓢRelDbIDExecuteded**BL**

B.2 Relational Database Static Behaviours

RelDb.StaticBehaviors ≜ RelBb.Execute (<I :: SQLCmd**S**>;
<O :: RecordSet**ST**, ⓢRelDbIDComplete**BL**>)
|| RelDb.Connect (<I :: ConnectString**S**>;
<O :: ⓢRelDbIDConnect**BL**>)
|| RelDb.Disconnect (<I :: (),
<O :: ⓢRelDbIDConnectBL >;

B.2.1 Connect to Relational Database

RelBb.Connect(<I :: Connection**S**>; <O :: ⓢRelDbIDConnected**BL**>) ≜
{
↳ ConnectDatabase(Connection**S**, ⓢConnected**BL**)
→ (? ⓢConnected**BL** = **T**
→ ⓢRelDbIDConnected**BL** := **T**
|?~
→ ⓢRelDbIDConnected**BL** := **F**
)
}

B.2.2 Disconnect to Relational Database

RelBb.Disconnect(<I :: ()>; <O :: ElementST, ⓢRelDbIDconnected**BL**>) ≜
{
↳ DisonnectDatabase(Connection**S**,)
→ ⓢRelDbIDConnected**BL** := **F**
}

B.2.3 Execute a SQL Command

RelBb.Execute(<I :: SQLCmd**S**>; <O :: RecordSet**ST**, ⓢRelDbIDconnect**BL**>)
≜
{
→ RelRecordSetID**ST** := ExecuteDatabase(SQLCmd**S**, ⓢExecuted**BL**)
→ ⓢRelDbIDExecuted**BL** := ⓢExecuted**BL**
}

B.3 Relational Database Dynamic Behaviours

RelDb.DynamicBehaviors ≜
{ (@Execute**S** ↳ RelDb.Execute (<I :: SQLCmd**S**>; <O :: Relreord-Set**ST**, ⓢRelDbIDExist**BL**>)
| @Connect**S** ↳ RelDb.Connect (<I :: Connection**S**>; <O :: ⓢRelDbIDConnected**BL**>)
| @DisconnectS ↳ RelDb.Disconnect (<I :: ()>; <O :: ⓢRelDbIDConnected**BL**>)
) → ⊗
}

B.4 Addendum

An application may want to store the results of multiple record sets in memory for the duration of the run and at the end update any changes back to the database. Additional behaviors can be added to the database class:

Add(<I::DataSet**DS**, SQLCmd**S**>; <O::@addSuccess**BL**>)

This "executes" the select query and "adds" the record set to the data set which is defined as:

$$DS = \underset{i=1}{\overset{NofreccordSets}{R}} \text{Name}(i\mathbf{N}) : \mathbf{S}, \text{NofRows}(i\mathbf{N}): \mathbf{N} \underset{j=1}{\overset{NofRows}{R}} \text{RowNo}(j\mathbf{N}) : \mathbf{N} \underset{k=1}{\overset{NofColumns}{R}} \text{Name}(k\mathbf{N}) : \mathbf{S}, \text{Type}(k\mathbf{N}): \mathbf{S}, \text{Value}(k\mathbf{N}) : \mathbf{RT}$$

Where $\mathbf{RT} \in \{\mathbf{N, Z, B, S, D, DT, TI}\}$

The application can then access these multiple matrices of data by the use of a DataView class. A DataView is defined by referencing part of the data set using either the index or the record set name.

DataView1 = DataView(DataSet.Tables("RecordSetName"))

A DataView is defined similar to a record set:

$$\mathbf{DV} = \underset{i=1}{\overset{NofRows}{R}} \text{Index}(i\mathbf{N}):\mathbf{N}, \text{NofColumns}(i\mathbf{N}):\mathbf{N} \underset{j=1}{\overset{NofColumns}{R}} \text{Name}(j\mathbf{N}):\mathbf{S}, \text{Type}(j\mathbf{N}):\mathbf{S}, \text{Value}(j\mathbf{N}): \mathbf{RT}$$

Where $\mathbf{RT} \in \{\mathbf{N, Z, B, S, D, DT, DI}\}$

A DataViewRow is then defined as a single row in a data view

$$\mathbf{RV} = \underset{j=1}{\overset{NofColumns}{R}} \text{Name}(j\mathbf{N}):\mathbf{S}, \text{Type}(j\mathbf{N}):\mathbf{S}, \text{Value}(j\mathbf{N}): \mathbf{RT}$$

Where $\mathbf{RT} \in \{\mathbf{N, Z, B, S, D, DT, DI}\}$

Thereafter, a specific value in a data set can be accessed after the data view row has be instantiated

DataViewRow1 = DataView(Index)
RowColumn1 = DataViewRow1("ColumnName") or DataViewRow1(Index)

Appendix C
RTPA Notation System

No.	Meta Process	Notation	Syntax
1	Assignment	$:=$	$y\mathbb{T} := x\mathbb{T},\ \mathbb{T} \in \mathcal{T}$
2	Evaluation	$\blacklozenge$	$\blacklozenge exp\mathbf{BL} \in \{\mathbf{T}, \mathbf{F}\},\ \blacklozenge exp\mathbf{N} \in \mathbb{P}\mathbf{N}$
3	Addressing	$\Rightarrow$	$id\mathbb{T} \Rightarrow \mathrm{MEM}[ptr\mathbf{P}]\ \mathbb{T}$
4	Memory allocation	$\Leftarrow$	$id\mathbb{T} \Leftarrow \mathrm{MEM}[ptr\mathbf{P}]\ \mathbb{T}$
5	Memory release	$\nLeftarrow$	$id\mathbb{T} \nLeftarrow \mathrm{MEM}[\perp]\mathbb{T}$
6	Read	$\gtrdot$	$\mathrm{MEM}[ptr\mathbf{P}]\mathbb{T} \gtrdot x\mathbb{T}$
7	Write	$\lessdot$	$x\mathbb{T} \lessdot \mathrm{MEM}[ptr\mathbf{P}]\mathbb{T}$
8	Input	$\mid\gtrdot$	$\mathrm{PORT}[ptr\mathbf{P}]\mathbb{T} \mid\gtrdot x\mathbb{T}$
9	Output	$\mid\lessdot$	$x\mathbb{T} \mid\lessdot \mathrm{PORT}[ptr\mathbf{P}]\mathbb{T}$
10	Timing	$\underline{\underline{@}}$	$@t\mathbf{TM}\ \underline{\underline{@}}\ \S t\mathbf{TM}$ $\mathbf{TM} = \mathbf{yy{:}MM{:}dd}$ $\mid \mathbf{hh{:}mm{:}ss{:}ms}$ $\mid \mathbf{yy{:}MM{:}dd{:}hh{:}mm{:}ss{:}ms}$
11	Duration	$\triangleq$	$@t_{\square}\mathbf{TM}\ \underline{\underline{\Delta}}\ \S t_{\square}\mathbf{TM} + \Delta n\mathbf{TM}$
12	Increase	$\uparrow$	$\uparrow(n\mathbb{T})$
13	Decrease	$\downarrow$	$\downarrow(n\mathbb{T})$
14	Exception detection	!	$!\ (@e\mathbf{S})$
15	Skip	$\otimes$	$\otimes$
16	Stop	$\boxtimes$	$\boxtimes$
17	System	$\S$	$\S(SysID\mathbf{S})$

© Dr. Y. Wang (2001)

Fig. C.1 RTPA meta processes

Y. Tu and P. Dean, *One-of-a-Kind Production*,
DOI: 10.1007/978-1-84996-531-6, © Springer-Verlag London Limited 2011

No.	Process Relation	Notation	Syntax
1	Sequence	$\rightarrow$	$P \rightarrow Q$
2	Jump	$\curvearrowright$	$P \curvearrowright Q$
3	Branch	$\mid$	$\blacklozenge exp\mathbf{BL} = \mathbf{T} \rightarrow P$ $\mid \blacklozenge\sim \rightarrow Q$
4	Switch	$\mid \ldots$ $\ldots$ $\mid \ldots$	$\blacklozenge exp_i\mathbb{T} = i$ $\rightarrow P_i$ $\mid \sim \rightarrow \otimes$ where $\mathbb{T} \in \{\mathbf{N}, \mathbf{Z}, \mathbf{B}, \mathbf{S}\}$
5	While-loop	R^{*}	$\underset{exp\mathbf{BL}=\mathbf{T}}{\overset{\mathbf{F}}{R}} P$
6	Repeat-loop	R^{+}	$P \rightarrow \underset{exp\mathbf{BL}=\mathbf{T}}{\overset{\mathbf{F}}{R}} P$
7	For-loop	R^{i}	$\underset{i\mathbf{N}=1}{\overset{n\mathbf{N}}{R}} P(i\mathbf{N})$
8	Recursion	$\circlearrowleft$	$\underset{i\mathbf{N}=n\mathbf{N}}{\overset{0}{R}} P^{i\mathbf{N}} \circlearrowleft P^{i\mathbf{N}-1}$
9	Function call	$\rightarrowtail$	$P \rightarrowtail F$
10	Parallel	$\parallel$	$P \parallel Q$
11	Concurrence	$\oiint$	$P \oiint Q$
12	Interleave	$\mid\mid\mid$	$P \mid\mid\mid Q$
13	Pipeline	»	P » Q
14	Interrupt	↯	P ↯ Q
15	Time-driven dispatch	$\hookrightarrow$	$@t_i\mathbf{TM} \hookrightarrow P_i$
16	Event-driven dispatch	$\hookrightarrow$	$@e_i\mathbf{S} \hookrightarrow P_i$
17	Interrupt-driven dispatch	$\hookrightarrow$	$@int_i\odot \hookrightarrow P_i$

© Dr. Y. Wang (2001)

Fig. C.2 RTPA process relations

No.	Primitive Type	Syntax for Variables	Syntax for Constants	Mathematical Domain
1	Natural number	**N**	**N***	{0, ..., +∞}
2	Integer	**Z**	**Z***	{-∞, ..., +∞}
3	Real	**R**	**R***	{-∞, ..., +∞}
4	String	**S**	**S***	{0, ..., #(**S**)}
5	Boolean	**BL**	**BL***	{**T**, **F**}
6	Byte	**B**	**B***	{0, ..., 256}
7	Hexadecimal	**H**	**H***	{0, ..., +∞}
8	Pointer	**P**	**P***	{0, ..., +∞}
9	Time	**TI** = **Hh:mm:ss:ms**	**TI*** = **hh:mm:ss:ms***	**hh** ∈ {0, ..., 23}, **mm**, **ss** ∈ {0, ..., 59}, **ms** ∈ {0, ..., 999}
10	Date	**D** = **Yy:MM:dd**	**D*** = **yy:MM:dd***	**yy** ∈ {0, ..., 99}, **MM** ∈ {1, ..., 12}, **dd** ∈ {1, ..., 31}
11	Date/Time	**DT** = **yyyy:MM:dd: hh:mm:ss:ms**	**DT*** = **yyyy:MM:dd: hh:mm:ss:ms***	**yyyy** ∈ {0, ..., 9999}, **MM** ∈ {1, ..., 12}, **dd** ∈ {1, ..., 31}, **hh** ∈ {0, ..., 23}, **mm**, **ss** ∈ {0, ..., 59}, **ms** ∈ {0, ..., 999}
12	Run-time determinable type	**RT**	–	–
13	System architectural type	**ST**	–	–
14	Event	@*e***S**	–	@*e***S** ∈ §
15	Timing	@*t***TM**	–	@*t***TM** ∈ §
16	Interrupt	@*int*⊙	–	@*int*⊙ ∈ §
17	Status	ⓢs**BL**	–	{**T**, **F**}

© Dr. Y. Wang (2001)

Fig. C.3 RTPA primitive types

Appendix D
RTPA Specification of Architectural Design of the Information Generation Module (IGM) and its Major Classes

D.1 Information Generation Module

§IGM ≜ IGM.Architecture
|| IGM.StaticBehaviors
|| IGM.DynamicBehaviors

D.1.1 IGM System Architecture

IGM.Architecture ≜ CLM : **ST** // Relational Database
|| Configuration DB : **ST** | [1]
|| InputDemand : **ST** | [1..*]
|| InputItem : **ST** | [1]
|| DemandDB : **ST** | [1]
|| Request : **S** | [1]
|| Response : **S** | [1]
|| OutputInfor : **ST** | [1..*]
|| ConfigModel : **ST** | [1..*]
|| Events : **ST**
|| Status : **ST**

IGM.Architecture.Events**ST** ≜ @ServerInitial**S**
|| @ShutDownServer**S**

IGM.Architecture.Status**ST** ≜ ⓢConfigModelsCreated**BL**
|| ⓢValidRequest**BL**
|| ⓢBOOGenerated**BL**
|| ⓢBOOOutputCompleted**BL**
|| ⓢImportDemandCompleted**BL**
|| ⓢUpdateDemandCompleted**BL**
|| ⓢItemInterface**BL**
|| ⓢServerShutDown**BL**

D.1.1.1 System CLM Schemas/Objects

IGM.Architecture.ConfigurationDB**ST** ≜ ConfigurationDB**S** :
(< TaskList : **ST** | [1..*] >,
< AttributesVariables : **ST** | [1..*] >,

Y. Tu and P. Dean, *One-of-a-Kind Production*,
DOI: 10.1007/978-1-84996-531-6, © Springer-Verlag London Limited 2011

< VariableList : **ST** | [1..*] >,
< GlobalVariables : **ST** | [1..*] >,
< WorkCenterOperations : **ST** | [1..*] >
)

IGM.Architecture.ConfigurationDB.TaskList**ST** ≙ TaskList**S** :
(< ProductID : **S** >,
< ConfigID : **S** >,
< AttributeID : **S** >,
< isBase : **BL** >,
< isSelected : **BL** >,
< WorkCenterID : **S** >,
< OperationID : **S** >,
< OperationCondition : **S** >,
< TaskSequence : **N** | 1 ≤ TaskSequence**N** ≤ 9999 >,
< TaskID : **S** >,
< TaskCondition : **S** >,
< TaskExpression : **S** >
)

IGM.Architecture.ConfigurationDB.AttributeVariables**ST** ≙ AttributeVariables**S** :
(< ProductID : **S** >,
< ConfigID : **S** >,
< AttributeID : **S** >,
< VariableID : **S** >,
< Type : **S** >,
< Value : **RT** >,
< Default : **RT** >,
< Min : **N** | 1 ≤ Min**N** ≤ 9999 >,
< Max : **N** | 1 ≤ Max**N** ≤ 9999 >
)

IGM.Architecture.ConfigurationDB.VariableList**ST** ≙ VariableList**S** :
(< ProductID : **S** >,
< ConfigID : **S** >,
< AttributeID : **S** >,
< VariableID : **S** >,
< ListValue : **S** >
)

IGM.Architecture.ConfigurationDB.GlobalVariables**ST** ≙ GlobalVariables**S** :
(< GlobalVariableID : **S** >,
< Type : **S** >,
< Value : **S** >
)

IGM.Architecture.ConfigurationDB.WorkcenterOperations**ST** ≙ WorkCenterOperations**S** :
(< WorkCenterID : **S** >,
< OperationID : **S** >,
< Min : **N** | 1 ≤ Min**N** ≤ 9999 >,
< Max : **N** | 1 ≤ Max**N** ≤ 9999 >
)

IGM.Architecture.InputItem**ST** ≙ InputItem**S** (< ItemID : **S** >) :
(< ItemConfigID : **S** >,
< ItemConfigString:**S**|

$$\mathop{R}\limits_{i=1}^{NofAttributes\mathbf{N}} [ATT(name(i\mathbf{N}), \quad \mathop{R}\limits_{j=1}^{NoVariables\mathbf{N}} [Var(name(j\mathbf{N}) = Value(j\mathbf{N})]|] >$$

)

IGM.Architecture.InputDemand**ST** ≜ InputDemand**S** (< OrderNo : **N** | 1 ≤OrderNo**N** ≤999999 >, < LineNo : **N** | 0 ≤ LineNo**N** ≤ 999 >) :
(< ConfigID : **S** >,
<ProductionLine : **S**>
< ScheduleDate : **D** >,
< SequenceNo : **N** | 0 ≤ SequenceNo**N** ≤ 9999 >,
< CompletedDate : **DT** >,
< ConfigString : **S** |
$\mathop{R}\limits_{i=1}^{NofAttributes\mathbf{N}}$ [ATT(name(i**N**), $\mathop{R}\limits_{j=1}^{NoVariables\mathbf{N}}$ [Var(name(j**N**) = Value(j**N**)]|] >
)

IGM.Architecture.DemandDB**ST** ≜ DemandDB**S** (< OrderNo : **N** | 1 ≤Order-No**N** ≤999999 >, < LineNo : **N** | 0 ≤ LineNo**N** ≤ 999 >) :
(< ConfigID : **S** >,
<ProductionLine : **S**>,
< ScheduleDate : **D** >,
< SequenceNo : **N** | 0 ≤ SequenceNo**N** ≤ 9999 >,
< CompletedDate : **DT** >,
< StdMins : **N** | 0 ≤ StdMins**N** ≤ 9999 >,
< ConfigString : **S** |
$\mathop{R}\limits_{i=1}^{NofAttributes\mathbf{N}}$ [ATT(name(i**N**), $\mathop{R}\limits_{j=1}^{NoVariables}$ [Var(name(j**N**) = Value(j**N**)]|] >
)

IGM.Architecture.Request**ST** ≜ Request**S** (< RequestType : **S** | Request**S** = "Import" ∧ Request**S** = "Generate" ∧ Request**S** = "Input" ∧ Request**S** = "Stop">) :
(<Parameters : **S** | $\mathop{R}\limits_{i=1}^{NofParameters\mathbf{N}}$ (ParameterName**S** = Value**S** >
)

IGM.Architecture.Response**ST** ≜ Response**S** : (< Response: **S** >)

IGM.Architecture.OutputInfo**ST** ≜ OutputBOO**S** (< OrderNo : **N** | 1 ≤Order-No**N** ≤999999 >, < LineNo : **N** | 0 ≤ LineNo**N** ≤ 999 >) :
(< ProductionLine : **S** >,
< ScheduledDate : **DT** >,
< SequenceNo : **N** | 0 ≤ SequenceNo**N** ≤ 9999 >,
< StdMins : **N** | 0 ≤ StdMins**N** ≤ 9999 >,
< WorkCenterID : **S** >,
< OperationID : **S** >,
< TaskID : **S** >,
< TaskStd : **R** | 0 TaskStd**R** 9999.99 >
)

IGM.Architecture.ConfigModel**ST** ≜ ConfigModel: **ST** |[1..*]>
||< Attribute: **ST** | [1..*]>
||< Component: **ST** | [1..*]>
||< RawMaterial: **ST** | [1..*]>
||< Constraint: **ST** | [1..*]>
||< Variable: **ST** | [1..*]>

||< Property: **ST** | [1..*]>
||< ProductionLine: **ST** | [1..*]>
||< WorkCenter: **ST** | [1..*]>
||< Operation: **ST** | [1..*]>
||< Task: **ST** | [1..*]>
||< Resource: **ST** | [1..*]>
||< Equipment: **ST** | [1..*]>
||< Storage: **ST** | [1..*]>
||< Report: **ST** | [1..*]>
||< DataField: **ST** | [1..*]>

IGM.Architecture.Configuration**ST** ≜ ConfigID**S**:
(<ProductID:**S**>,
<ProductName:**S**>
<ConfigID:**S**>,
<ConfigName:**S**>
< Attributes:**S** | $\underset{i=1}{\overset{NofAttributes\mathbf{N}}{R}}$ <Attribute(i**N**)**ST**>
)

IGM.Architecture.Attributes**ST** ≜ AttributeIDS:
(<Name:**S**>
<IsSelected:**BL** | T=Selected ^ F=Not Selected |default=FALSE>
< Variables:**ST** | $\underset{i=1}{\overset{NofVariables\mathbf{N}}{R}}$ <Variable(i**N**)>
< Operations:**ST** | $\underset{i=1}{\overset{NofOperations\mathbf{N}}{R}}$ <Operation(i**N**)>
< Components:**ST** | $\underset{i=1}{\overset{NofComponents\mathbf{N}}{R}}$ <Component(i**N**)>
)

IGM.Architecture.Component**ST**≜ ComponentIDS:
(<Name:**S**>
< Components:= $\underset{i=1}{\overset{NofComponents\mathbf{N}}{R}}$ <Component(i**N**)>
< Properties:= $\underset{i=1}{\overset{Nof\,\mathrm{Pr}operty\mathbf{N}}{R}}$ <Property(i**N**)>
< RawMaterials:= $\underset{i=1}{\overset{NofRawMaterial\mathbf{N}}{R}}$ <RawMaterial(i**N**)>
<StorageID:**S**>
)

IGM.Architecture.RawMaterial**ST** ≜ RawMaterialID**S**:
(<Name:**S**>
<PartNumber:**S**>
< Properties:= $\underset{i=1}{\overset{Nof\,\mathrm{Pr}operty\mathbf{N}}{R}}$ <Property(i**N**)>

< Reports:= $\mathop{R}\limits_{i=1}^{NofReports\mathbf{N}}$ <Report(i**N**)>

)

IGM.Architecture.Constraint**ST** ≙ ConstraintID**S**:

(<Name:**S**>

<ConfigID:**S**>

<Condition:**S**>

<Result:**S**>

< Variables:= $\mathop{R}\limits_{i=1}^{NofVariables\mathbf{N}}$ <Variable(iN)>

)

IGM.Architecture.Variable**AB** ≙ VariableID**S**: //Abstract class

(<Name:**S**>

<Type:**S** | “value” ∧ “list” ∧ “Boolean”>

<Value:**O** > // Object depending its type

)

IGM.Architecture.ValueVariable**ST** ≙ VariableIDS: //Inherits Variable

(<Value:**R**>

<DefaultValue:**R**>

<Minimum:**R**>

<Maximum:**R**>

)

IGM.Architecture.ListVariable**ST**≙ VariableID**S**: //Inherits Variable

(<Value:**S**>

<DefaultValue:**S**>

< ListValues:= $\mathop{R}\limits_{i=1}^{NofListValues\mathbf{N}}$ <ListValue(i**N**)**S**>

)

IGM.Architecture.BooleanVariable**ST** ≙ VariableID**S**: //Inherits Variable

(<Value:**BL**>

<DefaultValue:**BL**>)

IGM.Architecture.Property**ST** ≙ PropertyID**S**:

(<Name:**S**>

<ConfigID:**S**>

<Condition:**S**>

<Result:**S**>

< Variables:= $\mathop{R}\limits_{i=1}^{NofVariables\mathbf{N}}$ <Variable(i**N**)>

)

IGM.Architecture.ProductionLine**ST** ≙ ProductionLineID**S**:

(<Name:**S**>

<PerformanceFactor:**R**>

< WorkCenters:= $\mathop{R}\limits_{i=1}^{NofWorkCenters\mathbf{N}}$ <WorkCenter(i**N**)>)

IGM.Architecture.WorkCenter**ST** ≙ WorkCenterID**S**:

(<Name:**S**>

$$< \text{Operations} := \mathop{R}\limits_{i=1}^{NofOperations\mathbf{N}} <\text{Operation}(i\mathbf{N})>$$

$$< \text{Resources} := \mathop{R}\limits_{i=1}^{NofResources\mathbf{N}} <\text{Resource}(i\mathbf{N})>$$

$$< \text{Equipments} := \mathop{R}\limits_{i=1}^{NofEquipment\mathbf{N}} <\text{equipment}(i\mathbf{N})>$$

$$< \text{Reports} := \mathop{R}\limits_{i=1}^{NofReports\mathbf{N}} <\text{Report}(i\mathbf{N})>$$

)

IGM.Architecture.Operation**ST** ≙ OperationID**S**:
(<Name:**S**>
<Condition:**S**>
<SetupTime:**R**>
<MinimumOps:**N**>
<MaximumOps:**N**>
<AttributeID:**S**>
<ResourceID:**S**>

$$< \text{Tasks} := \mathop{R}\limits_{i=1}^{NofTasks\mathbf{N}} <\text{Task}(i\mathbf{N})>$$

)

IGM.Architecture.Task**ST** ≙ TaskID**S**:
(<Name:**S**>
<Sequence:**N**>
<Condition:**S**>
<StdMins:**R**> // Cycle Time
<Expression:**S**>

$$< \text{Equipments} := \mathop{R}\limits_{i=1}^{NofEquipment\mathbf{N}} <\text{Equipment}(i\mathbf{N})>$$

$$< \text{RawMaterials} := \mathop{R}\limits_{i=1}^{NofRawMaterials\mathbf{N}} <\text{RawMaterials}(i\mathbf{N})>$$

)

IGM.Architecture.Resource**ST** ≙ ResourceID**S**:
(<Name:**S**>
<HourlyRate:**R**>

$$< \text{Reports} := \mathop{R}\limits_{i=1}^{NofReports\mathbf{N}} <\text{Report}(i\mathbf{N})>$$

)

IGM.Architecture.Equipment**ST** ≙ EquipmentID**S**:
(<MachineNo:**S**>
<IsAutomated:**BL** | **T**=Automated ∧ **F** = Manual>
<SetupTime:**R**>
<MachineParameters:**S**>

< Reports:= $\overset{Nof\,Reports\mathbf{N}}{\underset{i=1}{R}}$ <Report(i**N**)>

)

IGM.Architecture.Storage**ST** ≜ StorageID**S**:
(<Sequence:**N**>
<FromLoaction:**S**>
<ToLocation:**S**>
<Type:**S** | Cart ∧ Buffer ∧ Kanban ∧ Storage ∧ Conveyer>
<Size:**N** | default = 1>
<Minimum:**N**>
<Maximum:**N**>
<BinNumber:**N**>

< Components:= $\overset{NofComponents\mathbf{N}}{\underset{i=1}{R}}$ <Component(i**N**)>

< RawMaterials:= $\overset{NofRawMaterials\mathbf{N}}{\underset{i=1}{R}}$ <RawMaterial(i**N**)>

Reports:= $\overset{Nof\,Reports\mathbf{N}}{\underset{i=1}{R}}$ <Report(i**N**)>

)

IGM.Architecture.Report**ST** ≜ ReportID**S**:
(<Name:**S**>
<Format:**S**>

< DataFields:= $\overset{NofDataFields\mathbf{N}}{\underset{i=1}{R}}$ <DataField(i**N**)>

)

IGM.Architecture.DataField**ST**≜ DataFieldID**S**:
(<Name:**S**>
<Type:**S**> // All values converted to string format
<Value:**S**>

D.1.2 IGM Static Behaviours

IGM.StaticBehaviors ≜ ServerInitial (< **I**::( ) >, < **O**::( ) >)
|| ShutDownServer (< **I**:: Response**S** >, < **O**::( ) >)
|| ProcessRequest (<**I** :: Request**S**>, <O :: Response**S**>)
|| ValidateRequest (<**I** :: Request**S**>, <O :: ⓢValidParameters**BL**>)
|| CreateConfigModels (< **I**::( ) >, < **O**:: ⓢCreatedConfigModels**BL** >)
|| ImportDemand (< **I**:: Request**S** >, < **O**:: Response**S**, ⓢImportDemandCompleted**BL** >)
|| GenerateBOO (< **I**:: Request**S** >, < **O**:: Response**S**, ⓢBOOGenerated**BL** >)
|| ItemInterface (< **I**:: Request**S** >, < **O**:: Response**S** >)

D.1.2.1 Server Initialization

```
IGM.ServerInitial ( < I::( ) >, < O::( ) >) ≙
{
↳ CreateConfigModels ( < I::( ) >, < O:: ⓈCreatedConfigModelsBL >)
→ (?ⓈCreatedConfigModelsBL = F
        →ResponseS := "Invalid Models"
        →ⓈShutDownServerBL = T
        ↳ StopServer (< I::ResponseS >, < O::( ) >)
  |?~
        →ⓈShutDownServerBL = F
        Exp:ST := ExpressionST( )        // Instantiates an Expression object to
    evaluate expressions
  )
}
```

D.1.2.2 Shut Down Server

```
IGM.ShutDownServer ( < I::ResponseS >, < O::( ) >) ≙
{
→OutputDevice.write ( ResponseS )
→⊠
}
```

D.1.2.3 Process Request

```
IGM.ProcessRequest ( < I:: RequestS >, < O:: ResponseS >) ≙
{
→ (? RequestS = "Import"
        ↳ ImportDemand (< I:: RequestS >, < O:: ResponseS >)
   |? RequestS = "Generate"
        ↳ GenerateBOO (< I:: RequestS >, < O:: ResponseS >)
   |? RequestS = "Input"
        ↳ ItemInterface (< I:: RequestS >, < O:: ResponseS >)
   |? RequestS = "Stop"
        →ResponseS := "User request to shut down server"
        ↳ ShutDownServer (< I:: ResponseS >, < O:: ( ) >)
   |?~
        →ResponseS := "Invalid request from user"
   )
}
```

D.1.2.4 Validate Request

```
IGM.ValidateRequest ( < I:: RequestS >, < O:: ⓈValidRequestBL >) ≜
{
→ (?RequestTypeS = "Import"            // Import demand Items for a specific
date
        → (? ParametersS(1).Type = Date
                →ImportDateD := CDate(ParametersS(1))
                →ⓈValidRequestBL = T
          |?~
                →ⓈValidRequestBL = F
                →⊗
        )
  |? RequestS = "Generate"         // Generate BOO for a set of items for a speci-
                                     fied date and Sequence number range
        → (? ParametersS(1).Type = Date
                →ImportDateD := CDate(ParameterS(1))
                →ⓈValidrequestBL = T
              |?~
                    →ⓈValidRequestBL = F
                    →⊗
            )
            → (? ParametersS(2).Type = Integer
                    →StartSeqNoN := CInt(ParametersS(2))
                    →ⓈValidRequestBL = T
              |?~
                    →ⓈValidRequestBL = F
                    →⊗
              )
            → (? ParametersS(3).Type = Integer
                    →endSeqNoN := CInt(ParametersS(3))
                    →ⓈValidRequestBL = T
              |?~
                    →ⓈValidRequestBL = F
                    →⊗
              )
      |? RequestS = "Input"            // Generate Interface to test configuration mod-
  els with an input item
            →ⓈValidRequestBL = T
            →⊗
      |? RequestS = "Stop"             // A request to shut down the server
            →ⓈValidRequestBL = T
            →⊗
      |? ~                             // Invalid Request Parameter
            →ⓈValidRequestBL = F
            →⊗
      )
    }
```

D.1.2.5 Create Configuration Models

```
IGM.CreateConfigModel ( < I:: ( ) >, < O:: ⓢCreatedConfigModelsBL >) ≙
{
→ConfigDB = RelDataBase( < I::( ) >, < O::( ) >)
→ConfigDB.Connect( < I::("Configuration.mdb" ) >, < O:: ⓢRelDBConnectedBL >)
→ (? ⓢRelDBConnectedBL = F
        → ⓢCreatedConfigModelsBL:= F
        →ResponseS := "Failed to connect to Configuration Database"
        →⊗              // Exit Configuration Model Function
    )
    →SQLCmdS := "Select * From TaskList"
    →TaskListRS := ConfigDB.Execute( < I::SQLCmdS >, < O:: RecordSetRS,
                                      ⓢRelDBExecutedBL >)
    → (? ⓢRelDBExecutedBL = F
            → ⓢCreatedConfigModelsBL:= F
            →ResponseS := "Failed to access Configuration Database - TaskList"
            →⊗              // Exit Configuration Models Function
    )
    // TaskList is sorted by ProductID, ConfigID, WorkCenterID, OperationID and TaskSequenceNo
    // The tree structure of each configuration is now built, add attributes, operations and tasks in this
    // part of the Framework application
    → ConfigsST := CollectionST( ) (or ArrayListST)
         NofTaskListRecordsN
    →        R          (→ ConfigKeyS := TaskList(iN)RS.ProductIDS & TaskList(iN)RS.ConfigIDS
            i=1
                           (? ConfigKeyS ≠ LastConfigKeyS
                                 → c:ST := ConfigST( )
                                 → cST.ProductIDS := TaskList(iN)RS.ProductIDS
                                 → cST.ConfigIDS := TaskList(iN)RS.ConfigIDS
                                 → ConfigsST.add(cST, ConfigKeyS)
                                 → LastConfigKeyS := ConfigKeyS
                           |?~
                                 → cST := ConfigsST(ConfigKeyS)
                           )
                          → AttKeyS := TaskList(iN)RS.AttributeIDS
                           (? AttKeyS ≠ LastAttKeyS
                                 → a:ST := AttributeST( )
                                 → aST.AttributeIDS := TaskList(iN)RS.AttributeIDS
                                 → aST.isSelectedBL := TaskList(iN)RS.isSelectedBL
                                 → cST.addAttribute(aST, AttKeyS)
                                 → LastAttKeyS := AttKeyS
                           |?~
                                 → aST := cST.Attributes(AttKeyS)
                           )
```

→ OpKey**S** := TaskList(i**N**)**RS**.WorkCenterID**S** & TaskList(i**N**)**RS**.OperationID**S**
(? OpKey**S** ≠ LastOpKey**S**
→ Op:**ST** := Operation**ST**( )
→ Op**ST**.OperationID**S** := TaskList(i**N**)**RS**.OperationID**S**
→ Op**ST**.Condition**S** := TaskList(i**N**)**RS**.OperationCondition**S**
→ a**ST**.addOperation(Op**ST**, OpKey**S**)
→ LastOpKey**S** := OpKey**S**
|?~
→ Op**ST** := a**ST**.Operation(OpKey**S**)
)
→ TaskKey**S** := TaskList(i**N**)**RS**.Sequence**N** & TaskList(i**N**)**RS**.TaskID**S**
(? TaskKey**S** ≠ LastTaskKey**S**
→ Tk:**ST** := Task**ST**( )
→ Tk**ST**.TaskID**S** := TaskList(i**N**)**RS**.TaskID**S**
→ Tk**ST**.Sequence**N** := TaskList(i**N**)**RS**.Sequence**N**
→ Tk**ST**.Condition**S** := TaskList(i**N**)**RS**.TaskCondition**S**
→ Tk**ST**.Expression**S** := TaskList(i**N**)**RS**.Expression**S**
→ Op**ST**.addTask(Tk**ST**, TaskKey**S**)
→ LastTaskKey**S** := TaskKey**S**
)
)
// The variables are added to the Attributes (along with the Operations)
→SQLCmd**S** := "Select * From Variables"
→Variable**RS** := ConfigDB.Execute(SQLCmd**S**)**RS**
→ (? ⓢRelDBExecuted**BL** = **F**
→ ⓢCreatedConfigModels**BL**:= **F**
→ResponseS := "Failed to access Configuration Database - Variables"
→⊗ // Exit Configuration Models Function
)
→ $\overset{NofVariableRecords\mathbf{N}}{\underset{i=1}{R}}$ (→ c:**ST** := Configs(Variable(i**N**)**RS**.ProductID**S** & Variable(i**N**)**RS**.ConfigID**S**)**ST**
→ a:**ST** := c**ST**.Attributes(Variable(i**N**)**RS**.AttributeID**S**)**ST**

→ v:**ST** := Variables**ST**
→ v**ST**.Name**S** := Variable(i**N**)**RS**.VariableID**S**
→ v**ST**.Type**S** := Variable(i**N**)**RS**.Type**S**
→ v**ST**.Value**RT** := Variable(i**N**)**RS**.Value**RT**
→ v**ST**.Default**RT** := Variable(i**N**)**RS**.Defaul**RT**
→ v**ST**.Min**N** := Variable(i**N**)**RS**.Max**N**
→ v**ST**.Max**N** := Variable(i**N**)**RS**.Min**N**
→ a**ST**.addVariable(v**ST**, v**ST**.Name**S**)
)
// The variable Lists are added to the Variables of type LIST
→SQLCmd**S** := "Select * From VariableList"
→VarList**RS** := ConfigDB.Execute(SQLCmd**S**)**RS**
→ (? ⓢRelDBExecuted**BL** = **F**
→ ⓢCreatedConfigModels**BL**:= **F**
→ResponseS := "Failed to access Configuration Database – Variable List"
→⊗ // Exit Configuration Models Function
)
→ $\overset{NofVariableRecords\boldsymbol{N}}{\underset{i=1}{R}}$ (→ c:**ST** := Configs(VarList(i**N**)**RS**.ProductID**S** & VarList(i**N**)**RS**.ConfigID**S**)**ST**
→ a:**ST** := c**ST**.Attributes(VarList(i**N**)**RS**.AttributeID**S**)**ST**
→ lv:**ST** := a**ST**.Variable(VarList(i**N**)RS.VarID**S**)**ST**
→ lv**ST**.add(VarList(i**N**)**RS**.VarListValue**S**)
)
→ ⓢCreatedConfigModels**BL**:= **T**
}

D.1.2.6 Import Demand Items

```
// It is assumed that the demand items have been transferred from the ERP system
// to the Demand Database by a SQL command or stored procedure
// This functions selects the demand items transferred and created a collection of Items
// which it then calculates the standard time and updates the Demand database
IGM.ImportDemand ( < I:: RequestS >, < O:: ResponseS, ⓢImportDemandCompletedBL >) ≜
{
↳ ValidateRequest ( < I:: RequestS >, < O:: ⓢValidRequestBL >)
→ (?ⓢValidRequestBL = F
        →ⓢImportDemandCompletedBL = F
        →ResponseS := "Invalid Request Parameters"
        →⊗            // Exit Import Function
  )
→ Id:ST := ImportData( < I:: ( ) >, < ⓢRelDBConnectedBL >)
→SQLCmdS := "Select * From Items where Items.ScheduleDate =" & ImportDateD
→ (? ⓢRelDBExecutedBL = T
        →DemandItemRS := IdST.Execute( < I::SQLCmdS >, < O:: RecordSetRS,
                          ⓢRelDBexecutedBL >)
        → (? ⓢRelDBExecutedBL = F
                →ⓢImportDemandCompletedBL = F
                →ResponseS := "Failed to access Demand Database"
                →⊗            // Exit Import Function
          )
          NofItemsN
        →    R      (→ c:ST = config(DemandItemRS(iN).ProductIDS &
            i=1
                          DemandItemRS(iN).ConfigIDS)
                  → i:ST := ItemST(DemandItemRS(iN).OrderNoN & DemandItemRS(iN).LineNoN)
                  → iST.copyConfig(cST)
                  → iST.ApplyConstraints(DemandItemRS(iN).ConfigStringS)
                  → iST. ConfigStd( )
                  → ItemsST.add(iST)
                → ⓢImportDemandCompletedBL := T
                )
→ (? ⓢImportDemandCompletedBL = T
        ↳ Id.UpdateDemand ( < I:: ItemsST >, < O:: ⓢDemandUpdateCompletedBL >)
  |?~
        →ⓢUpdateDemandUpdatedBL = F
  )
}
```

D.1.2.7 Generate BOO

// Generate BOO for items within a specified sequence range

IGM.GenerateBOO (< **I::** Request**S** >, < **O::** Response**S**, ⓢBOOGenerated**BL** >) ≙
{
↳ ValidateRequest (< **I::** Request**S** >, < **O::** ⓢValidRequest**BL** >)
→ (?ⓢValidRequest**BL** = **F**
→ⓢGeneratedBOO**BL** = **F**
→Response**S** := "Invalid Request Parameters"
→⊗ // Exit Import Function
)
→ Id:**ST** := ImportData(< **I::** ( ) >, < ⓢRelDBConnected**BL** >)
→ SQLCmd**S** = "Select * From Items Where Items.ScheduleDate = " & ImportDate**D** & " and Items.SequeneNo Between " & StartSeqNo**N** & " and " & EndSeqNo**N**
→ (? ⓢRelDBExecuted**BL** = **T**
→DemandItem**RS** := Id**ST**.Execute(< **I::**SQLCmdS >, < **O::** RecordSet**RS**, ⓢRelDBexecuted**BL** >)
→ (? ⓢRelDBExecuted**BL** = **F**
→ⓢGeneratedBOO**BL** = **F**
→ResponseS := "Failed to access Demand Database"
→⊗ // Exit Import Function
)
→ $\overset{NofItems\mathbf{N}}{\underset{i=1}{R}}$ (→ c:**ST** = config(DemandItem**RS**(i**N**).ProductID**S** & DemandItem**RS**(i**N**).ConfigID**S**)
→ i:**ST** := Item**ST**(DemandItem**RS**(i**N**).OrderNo**N** & DemandItem**RS**(i**N**).LineNo**N**)
→ i**ST**.copyConfig(c**ST**)
→ i**ST**.ApplyConstraints(DemandItem**RS**(i**N**).ConfigString**S**)
→ i**ST**. ConfigStd( )
→ Items**ST**.add(i**ST**)
→ ⓢImportDemandCompleted**BL** := **T**
)
→ (? ⓢImportdemandCompleted**BL** = **T**
↳ Id.OutputBOO (< **I::** Items**ST** >, < **O::** ⓢBOOGenerated**BL** >)
|?~
→ⓢBOOGenerated**BL** = **F**
)
}

D.1.2.8 Item Interface

// An interface generated based on the selected configuration, used to test configuration models

IGM.ItemInterface (< I:: Request𝐒 >, < O:: Response𝐒 >) ≜

{

↳ ValidateRequest (< **I::** Request𝐒 >, < **O::** ⓈValidRequest𝐁𝐋 >)

→ (?ⓈValidRequest𝐁𝐋 = 𝐅

→ⓈItemInterface𝐁𝐋 = 𝐅

→Response𝐒 := "Invalid Request Parameters"

→⊗ // Exit Import Function

)

}

D.1.3 IGM Dynamic Behaviours

IGM.DynamicBehaviors ≜

{

(ServerInitial

|| ShutDownServer

|| ProcessRequest

|| ValidateRequest

|| CreateConfigModels

|| ImportDemand

|| GenerateBOO

|| ItemInterface

)

}

IGM.ProcessDeployment ≜

{ // Server in wait state

@ServerInitial𝐒

↳ (ServerInitial

$\overset{@ServerShutDown=\mathbf{T}}{\underset{\geq 1}{R}}$ ProcessRequest (< **I::** Request𝐒 >, < **O::** Response𝐒 >)

↳ ShutDownServer (< **I::** Response𝐒 >, < **O::** ( ) >)

)

|| // Interrupt Shut Down

↳ @Interrupt

→Response𝐒 := "Server interrupted – shutting down!"

→OutputDevice.write (Response𝐒)

→⊠

}

D.2 Item Class

§ITEM ≙ ITEM.Architecture
|| ITEM.StaticBehaviors
|| ITEM.DynamicBehaviors

D.2.1 Item Architecture

System and CLM Schemas/Objects

D.2.1.1 Item System Architecture

ITEM.Architecture ≙ CLM : **ST** // Item Class
|| Item:**ST**
|| Events : **ST**
|| Status : **ST**

ITEM.Architecture.Events**ST** ≙ NA
ITEM.Architecture.Status**ST** ≙ ⓢConfigCopied**BL**
|| ⓢStdTimeCalculated**BL**

D.2.1.2 System CLM Schemas/Objects

ITEM.Architecture.Item**ST** ≙ ItemID**S** (< OrderNo : **N** | 1 ≤OrderNo**N** ≤999999 >, < LineNo : **N** | 0 ≤ LineNo**N** ≤ 999 >) :
(<ProductID : **S**>
< ConfigID : **S** >,
<ProductionLine : **S**>,
< ScheduleDate : **D** >,
< SequenceNo : **N** | 0 ≤ SequenceNo**N** ≤ 9999 >,
< CompletedDate : **DT** >,
< StdMins : **R** | 0 ≤ StdMins**N** ≤ 9999.99 >,<Attributes : **ST** |
$\overset{NofAttributes\mathbf{N}}{\underset{i=1}{R}}$ Attribute**ST**
)

D.2.2 ITEM Static Behaviours

ITEM.StaticBehaviors ≜ New (< **I::** < OrderNo : **N** | 1 ≤OrderNo**N** ≤999999 >, < LineNo : **N** | 0 ≤ LineNo**N** ≤ 999 >, < **O::**( ) >)
|| AddAttribute (< **I::** Attribute**ST** >, < **O::**( ) >)
|| CopyConfig (<**I** :: ConfigID**S**>, <O :: ( )>)
|| ConvertToCfgString (< **I::** ( ), < **O::** <Config-String**S**>)
|| CalcStdTime (<**I** :: ( )>, <O :: StdTime**R**>)
|| ToDisplay (< **I::** ( ) >, < **O::** ( ) >)
|| ApplyConstraints (< **I::**( ) >, < **O::** ( )>
)

D.2.2.1 New

ITEM.NewItem (< **I::** < OrderNo : **N** | 1 ≤OrderNo**N** ≤999999 >, < LineNo : **N** | 0 ≤ LineNo**N** ≤ 999 >>, < **O::**( ) >) ≜
{
→ _itemOrderNo**N** := OrderNo**N**
→ _itemLineNo**N** := LineNo**N**
}

D.2.2.2 Add Attribute

ITEM.AddAttribute (< **I::** Attribute**ST** >, < **O::**( ) >) ≜
{
→ Attributes.add(Attribute**ST**, attributeKey**S**)
}

D.2.2.3 Copy Configuration

ITEM.CopyConfig (< **I::** Configuration**ST** >, < **O::** ( ) >) ≜
{
→ $\overset{NofAttributes\mathbf{N}}{\underset{i=1}{R}}$ AddAttributes(Configuration.Attribute(i**N**)**ST**, Configuration.AttributeKey**S**)
(? Exception = **T**
→ ⓢConfigCopied**BL** := **F**
|?~
→ ⓢConfigCopied**BL** := **T**
)
}

D.2.2.4 Set Variable Values in Item from the Configuration String

ITEM.ApplyConstraints (< **I::** ConfigString**S** >, < **O::** ( ) >) ≙
{
→ ConvertToAttributeVariables(ConfigString**S**, AttributeVariables**ST**)
// AttributeVariables is a collection of AttributeVariable(Att, ID, Name, Value)

$\rightarrow \overset{NofAttributeVariables\mathbf{N}}{\underset{i=1}{R}}$ (? AttributeVariable(i**N**).IDS = "ATT"

→ Atttribute (AttributeVariable(i**N**).Att**S**).isSelected**BL** := AttributeVariable(i**N**).Value**BL**

|?~

→ Atttribute (AttributeVariable(i**N**).Att**S**).Variable (AttributeVariable(i**N**).Name**S**).Value**RT** := AttributeVariable(i**N**).Value**RT**

)
}

D.2.2.5 Calculate Standard Time (minutes) for Item

ITEM.CalcStdTime (< **I::** ( ) >, < **O::** StdTime**R** >) ≙
{
→ StdTime**R** := 0
→ Exp.Clear ( )
→ Exp.setGlobals ( )

$\rightarrow \overset{NoAttributes\mathbf{N}}{\underset{i=1}{R}}$ (→ Exp.AddRange(Attributes(i**N**).Variables**ST**)

→ (? Attributes(i**N**).isSelected = **T**

$\rightarrow \overset{NoOperations\mathbf{N}}{\underset{j=1}{R}}$ StdTime**R** := stdTime**R** + Attributes(i**N**).Operation(j**N**).calcOpStd( )**R**

)

)

(? Exception = **T**

→ ⓢStdTimeCalculated**BL** := **F**

|?~

→ ⓢStdTimeCalculated**BL** := **T**

)

D.2.2.6 Output BOO to Display

ITEM.ToDisplay (< **I::** ( ) >, < **O::** StdTime**R** >) ≙

{

→ OutputDevice.WriteLine("Custom Product")

→ OutputDevice.WriteLine("Product Family" & ProductID**S**)

→ OutputDevice.WriteLine("Configuration" & ConfigID**S** & " Std = " & CalcStdTime( )**R**)

→ $\overset{NofAttributesN}{\underset{i=1}{R}}$ (→ OutputDevice.WriteLine("Attribute: " & Attribute(i**N**).AttID**S** & " Selected = " &
Attribute(i**N**).AttID**S** & " Std = " & CalcStdTime( )**R**)

→ $\overset{NofVariables\boldsymbol{N}}{\underset{j=1}{R}}$ OutputDevice.WriteLine("Variable: " &
Attribute(i**N**).Variable(j**N**).Name**S** & " = " & Attribute(i**N**).Variable(j**N**).Value**RT**)

→ $\overset{NofOperations\boldsymbol{N}}{\underset{m=1}{R}}$ (→ OutputDevice.WriteLine("Work Center: " &
Attribute(i**N**).Operation(m**N**).WCID**S** & " Operation: " & Attribute(i**N**).Operation(m**N**).OpID**S** & Condition = " & OpCond(Attribute(i**N**).Operation(m**N**).Cond**S**)**BL** & "Std = " & OpCalcOpStdMins(Attribute(i**N**).Operation(m**N**))**R**

→ $\overset{NofTasks\boldsymbol{N}}{\underset{n=1}{R}}$ (→ OutputDevice.WriteLine("Task: " &
Attribute(i**N**).Operation(m**N**).Task(n**N**).name**S** & " Condition = " & TaskCond(Attribute(i**N**).Operation(m**N**).Task(n**N**).Cond**S**)**BL** & " Std = " & TaskCond(Attribute(i**N**).Operation(m**N**).Task(n**N**).stdMins**R**

)

)

)

}

D.2.2.7 Convert Input Variables to Configuration String

//
|ATT(BASE=TRUE),VAR(FIRST=1),VAR(SECOND=2)|ATT(OPTION1=FALSE),VAR(THIRD=3)
ITEM.ConvertToCfgString (< **I::** ( ) >, < **O::** ConfigString**S** >) ≙
{
→ ConfigString**S** := " "
→ $\overset{NofAttributes\mathbf{N}}{\underset{i=1}{R}}$ (→ ConfigString**S** := ConfigString**S** & "ATT(" & Attribute(i**N**).AttID**S** & ")=" &
Attribute(i**N**).isSelected**BL**
→ $\overset{NofVariables\mathbf{N}}{\underset{j=1}{R}}$ ConfigString**S** := ConfigString**S** & ",VAR(" &
Attribute(i**N**).Variable(j**N**).Name**S** & ")=" &
Attribute(i**N**). Variable(j**N**).Value**RT**
)
}

D.2.3 ITEM Dynamic Behaviours—NA

D.3 Import Data Class

§ID ≙ ID.Architecture
|| ID.StaticBehaviors
|| ID.DynamicBehaviors

D.3.1 Architecture

System and CLM Schemas/Objects

D.3.1.1 Import Data System Architecture

ID.Architecture ≙ CLM : **ST** // Import Data Class
|| ImportData:**ST**
|| InputDemand: **ST**
|| Events : **ST**
|| Status : **ST**
ID.Architecture.Events**ST** ≙ NA
ID.Architecture.Status**ST** ≙ ⓢRelDBConnected**BL**
|| ⓢRelDBexecuted**BL**
|| ⓢImportDataCompleted**BL**
|| ⓢUpdateItemsCompleted**BL**
|| ⓢBOOGenerated**BL**

D.3.1.2 System CLM Schemas/Objects

ID.Architecture.ID**ST** ≙ ImportData**S**:
(<SQLCmd : **S**>,
<Response : **S**>,
<RecordSet : **ST** | $\overset{NofItems\mathbf{N}}{\underset{i=1}{R}}$ InputDemand**ST**>
)

ID.Architecture.InputDemand**ST** ≙ InputDemand**S** (< OrderNo : **N** | 1 ≤Order-No**N** ≤999999 >, < LineNo : **N** | 0 ≤ LineNo**N** ≤ 999 >) :
(< ConfigID : **S** >,
<ProductionLine : **S**>,
< ScheduleDate : **D** >,
< SequenceNo : **N** | 0 ≤ SequenceNo**N** ≤ 9999 >,
< CompletedDate : **DT** >,
< ConfigString : **S** |
$\overset{NofAttributes\mathbf{N}}{\underset{i=1}{R}}$ [ATT(name(i**N**), $\overset{NoVariables\mathbf{N}}{\underset{j=1}{R}}$ [Var(name(j**N**) = Value(j**N**)]|] >
)

D.3.2 Import Data Static Behaviours

ID.StaticBehaviors ≙ ImportData (< **I::** SQLCmd**S** >, < **O::** ⓢImportDataCompleted**BL**)
|| getRecordSet (< **I::** ( ) >, < **O::** RecordSet**RS** >)
|| UpdateItems (< **I::** ( ) >, < **O::** ⓢUpdateItemsCompleted**BL** >)
|| OutputBOO (<**I::** ( )>, <**O::** ⓢOutputBOOCompleted**BL** >)

D.3.2.1 New Import Data

ID.ImportData (< **I::** ( ) >, < **O::** ⓢRelDBConnected**BL** >) ≙
{
→DemandDB:**ST** = RelDataBase(< **I::**( ) >, < **O::**( ) >)
→DemandDB**ST**.Connect(< **I::**("Demand.mdb") >, < **O::** ⓢRelDBConnected**BL** >)
→ (? ⓢRelDBConnected**BL** = **F**
→ⓢImportDataCompleted**BL** = **F**
→Response**S** := "Failed to connect to Demand Database"
→⊗ // Exit Import Function
)
}

D.3.2.2 Get Record Set from Demand Database

```
ID.GetRecordSet ( < I:: SQLCmdS >, < O:: <RecordSetST>,
<ⓢImportDataCompletedBL >) ≙
  {
  →RecordSet:ST := DemandDBST.Execute( < I::SQLCmdS >, < O:: Record-
SetRS,
                              ⓢRelDBexecutedBL >)
  → (? ⓢRelDBExecutedBL = F
          →ⓢImportDataCompletedBL = F
          →ResponseS := "Failed to access Demand Database"
          →⊗            // Exit Import Function
    )
  → ⓢImportDataCompletedBL := T
  }
```

D.3.2.3 Update Items in Demand Database

```
  ID.UpdateItems ( < I:: ItemsST >, < O:: <ⓢUpdateItemsCompletedBL >) ≙
  {
     NofItemsN
  →     R      (→ SQLcmdS = "Update Items SET ItemStdMins = " &
       i=1
Items(iN)ST.ItemStdR &
                    " WHERE ItemOrderNo = '" & Items(iN)ST.ItemOrderNoN &
                 "' AND LineNo = '" & Items(iN)ST.LineNoN & "';"
                 →DemandDBST.Execute( < I::SQLCmdS >, <
       O::ⓢRelDBexecutedBL >)
                 → (? ⓢRelDBExecutedBL = F
                      →ⓢUpdateItemsCompletedBL = F
                      →ResponseS := "Failed to access Demand Database"
                      →⊗                  // Exit Import Function
                    )
                 )
  → ⓢUpdateItemsCompletedBL := T
  }
```

D.3.2.4 Output BOO to Demand Database

```
ID.OutputBOO ( < I:: ItemsST >, < O:: <ⓢBOOGeneratedBL >) ≙
{
  NofItemsN
→     R        (→OrderNoN := Item(iN).OrderNoN
     i=1
→OrderLineN := Item(iN).OrderLineN
→ProductionLineS := Item(iN).ProductionLineS
→ScheduleDateD := Item(iN).ScheduleDateD
→SequenceNoN := Item(iN).SequenceNoN
→StdMinsR := Item(iN).StdMinsR
  NofAttributesN
→        R         (? Items.Attribute(iN)ST.isSelectedBL = T
        i=1
              NofOperationsN=T
    →               R              (→  OperationIDS  :=  Attrib-
                   j=1
      ute(iN).Operation(jN).OpIDS
                                 →  WorkCenterIDS  :=  Attrib-
                  ute(iN).Operation(jN).OpIDS
                  NofTasksN=T
            →         R        (→    TaskIDS    :=    Attrib-
                     n=1
           ute(iN).Operation(jN).Task(nN).NameS
                              →    TaskStdMinsR    :=    Attrib-
                  ute(iN).Operation(jN).Task(nN).StdMinsR
                                 )
                               )
                 )
         )
→ SQLCmdS = "Insert into ItemBOO (OrderNo, LineNo, ProductionLine, ScheduleDate, SequenceNo, StdMins, WorkCenterOperation,Task, TaskStdMins) Values (" & OrderNoN & ", " & LineNoN & ", "' & ProductionLineS & "', '" & ScheduleDateD & "', '" & StdMinsR & "' , '" & WorkCenterIDS & "', '" & OperationIDS & "', '" & TaskIDS & "', " & TaskStdMinsR & ")"
         →DemandDBST.Execute( < I::SQLCmdS >, < O::ⓢRelDBexecutedBL >)
         → (? ⓢRelDBExecutedBL = F
              →ⓢBOOGeneratedBL = F
              →ResponseS := "Failed to access Demand Database"
              →⊗                    // Exit Import Function
            )
→ ⓢBOOGeneratedBL := T
}
```

D.3.3 Import Data Dynamic Behaviours—NA

D.4 Expression class

§Expression ≙ Expression.Architecture
|| Expression.StaticBehaviors
|| Expression.DynamicBehaviors

D.4.1 Architecture

System and CLM Schemas/Objects

D.4.1.1 Expression System Architecture

Expression.Architecture ≙ CLM : **ST** // Expretion Class
|| Exp:**ST**
|| ExpVariable : **ST**
|| Events : **ST**
|| Status : **ST**
Expression.Architecture.Events**ST** ≙ NA
Expression.Architecture.Status**ST** ≙ ⓢExpressionEvaluated**BL**

D.4.1.2 System CLM Schemas/Objects

Expression.Architecture.Exp**ST** ≙ Exp**S**:
(<Variables : **ST** | $\overset{NofVariables\mathbf{N}}{\underset{i=1}{R}}$ ExpVariable**ST**>
)
Expression.Architecture.ExpVariable**ST** ≙ ExpVariable**S**:
(< Variable : **S** >,
< Value : **RT** >
)

D.4.2 Expression Static Behaviours

Expression.StaticBehaviors ≙ Expression**S** (< **I::** ( ) >, < **O::** ( ) >)
|| Evaluate (< **I::** < Expression**S** >, < **O::** Value**RT** >)
|| AddVariable (< **I::** < ExpVariable**ST** >, < **O::**( ) >)
|| SetGlobals (<**I** :: ( ) >, <O :: ( )>)
|| AddRange (< **I::** VariableArray**ST** >, < **O::** ( )>)
|| Clear (< **I::** ( ) >, < **O::** ( ) >)

D.4.2.1 New Expression

ExpressionS (< **I::** < ( ) >, < **O::**( ) >) ≙
// Instantiates the new expression by creating an instance of the Evaluator class
// Evaluator class is a publicly available class that evaluates expression s and return either a
// Boolean value (True/False) or numeric value (decimal value, real, or double)
{
→ Evaluator1 := New Evaluator( )
}

D.4.2.2 Evaluate Expression

Evaluate (< **I::** Expression**S** >, < **O::** Value**RT** >) ≙
{
→ Value**RT** = Evaluator1.Eval(Expression**S**)**RT**
(? Exception = **T**
→ ⓢExpressionEvaluated**BL** = **F**
|?~
→ ⓢExpressionEvaluated**BL** = **T**
)
}

D.4.2.3 Add Variable

AddVariable (< **I::** Variable**ST** >, < **O::** ( ) >) ≙
{
→ ExpVar**ST** = new ExpVariable**ST**
→ ExpVar.Variable**S** = Variable.Name**S**
→ ExpVar.Value**RT** = Variable.Value**RT**
→ Variables.Add(ExpVar**ST**)

D.4.2.4 Add Global Variables

// Global variables are held in the Configuration database in a table
// called GlobalVariables
SetGlobals (< **I::** ( ) >, < **O::** ( ) >) ≙
{
→ GlobalVariables**RS** := ConfigDB.Execute("Select * from GlobalVariables")**RS**
→ $\overset{NofGlobalVariables\mathbf{N}}{\underset{i=1}{R}}$ (→ ExpVar**ST** = new ExpVariable**ST**
→ ExpVar.Variable**S** = GlobalVariables(i**N**).Name**S**
→ ExpVar.Value**RT** = GlobalVariables(i**N**).Value**RT**
→ Variables.Add(ExpVar**ST**)
)
}

D.4.2.5 Add Range of Variables (in an array)

AddRange (< **I::** VariableArray**ST** >, < **O::** ( ) >) ≙
{
→ $\overset{NofVariables\mathbf{N}}{\underset{i=1}{R}}$ (→ ExpVar**ST** = new ExpVariable**ST**
→ ExpVar.Variable**S** = Variable(i**N**).Name**S**
→ ExpVar.Value**RT** = Variable(i**N**).Value**RT**
→ Variables.Add(ExpVar**ST**)
)
}

D.4.2.6 Clear Variables

Clear( ) (< **I::** ( ) >, < **O::** ( ) >) ≙
{
→ Variables.Clear ( )
}

D.4.3 Expression Dynamic Behaviours—NA

D.5 Attribute Class

§ATT ≜ ATT.Architecture
|| ATT.StaticBehaviors
|| ATT.DynamicBehaviors

D.5.1 Architecture

Systems and CLM Schemas/Objects

D.5.1.1 ATT System Architecture

ATT.Architecture ≜ CLM : **ST** // ATT Class
|| ATT:**ST**
|| Events : **ST**
|| Status : **ST**

ATT.Architecture.Events**ST** ≜ NA
ATT.Architecture.Status**ST** ≜ NA

D.5.1.2 System CLM Schemas/Objects

ATT.Architecture.ATT**ST** ≜ AttributeIDS:
(<Name:**S**>
<IsSelected:BL | T=Selected ^ F=Not Selected |default=FALSE>
< Variables:= $\overset{NofVariables\mathbf{N}}{\underset{i=1}{R}}$ <Variable(i**N**)>
< Operations:= $\overset{NofOperations\mathbf{N}}{\underset{i=1}{R}}$ <Operation(i**N**)>
< Components:= $\overset{NofComponents\mathbf{N}}{\underset{i=1}{R}}$ <Component(i**N**)>)

D.5.2 ATT Static Behaviours

ATT.StaticBehaviors ≙ AddVariable (< **I::** Variable**ST** >, < **O::**( ) >)
|| AddOperation (<**I ::** Operation**ST**>, <O :: ( )>)
|| AddComponent (< **I::** Component**ST**>, < **O::** ( )
>)

D.5.2.1 Add Variable

AddVariable (< **I::** < Variable : **ST** >, < **O::**( ) >) ≙
{
→ Variables.add(Variable**ST**, variableKey**S**)
}

D.5.2.2 Add Operation

AddOperation (< **I::** < Variable : **ST** >, < **O::**( ) >) ≙
{
→ Operation.add(Operation**ST**, operationKey**S**)
}

D.5.2.3 Add Component

AddComponent (< **I::** < Variable : **ST** >, < **O::**( ) >) ≙
{
→ Componets.add(Componet**ST**, componentKey**S**)
}

D.5.3 ATT Dynamic Behaviours—NA

D.6 Operation Class

§OPER ≙ Operation.Architecture
|| Operation.StaticBehaviors
|| Operation.DynamicBehaviors

D.6.1 Architecture

Systems and CLM Schemas/Objects

D.6.1.1 Operation System Architecture

OPER.Architecture ≙ CLM : **ST** // OP Class
|| OP:**ST**
|| Events : **ST**
|| Status : **ST**
OPER.Architecture.Events**ST** ≙ NA
OPER.Architecture.Status**ST** ≙ NA

D.6.1.2 System CLM Schemas/Objects

OPER.Architecture.CLM.Operation**ST** ≙ OperationID**S**:
(<Name:**S**>
<Condition:**S**>
<SetupTime:**R**>
<MinimumOps:**N**>
<MaximumOps:**N**>
<AttributeID:**S**>
<ResourceID:**S**>
< Tasks := $\overset{NofTasks\mathbf{N}}{\underset{i=1}{R}}$ <Task(i**N**)>
)

D.6.2 Operation Static Behaviours

OPER.StaticBehaviors ≙ AddTask (< **I**:: Task**ST** >, < **O**::( ) >)
|| OpCaclStdMins (<**I** :: Operation**ST**>, <O :: ( )>)
|| OpCond (< **I**:: Condition**S**>, < **O**:: Value**BL** >)

D.6.2.1 Add Task

OPER.AddTask (< **I**:: Task**ST** >, < **O**::( ) >) ≙
{
→ Tasks.add(Task**ST**, taskKey**S**)
}

D.6.2.2 Calculate the Standard Time for the Operation

OPER.OpCalcStdMins (< **I::** Variable**ST** >, < **O::**OpStdMins**R** >) ≙
{
→ OpStdMins**R** := 0
→ (? OpCond(Condition**S**) = **T**

$\rightarrow \mathop{R}\limits_{i=1}^{NofTasks\mathbf{N}}$ (? TaskCond(Task(i**N**).Condition**S**) = **T**

→ OpStdMins**R** := OpStdMins**R** + Exp.Evaluate(Task(i**N**).TaskExpresion**S**)**R**
)
}

D.6.2.3 Evaluate the Condition

OPER.OpCond (< **I::** Condition**S** >, < **O::** Value**BL** >) ≙
{
→ Value**BL** := Exp.Evaluate(Condition**S**)**BL**
}

D.6.3 Operation Dynamic Behaviours—NA

D.7 Task Class

§TASK ≙ TASK.Architecture
|| TASK.StaticBehaviors
|| TASK.DynamicBehaviors

D.7.1 Architecture

Systems and Schemas/Objects

D.7.1.1 TASK System Architecture

TASK.Architecture ≙ CLM : **ST** // TASK Class
|| TASK:**ST**
|| Events : **ST**
|| Status : **ST**
TASK.Architecture.Events**ST** ≙ NA
TASK.Architecture.Status**ST** ≙ NA

D.7.1.2 System CLM Schemas/Objects

TASK.Architecture.CLM.Task**ST** ≙ TaskID**S**:
(<Name:**S**>
<Sequence:**N**>
<Condition:**S**>
<StdTime:**R**> // Cycle Time
<Expression:**S**>
< Equipments:= $\overset{NofEquipment\mathbf{N}}{\underset{i=1}{R}}$ <Equipment(i**N**)>
< RawMaterials:= $\overset{NofRawMaterials\mathbf{N}}{\underset{i=1}{R}}$ <RawMaterials(i**N**)>)

D.7.2 TASK Static Behaviours

TASK.StaticBehaviors ≙ AddEquipment (< **I::** Task**ST** >, < **O::**( ) >)
|| AddRawMaterials (<**I** :: TASKeration**ST**>, <O :: ( )>)
|| TASKCond (< **I::** Condition**S**>, < **O::** Value**BL** >)

D.7.2.1 Add Task

AddTask (< **I::** Task**ST** >, < **O::**( ) >) ≙
{
→ Tasks.add(Task**ST**, taskKey**S**)
}

D.7.2.2 Calculate the Standard Time for the TASK

AddTASK (< **I::** Variable**ST** >, < **O::**TASKStdMins**R** >) ≙
{
→ TASKStdMins**R** := 0
→ (? TASKCond(Condition**S**) = **T**
→ $\overset{NofTasks\mathbf{N}}{\underset{i=1}{R}}$ (? TaskCond(Task(i**N**).Condition**S**) = **T**
→ TASKStdMins**R** := TASKStdMins**R** + Exp.Evaluate(Task(i**N**).TaskExpresion**S**)**R**
)
}

D.7.2.3 Evaluate the Condition

TASKCond (< **I::** Condition**S** >, < **O::** Value**BL** >) ≜
{
→ Value**BL** := Exp.Evaluate(Condition**S**)**BL**
}

D.7.3 TASK Dynamic Behaviours—NA

Appendix E
Industrial Case Study

The following cases are collected from industries. These cases are used for readers to understand how the principles, concepts, methods and the OKP (one-of-a-kind production) management and control software (OKPMCS) system©, which formulate the OKP strategy and technology for mass-producing customized products are applied in industries. Through these case studies, readers may get an insight into the problems in OKP companies and a better understanding of the OKP strategy and technology as presented in this book. For commercial reasons, we do not provide the names of these companies who provide the cases for us. We also modified the original data collected from the companies. However, these modifications will not affect readers to use these cases to observe the problems and to study the OKP strategy and technology as presented in this book.

To start modelling an OKP production of a product, the following data about the product need to be collected:

1. BOM (bill-of-materials) of the product, including cost of all the parts.
2. Specification of skills/trades and hourly rates that will be needed for the project.
3. Specification of work centres, i.e. brief summary of the usage of these work centres in production.
4. Workflow chart of the production of the product, including time estimate (or work efforts) and task definition, i.e. activities, the number of workers or a work-team, the worker skills/trades, as well as the work centres in which the operations are carried out.
5. Profile of suppliers, i.e. company name and post address, phone and fax numbers, e-mail address of the contact person, and the parts to supply.
6. Profile of subcontractors, i.e. company name and post address, phone and fax numbers, e-mail address of the contact person, and the work to be taken.

In OKP, due to the customer involvement and unexpected uncertainties, the data as mentioned above may be updated according to the changes in production.

Y. Tu and P. Dean, *One-of-a-Kind Production*,
DOI: 10.1007/978-1-84996-531-6, © Springer-Verlag London Limited 2011

By using the OKPMCS system, these data can be changed or updated easily. In early design and production stages, these data are set or estimated according to the prototype, i.e. the PPS of a similar product which was made before in the company (see Chap. 3), and the personal experiences in the company.

E.1 Case 1—A Compressor Suction Scrubber

A compressor suction scrubber is a high pressure vessel and it is a major part of a compressor in oil/gas production. It is used to remove any free liquids from the incoming gas stream so that the compressor will not be overworked, leading to high compression temperatures, mechanical wear and overall machine deterioration. Propane refrigerant is the liquid present in this system and the vessel will be operating at 37 degree Fahrenheit and 52 psi gauge. The inlet to the vessel is the two-phase flow (liquid and vapour) of propane. The outlets are the relatively "liquid free" vapour to the compressor, and the liquid propane that is drained from the bottom of the vessel and sent to the propane liquid transfer pump. Since a compressor is produced according to the particular requirements of a customer, the vessel is thereafter customized according to the specific order of the compressor. Therefore, the vessel can be viewed as a typical OKP product. A compressor suction scrubber or vessel is shown in Fig. E.1. The cost in this case is in Canadian dollars or $.

Fig. E.1 A compressor suction scrubber

The data for producing the vessel as shown in Fig. E.1 are listed as follows:

E.1.1 BOM (bill-of-materials)

Item #	Quantity	Description	Cost	
1	1	1/2" Thick Plate Rolled To 36" x 8' S/S for Shell	$	750.00
2	2	(11.1 Min.) Gauge (12.7 Nom.) Thick x 36" O.D. SE Head c/w 2" STR. Flange	$	800.00
3	1	1/2" Thick Plate Rolled to 36" x 21.3" for Skirt	$	430.00
4	1	1/2" Thick Plate x 42" O.D x 22" I.D. Base Ring	$	415.00
5	1	Namplate and Namplate Bracket	$	112.00
6	2	1/2" Thick Plate 305 x 102 Lifting Lugs	$	80.00
7	1	1/2" Thick Plate x 1066 O.D. X 920 I.D. For Insulation Ring	$	825.00
8	5	Flanges c/w Bolts and Nuts	$	435.00
9	6	Couplings and Fittings	$	320.00
10	1	8' Schedule 40 Pipe for creating the pipe pups for nozzles	$	125.00
11	8	Drain, Vent and Isolation Valves (Ball and Gate)	$	550.00

Fig. E.2 BOM of case 1

E.1.2 Specification of Skills/Trades

Eng: Engineering; Rate: \$75/h

An engineer or technologist is responsible for product design, drafting and documentation.

QC: Quality Control; Rate: \$45/h

A technician is responsible for inspecting the stages of vessel production that are outlined in the workflow chart as shown in Fig. E.3a, and E.3b.

BW: B Welding; Rate: \$50/h

A class B welder is responsible for welding operations as outlined on the workflow chart as shown in Fig. E.3a and b.

VF: Vessel Fitting; Rate: \$40/h.

A vessel fitter is responsible for installing the fittings, nozzles and heads on to the vessel.

PM: Purchasing Manager; Rate: \$50/h.

A manager is responsible to purchase the necessary materials as listed in the BOM as shown in Fig. E.2.

E.1.3 Specification of Work Centres

WC001: Engineering/Drafting

This is where the documentation and drawings are come from. The operations as indicated by ABSA on the workflow chart as shown in Fig. E.3a and E.3b are also carried out in work centre WC001.

WC002: Procurement Department

This is where the required materials are ordered and received.

WC003: Vessel manufacturing unit

All the in-house manufacturing and quality control operations, i.e. shop floor operations as defined in Sect. 3.4.1, are carried out in work centre WC003.

WC004: Final assembly work centre

The finished vessel needs to be moved into work centre WC004 for the final assembly of the compressor.

E.1.4 Workflow Chart

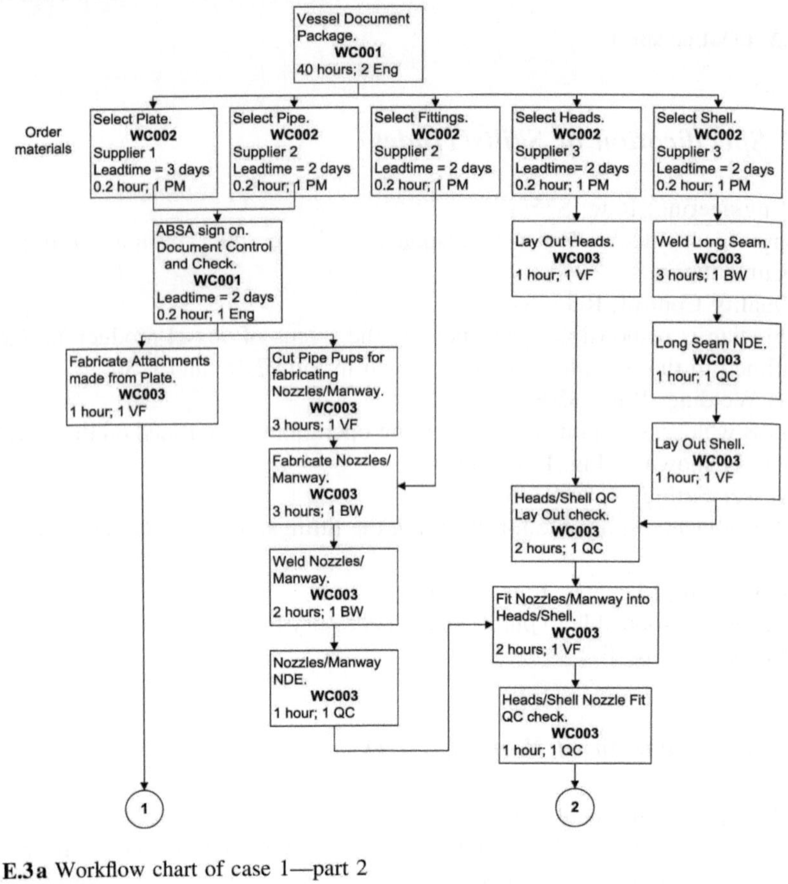

Fig. E.3a Workflow chart of case 1—part 2

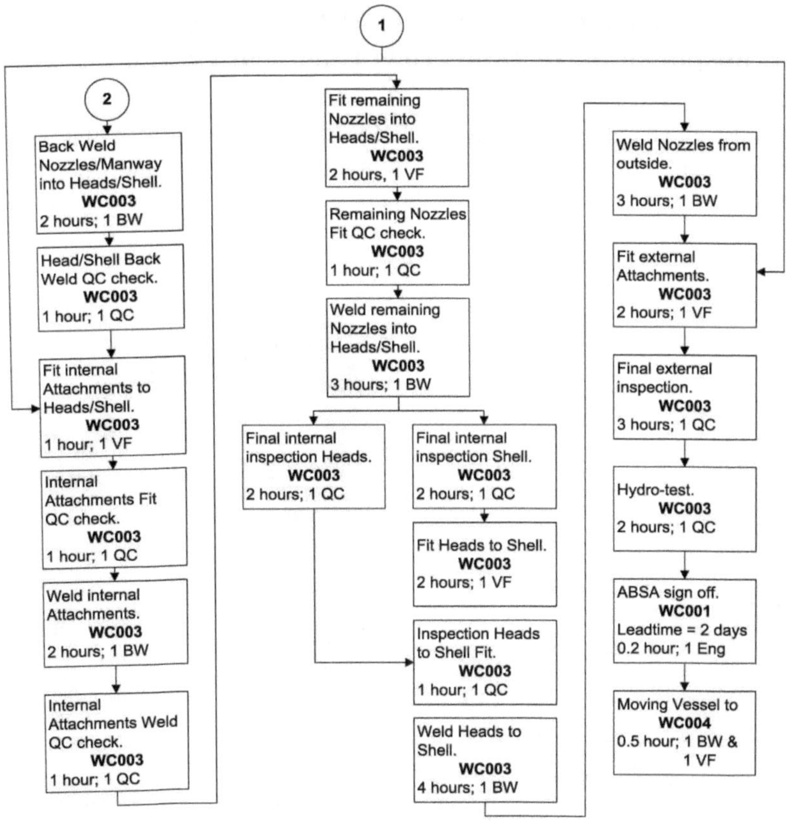

Fig. E.3b Workflow chart of case 1—part 2

E.1.5 Profile of Suppliers

The materials as listed in the BOM as shown in Fig. E.2 are purchased from three suppliers. For commercial reasons, we omit the names and the details of these suppliers, i.e. contact person, post address, phone and fax numbers, etc. We simply call these suppliers as: Supplier 1, 2 or 3. Supplier 1 provides steel plates, Supplier 2 provides all kinds of pipes and pipe fittings, and Supplier 3 provides heads and shell for the vessel.

E.1.6 Profile of Subcontractors

No subcontractor is involved for producing the vessel.

E.2 Case 2—a High Pressure Vessel for Storage and Transportation of Dangerous Chemical Liquid

A high pressure vessel as shown in Fig. E.4 is designed and produced according to the requirements of a customer. It is used for storage and long distance transportation of dangerous chemical liquid. The cost in this case is in Chinese Yuan or ¥.

Fig. E.4 A high pressure vessel for storage and transportation of dangerous chemical liquid

E.2.1 BOM (bill-of-materials)

The company is one of the main suppliers for customized high pressure vessels. It orders the commonly used parts, materials and utilities in large batches or amount from the suppliers. Therefore, in the workflow chart as shown in Fig. E.10, only these specially ordered parts are displayed. The common parts and materials are considered in the warehouse. The sum of "others" costs as shown in the BOM or Fig. E.5, i.e. 500 + 4,000 + 100 = ¥4600, can be input into "Adding Cost" of the last shop floor operation in the PPS.

Main	Description	Material	Specification	Quantity	Provider	Cost	Usage
1	Head	Carbon Steel	2400X2400X20	2	Subcontractor 1	¥6000	Heads
2	Shell Course	Carbon Steel	2300X7600X20	2	Supplier 1	¥9000	Shell Courses
3	Vent	Carbon Steel		1	Supplier 2	¥1000	Shell
4	Rings	Carbon Steel	7000X250X8	2	Supplier 3	¥5000	Final Assembly
5	Flange	Carbon Steel		3	Supplier 2	¥500	Pre-making Heads
6	Pipe 1	Carbon Steel		6	Supplier 4	¥200	Pre-making Heads
7	Heat Process					¥7000	Heat Process
8	Pipe 2	Carbon Steel		3	Supplier 4	¥800	Installing and welding pipes onto Shell
9	Internal Process					¥5000	Internal Process
10	Grinding & Painting					¥1000	Grinding & Painting
11	Valve				Supplier 5	¥8000	Finishing
12	Other Parts				Supplier 6	¥1000	Finishing
13	Steel Frames	Carbon Steel			Supplier 7	¥5000	End-holding Frames
Welding	Type 1				Supplier 8	¥200	Pre-making Heads
Material	Type 2				Supplier 8	¥500	Shell Courses
	Type 3				Supplier 8	¥300	Final Assembly
	Type 4				Supplier 8	¥150	Installing & welding pipes onto Shell
Others	Gas				Supplier 9	¥500	
	Electricity				Supplier 10	¥4000	
	Water				Supplier 11	¥100	Pressure Test

Fig. E.5 BOM of case 2

E.2.2 Specification of Skills/Trades

Rv: Riveting; Rate: ¥16/h.

A riveter is responsible for marking a plate and rolling the plate into a shell course.

Wd: Welding; Rate: ¥16/h.

A welder is responsible for welding the shell courses and final assembly welding.

RW: Riveting and Welding; Rate: ¥16/h.

A riveting and welding worker is responsible for installing and welding the pipes, fittings, and heads on to the shell.

Hp: Heat processing; Rate: ¥16/h

A heat processing worker is responsible for heat processing.

GP: Grinding and Painting; Rate: ¥16/h

A grinding and painting worker is responsible for grinding and painting the vessel.

Vf: Fitting; Rate: ¥16/h

A vessel fitter is responsible for pressure test of the vessel.

Gc: General category; ¥16/h

A worker under the general category skill is responsible for internal process and finishing process of the vessel.

Pm: Purchasing manager; Rate: ¥30/h.

A purchasing manager is responsible to purchase the specially ordered parts as shown on the workflow chart (Fig. E.10).

E.2.3 Specification of Work Centres

CaiGou: Procurement department and warehouse

This is where the required materials and parts are ordered and received.

Tong-1: Shell courses rolling and welding work centre

In work centre Tong-1, the three operations as shown on the workflow chart (Fig. E.10) are carried out. (1) Marking and rolling the shell courses. Two shell courses are needed for the case product as shown in Fig. E.4, (2) Welding the shell courses, (3) Assembling and welding the shell, i.e. to assemble and weld the two shell courses into the shell. Figure E.6 shows a shell course is rolled and welded in work centre Tong-1.

Fig. E.6 Rolling and welding a shell course in work centre Tong-1

Tou-0: Heads manufacturing work centre

The heads are formed or shaped by Subcontractor 1. In work centre Tou-0, the pipes and fittings are installed and welded onto the shaped heads, i.e. operation "Pre-making heads" as shown in the workflow chart. Figure E.7 shows a pre-made head in work centre Tou-0.

Fig. E.7 A pre-made head in work centre Tou-0

Tong-2: Vessel body making work centre

In work centre Tong-2, the three operations as shown on the workflow chart (Fig. E.10) are carried out: (1) Assembling and welding the heads onto Shell, (2) Heat Process, (3) Installing and welding pipes onto Shell. Figure E.8 shows the vessel body in work centre Tong-2.

Fig. E.8 The vessel body in work centre Tong-2

DK-0: End-holding frame fabrication work centre

As shown in Fig. E.4, two end-holding frames are needed for a vessel. In work centre DK-0, operation "Fabricating Frames" as shown on the workflow chart (Fig. E.10) is carried out.

ZZ-0: Vessel assembling work centre

In work centre ZZ-0, the five operations as shown on the workflow chart (Fig. E.10) are carried out. (1) Final Assembly welding, (2) Pressure Test, (3) Internal Process, (4) Grinding and Painting, (5) Finishing. Figure E.9 shows a vessel is assembled in work centre ZZ-0.

Fig. E.9 Assembling a vessel in work centre ZZ-0

E.2.4 Workflow Chart

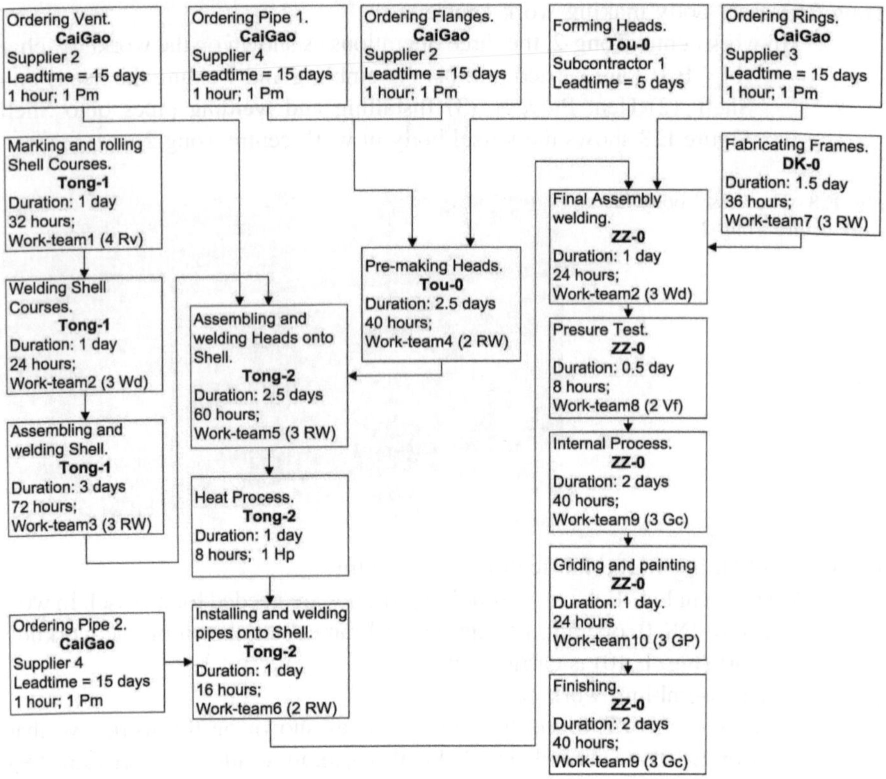

Fig. E.10 Workflow chart of case 2

E.2.5 Profile of Suppliers

The materials as listed in the BOM are purchased from 11 suppliers. For commercial reasons, we omit the names and the details of suppliers, i.e. contact person, post address, phone and fax numbers, etc., and simply call these suppliers as: Suppliers 1 through 11. For Case 2, the BOM as shown in Fig. E.5 provides the information on what the suppliers supply. Only the three suppliers, i.e. Suppliers 2, 3 and 4 who provide specially ordered parts according to customer requirements, are shown on the workflow chart in Fig. E.10.

E.2.6 Profile of Subcontractors

In Case 2, a subcontractor provides shaped or formed heads at a cost of ¥6000/head or ¥12,000 for two heads. We call this subcontractor as Subcontractor 1.

E.3 Modelling and Simulating the Case Product Production

Based on the data collected from the two companies, we can model the product production structures (PPSs) for the two cases. The PPSs of the two industrial cases have been modelled in databases and can be downloaded from webpage www.cimsys.org. If a reader is going to build his/her own PPSs based on the data as provided in Sects. E.1 and E.2, he/she may take the following suggestions:

Step 1: Model the cost of purchasing manager's time in both cases in the Resource Cost of the outsourcing operations in the PPSs (see Fig. 4.9).
Step 2: By the user definition interface (see Fig. 8.4), define users in Case 1 as:

2 Eng
1 QC
1 BW
1VF
1 PM
Supplier 1, Supplier 2 and Supplier 3.
Define users in Case 2 as:
Work-team 1 that consists of 4 Rv
Work-team 2 that consists of 3 Wd
Work-team 3 that consists of 3 RW
Work-team 4 that consists of 2 RW
Work-team 5 that consists of 3 RW
Work-team 6 that consists of 2 RW
Work-team 7 that consists of 3 RW

Work-team 8 that consists of 2 Vf
Work-team 9 that consists of 3 Gc
Work-team 10 that consists of 3 GP
1 Hp
1 Pm
Subcontractor 1, Suppliers 2, 3 and 4.

Since the commonly used parts are assumed in the warehouse or work centre CaiGao in Case 2, only Suppliers 2, 3 and 4 will be modelled in the PPS. Furthermore, set e-mail address of all the suppliers and subcontractor in the PPSs of both cases to your e-mail address so that you can emulate the production progress.

Please note that the company in Case 2 produces customized vessels on a flow production line which consists of 4 work centres, i.e. Tong-1, Tou-0, Tong-2 and ZZ-0, whereas the company in Case 1 produces a compressor suction scrubber in one work centre, i.e. W003. From these two cases, it is obvious that different companies employ different methods to produce the products, which are reflected by the different workflow charts, resource allocation plans and production system configurations of the two cases. These differences can be flexibly or easily modelled by in a PPS using the OKPMCS system. After the PPS is modelled, the OKPMCS system schedules and controls the production of the product according to the production method of a company.

Step 3: A reader should adopt a coding system to identify products and operations so that the PPSs of similar products and similar operations in PPSs can be searched by a computer module from the company database.

A simple coding system, i.e. segment code system is usually used in communication, is applied. A code in this coding system composes several sections from left side to the right side and each of these sections has a particular meaning. By this coding system, for instance, we can use the 1st section (from the left side of a code) to indicate the product family, the 2nd section to indicate the product group, and the 3rd section to indicate the serial number of the product. Using this coding system to code the identification numbers of PPSs, the similar PPSs can be identified by the 1st section and/or the 2nd section of the PPS identification numbers. Likewise, we can identify similar operations by this coding system.

Step 4: Insert transformation states or boxes in a PPS (see Sect. 3.4.1) so that the production progress and intermediate product states can be monitored and controlled during the production.

For Case 1, the proper place to insert a transformation state or box is where a QC (quality control) operation is taken place on the workflow chart. For Case 2, insert a transformation state or box after each of the operations on the workflow chart.

To simulate the production procedure, a reader can first schedule the production by using either Forward Scheduling or Backward Scheduling function (see Fig. 8.6) of the OKPMCS system. Then by changing the time setting of a computer as shown in Fig. E.11 hourly or daily from the start time towards the end time of the production, a reader can use PPSMaint, PPSManager and PPSShopFloor modules to simulate the production procedure of a case and to control the production.

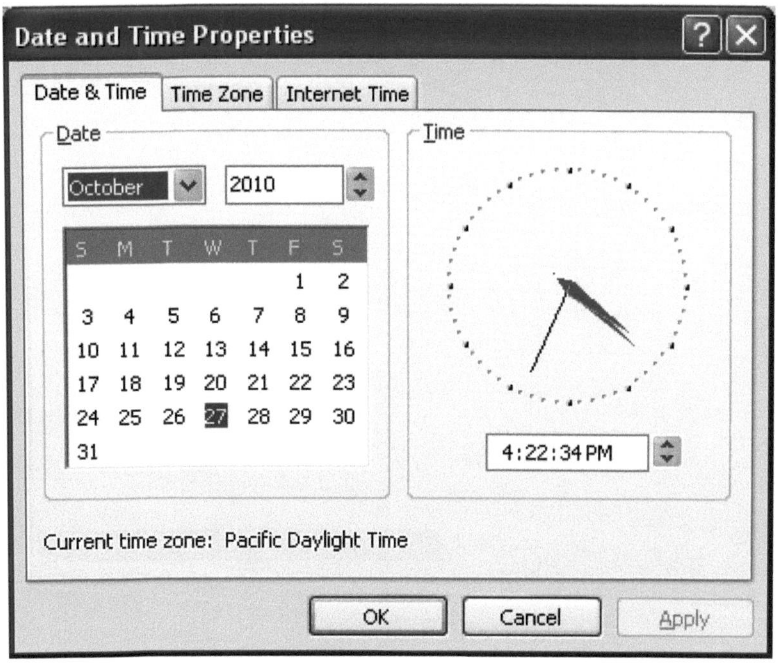

Fig. E.11 Changing a computer time setting

During the simulation of the production procedure and control, the reader can change the production data under the following hypothesis scenarios: production routine change (i.e. workflow change), product design change (i.e. BOM change), supply change (i.e. lead-time change), operation time (or work effort) change, change of the total number of workers, etc. To cope with these changes, the reader can apply the methods as presented in this book or the methods from the other references.

The case study as presented in this Appendix may be used by an instructor to teach a course on project management, operations management, one-of-a-kind production or rapid product development.

Index

A

Abstract data types, 36–37
Actual cost (AC), 44, 46, 58–59, 97–98
Adaptive production scheduling and control, 21, 77, 113, 182–185, 220
ADT, 36, 77
A-floor, 77
Agile manufacturing, 10, 62–63, 65
Agile unified process, 39
AUP, 39

B

Backward Scheduling, 208, 221, 233, 296
Base product, 17–19, 40, 45, 47, 49, 53, 85, 117, 120, 122–123, 125, 127, 169
Batch production, 3, 17–18, 77, 117
B-floor, 77
Bill of operations, 3, 6–7, 12, 19, 40
Bill of information, 27, 55, 59, 117, 133, 149
Bill of material, 67, 70, 73, 101, 117, 171, 211, 215–216, 218, 224
Bill of materials, 3, 6–7, 19, 40, 285, 287
Bill of operation, 45, 73, 117, 119, 149, 171, 211, 215–216, 218
BOI, 27, 29, 55, 117, 120–122, 133, 142, 144, 149, 153, 155, 167, 219
BOI template, 52, 120, 122, 144
BOM classes, 53–54
BOM template, 47–49, 122–126, 132–133
BOO classes, 54–55
BOO template, 48
Burgess and Kilebrew heuristic, 195–196, 198, 208

C

CAD, 3, 10, 19, 23, 45, 65, 87, 91
CAD/CAM, 11, 48
CAD/CAPP/CAM, 64
CAM, 10
Canadian Manufacturers and Exporters, 1
CAPP, 61–62, 64–65, 67, 87
CDS heuristic, 172–175, 186–189
C-floor, 77
CIM, 3–4, 25, 29, 65
CLM, 36, 144, 253, 268, 272–273, 276, 279, 281, 283
CME, 1
CNC, 2–4, 8, 13, 18, 48, 52, 53, 56, 107, 119, 120, 128, 170
Complicated project, 231
Component logical model, 36
Computational complexity, 174–176, 181–182, 192, 211
Computer aided process planning, 61, 63–64
Computer integrated manufacturing, 3, 25, 63–64
Computer numerical control, 2–3, 18, 170
Computer-aided cost estimate (CACE), 87
Computer-aided design, 45, 87
Computer-aided manufacturing, 10, 70
Configuration database, 143, 145
Cost Analysis Module, 66–69, 88, 90–95, 102, 106–107, 110–111, 113
Cost estimation and optimization, 93, 96, 109, 112
Cost index hierarchy, 90
Cost index structure, 88–91, 111, 219
Cost Performance Index (CPI), 97, 100
Cost Variance (CV), 87, 100
CPI, 97–98, 100, 109, 110, 220
Current production control phase, 75

Y. Tu and P. Dean, *One-of-a-Kind Production*,
DOI: 10.1007/978-1-84996-531-6, © Springer-Verlag London Limited 2011

Customer interface, 14, 23, 27, 42, 45, 67, 72, 123, 132, 147, 152–153, 167
CV, 97–98, 100, 109–110, 220
Cycle time, 13–14, 20, 34–35, 40, 43, 48–49, 84–85, 112, 127, 131, 163–164

D
Data Interchange Schema, 11
Data modeling module, 145
DDIS, 11
Demand database, 145, 148–152, 161, 165, 274–275
Design for cost (DFC), 86
D-floor, 77
Direct subcontracting strategy, 16, 26, 41
DIS, 11
DMM, 46, 143, 145, 153
DSS, 16, 19, 25–26, 41
Dynamic Data Interchange Schema, 11
Dynamic programming, 95, 108

E
Economies of scale, 2, 4, 12, 21
EDI, 19
EEC, 17
eGantt chart, 22, 172, 193–196, 198, 203–209, 211, 217, 222, 234
Electronic data interchange, 19
Electronic Gantt chart (eGantt chart), 172
Electronic Kanban, 217, 218, 223, 230
Engineer-to-order, 11, 21, 45, 112
Enterprise Production Data, 66–67, 204, 207, 219, 226
Enterprise resource planning (ERP), 2–3, 44, 69, 118, 215
EPD, 66–68, 204, 219, 221, 226
ESPRIT, 17, 64
ETO, 21, 45
European Economic Community, 17

F
Factory of Future (FOF), 17
FB, 98, 109
Flexible manufacturing systems, 29
Flow production lines, 4, 8
Flow shop, 56–57, 171–173, 175, 177, 185–188, 193
FMS, 4, 18
FMSs, 4, 29
FOF, 17
Forecasted Budget (FB), 98
Forward Scheduling, 233–234, 296

G
GBOMO, 12–13
GBOMs, 6, 12
GBOOs, 6
Generated Information, 149
Generative cost estimate method, 91, 93, 105
Generic algorithm, 147
Generic bill of materials, 6, 12
Generic bill of operations, 6
Generic bill-of-material-and-operation, 12
Generic process structure, 12
Generic product and process structure, 12
Generic product structure, 12
Generic product production structures, 6
GPcS, 12
GPdS, 12
GPPS, 12–13
GPPSs, 6, 14
Group technology (GT), 63, 67, 93, 228

H
HFS, 172–174, 176, 181–182, 187
High pressure vessel, 218, 286, 290
Hybrid flow shop (HFS), 172

I
IDPP, 67–69, 78–79
IGM, 46, 143–145, 147–151, 155, 158, 160–161, 167, 170, 253, 259, 267
Incremental Design and Process Planning, 62, 67–69, 79
Information classes, 53–55
Information generation module, 46, 55–56, 143, 167, 253
Information technology (IT), 9, 33, 135
Input Demand Items, 145–146, 150
Input Items, 145, 147

J
Job shops, 7
Job-shop, 4, 19, 27
Johnson's algorithm, 173–175, 192–193

K
Knowledge management, 24, 112
Knowledge mapping, 117, 133, 141
Knowledge mapping method, 117, 143
Knowledge mapping process , 117–118, 133–134, 136, 145, 167

L
LD heuristic, 197–198
Lean production, 25, 42–43, 62–63, 170, 215, 222
Least slack heuristic, 194, 222
Lever concept, 176–179, 186, 192
Longest duration heuristic, 194
LS heuristic, 198–200, 208
LWBJD (least weighted between jobs delay), 174

M
Make-to-order, 11, 21, 71, 112
Manufacturing Plant Model, 78
Manufacturing resource planning, 2, 44, 215
Mass customization, 2, 4–6, 8–11, 13
Mass Customization Information System, 10
Mass production, 1, 3–5, 7–8, 13, 19, 23–24, 33–34, 40
Material requirement planning, 170
MC, 4–6, 8–11, 13
MCIS, 10–11
Mechanism, 34–35
Minimum cost project crashing technique, 172
MP, 7, 33–34, 40
MRP I, 21, 26, 44–45, 70, 170
MRP II, 2–3, 12, 19, 22, 27–28, 44–45, 73, 141, 215, 217, 236
MTO, 21

N
National Industrial Bicycle Company, 5
NEH heuristic, 173–174
NIS heuristic, 172, 174–175, 187–188
NP-completeness theory, 173

O
Objective Management Group, 37
OKP, 2, 16
OKP (one-of-a-kind production) phenomenon, 17
OKP management and control software system©, 22, 24, 27–28, 33
OKP strategy, 33, 35, 41–43, 60, 66, 85, 113, 380
OKPMCS system, 59, 216, 239
OMG, 37
One-of-a-kind production, 4, 16
One-off, 4, 19, 27
One-piece flow production line, 142, 167
One-piece flow technology, 4, 42, 170
OP, 71, 92
OS, 71, 92

P
Pattern, 34–36
Pattern-Oriented Software Architecture, 36
PB, 91, 98, 100, 109–110, 220
PD, 83–84
PD chain, 83–84, 86, 89, 109
PD cost, 83–88, 90
PDM, 112, 219
PDR, 171, 174
Planned Budget (PB), 83, 97–98, 100, 109, 110
Planned Cost (PC), 97
POSA, 36
PPS, 71
PPSMaint module, 217
PPSManager module, 224
PPSs, 69, 93, 107–108, 110, 113, 197–198, 211, 220, 230
PPSShopFloor module, 229
Priority dispatching rules (PDRs), 171
Process Model, 36, 64, 78
Process planning, 3, 12, 24, 25, 61–68
Product data management, 219
Product development (PD) chain, 83
Product Model, 35, 64, 78
Product production structure, 71
Production calendar, 219, 221
Production estimation and plan phase, 75
Production planning and control structure, 76
Project management, 98, 171, 194, 200, 203, 209, 211, 216, 298
Project-based production, 24, 42, 171, 212, 216
Prototype, 45
Prototype-based incremental and concurrent product development method, 62, 68, 71, 79, 111, 133, 217, 219

Q
QFD (quality function deployment), 147

R
Real and Informational System (RIS), 74
Real and Physical System (RPS), 74
Real time computer aided process planning (RTCAPP), 64
Real-time process algebra, 36, 136, 243

R (*cont.*)
Relational database, 36–37, 89, 112, 118, 141, 162, 243
Requests for Information, 149
Resource levelling, 194–195, 222
Resource planning, 172, 193–194, 211
RTPA, 33, 36–37, 39–40, 136, 143–154, 249

S
SC, 71
SCM, 10–11
SFC, 47
Shop floor control, 47, 56, 65, 76, 79, 143, 149
Small or medium sized enterprices, 11, 43, 66, 215
SMEs, 1–4, 11, 16, 18, 22, 26, 29, 43, 66, 70, 215
SS heuristic, 171–172, 175–177, 179, 181–185, 187–193, 220
State space (SS) heuristic, 171, 176, 211
State space concept, 176, 179, 185, 192
STEP, 65
Stepwise concurrent product design and process planning management, 62, 66–67
Supply chain integration, 222
Supply chain management, 3, 10, 70, 87
Supply chain schedule, 222

T
Taillard's benchmarks, 185–189, 191, 211
Takt time, 20
Taxonomy, 9–10
TFS, 172–174, 177, 185, 192–193
THOCPN-CS, 183–184
Totally integrated manufacturing cost estimating system (TIMCES), 87
Traditional flow shop (TFS), 172
Transformation state, 71, 296

U
UML, 36–37
Unified modeling language, 37
Up-To-Date Expenditure (UTDE), 97
User interface, 35, 41, 151–152, 155
UTDE, 97–98, 100, 109

V
Variant cost estimate method, 93–94, 102
Virtual and Informational System (VIS), 74
Virtual and Physical System (VPS), 74
virtual manufacturing systems, 19, 75
VMS, 19
VMSs, 19, 26, 41

W
Whiteboard, 129, 218
WIP, 8, 20–21, 42, 44, 58, 148–149, 172, 176–177, 180
Workflow chart, 285, 287–290, 292–295
Workflow template, 120
Work-in-progress, 8, 42, 148

X
XML (extensible markup language), 89

MIX
Papier aus verantwortungsvollen Quellen
Paper from responsible sources
FSC® C105338

If you have any concerns about our products,
you can contact us on
ProductSafety@springernature.com

In case Publisher is established outside the EU,
the EU authorized representative is:
Springer Nature Customer Service Center GmbH
Europaplatz 3, 69115 Heidelberg, Germany

Printed by Libri Plureos GmbH
in Hamburg, Germany